From Industry 4.0 to Industry 6.0

From Industry 4.0 to Industry 6.0

Edited by

Carolina Machado
J. Paulo Davim

iSTE

WILEY

First published 2025 in Great Britain and the United States by ISTE Ltd and John Wiley & Sons, Inc.

ISTE Ltd
27-37 St George's Road
London SW19 4EU
UK

www.iste.co.uk

John Wiley & Sons, Inc.
111 River Street
Hoboken, NJ 07030
USA

www.wiley.com

Library of Congress Control Number: 2025932523

British Library Cataloguing-in-Publication Data
A CIP record for this book is available from the British Library
ISBN 978-1-83669-029-0

Contents

Preface

Industry 4.0 marked a revolution in industrial processes, introducing connectivity and digitalization as key elements to improve efficiency and productivity. However, technological evolution moves ever forward, and now the prospect of Industry 6.0 has emerged, a new era that promises to radically transform the way we view industrial production. Industry 4.0 was characterized by the integration of cyber-physical systems, the Internet of Things (IoT), artificial intelligence (AI) and data analysis to create smarter and more efficient factories. Connected sensors, autonomous machines and predictive analytics have become the norm, enabling process optimization and more informed decision-making. However, Industry 6.0 looks to go further, introducing revolutionary concepts that will redefine how we view industrial production. One of the pillars of Industry 6.0 is the complete integration of advanced artificial intelligence into all aspects of production. This implies systems that not only collect and analyze data but also continually learn and adapt. Autonomous machines will be able to not only follow predefined schedules but also make complex decisions in real time, adjusting to constantly changing scenarios. This will increase efficiency and reduce the need for human intervention in routine tasks. Another crucial element of Industry 6.0 is augmented reality (AR) and virtual reality (VR), which will be integrated more deeply into production processes. This will open doors to virtual work environments where engineers and operators can interact with complex machines and processes in an immersive and remote manner. Predictive maintenance will also benefit from augmented reality, allowing technicians to view crucial information about equipment in real time, facilitating faster diagnoses and repairs. Cybersecurity becomes even more pressing in the transition to Industry 6.0, as increasing interconnection and dependence on intelligent systems increase the risks of cyberattacks. New security protocols and advanced algorithms will be essential to protect sensitive data and maintain the integrity of industrial processes. Confidence in emerging technologies will largely depend on the ability to implement robust cybersecurity measures. Sustainability also becomes a central focus in Industry 6.0. The search for more ecological and efficient production processes is gaining prominence, driven by growing environmental awareness and stricter regulations. Technologies such as the Environmental Internet of Things (EIoT) and AI-based energy optimization will be key to reducing the environmental impact of industrial production, promoting more responsible and eco-efficient practices. Furthermore, Industry 6.0 not only transforms production processes, but also redefines the relationship between companies and consumers. Mass customization, driven by advanced data analysis algorithms, allows products to be tailored to individual consumer preferences. This not only meets the growing demand for personalized products but also promotes a more engaging and satisfying experience for customers. In short, the evolution from Industry 4.0 to Industry 6.0 represents a significant advancement in how we conceive production and the interaction between machines, humans and systems. Integrating advanced artificial intelligence, augmented reality, enhanced cybersecurity and a renewed commitment to sustainability redefines the limits of what is possible. As we embrace this new industrial era, it is imperative to not only keep up with technological innovations but also embrace a mindset of constant adaptation to maximize the benefits of this ever-evolving industrial revolution. This transition not only increases operational efficiency but also paves the way for new business models, improved sustainability and a more connected and aware society.

Conscious of this reality, *From Industry 4.0 to Industry 6.0* highlights the importance and impact this technological evolution will have on the way today's organizations develop into proactive, innovative and competitive agents.

Organized into eight chapters, this book looks to discuss in Chapter 1, *From Industry 4.0 Onward: Is There a Need for "Industry 6.0?"*, while Chapter 2 highlights *Industry 6.0 Transformation: Conceptual Transition Framework, Opportunities and a Research Agenda*. Chapter 3 focuses on the *Impact of Industry 6.0 on Human Cognitive Behavior*. Chapter 4 speaks about *Understanding the Metaverse – A Holistic Approach of a Rapprochement with the Marketing Domain*, Chapter 5 focuses on *Model-Based Management – A Safari of Essential Business Models*, Chapter 6 is entitled *Approaching the Portuguese Labor Market From a Gender and Generational Perspective in the Era of Industry 4.0, Robotization and Artificial Intelligence*, Chapter 7 covers *Add More Marketing to Marketing Doctoral Programs – Answering Hunt and Yadav's Calls* and, finally, Chapter 8 looks to discuss *Industry 6.0: Why Talk About It Now?*.

From Industry 4.0 to Industry 6.0 can be used by various potential stakeholders, not only academics and researchers, but managers, engineers, practitioners and other professionals who develop their professional activity in different areas of management and engineering. It is also relevant to emphasize that experts who contributed with a chapter were encouraged to identify the theoretical and practical implications of the different aspects that define the evolution from Industry 4.0 to

Industry 6.0 to provide a more effective understanding and implementation of these issues in different types of organizations.

The present book, organized in eight chapters, can serve as a valuable reference for academics, lecturers, researchers, graduated and postgraduate students, managers, engineers and other professionals in related matters with Industry 4.0 and 6.0.

The editors acknowledge their gratitude to ISTE-Wiley for this opportunity and their professional support. Finally, we would like to thank all chapter authors for their interest and availability to work on this project.

Carolina MACHADO
J. Paulo DAVIM
February 2025

1

From Industry 4.0 Onward: Is There a Need for "Industry 6.0"?

The process of industrialization – as a sustained trend in history and developing new socio-economic concepts – presuposes higher rates of growth of the economy and a structural change. It is on this basis diverse concepts of industrial revolutions have been accepted. These concepts have been coherent to reveal singular developments. Recently, the accepted concept of Industry 4.0 (i4.0) has shown some limitations regarding the need to develop automation technology in an anthropocentric orientation. This is why Industry 5.0 has two orientations: either the experiences that adopt this concept seek solutions to adapt the human factor to the features of the technology or, knowing the social and organizational requirements, look for solutions to develop this technology in accordance with these requirements. The debate is still developing. There is not yet conceptual maturity to propose a new topic that would be based in a new eventual structural change observed like a possible "Industry 6.0" era. Most arguments for an Industry 6.0 are still those which have been discussed for Industry 4.0, since they are mostly based on eventual not yet ready developments of technology. That is why we should continue the debate on the late developments of industrialization and its social and economic conditions.

1.1. Introduction

When invited to write about the concept "Industry 6.0", I was curious about its novelty. I have been working for more than a decade on the topic of Industry 4.0, and more recently on Industry 5.0. These last concepts integrate all of the dimensions related with the technological development in the manufacturing industry, and they especially integrate options on company management strategies.

Chapter written by António Brandão MONIZ.

The case for Industry 5.0 is even more advanced: it integrates the older debate on the new organizational options and design and sustainability criteria for management. "Integrates" means that, at the end of the day, the forefront manufacturing technology must be designed considering organizational criteria. Therefore, these criteria have to be consistent with the circular economy and sustainable factors. In this sense, new ideas are completely new according to scientific literature. They have been recently compartmented: technology development and its readiness levels, for one side, and new organizational design, social partners involvement and management strategies, for another side. Now, what is proposed, is for a new joint concept to become debated, leading to more advanced standards. The interdisciplinary scientific community has not yet completely involved with this discussion. There are still few case studies and successful benchmarks. We are still in the early phases of the debate.

To my surprise, some authors are starting to speak about Industry 6.0. My curiosity led me to research the relevant literature, wide discussions and experiments, but unfortunately, I did not find enough evidence. The published literature is mostly conjectural, based on impressive scenarios, and tries to push for a concept that is not yet consistent in terms of novelty and based on any evidence. It is also revealing that most scholars involved in the debate on Industry 4.0 and Industry 5.0 are not discussing Industry 6.0. Why? My answer would be that there is not yet conceptual maturity to propose a new topic that would be based in a new eventual structural change observed – even in few cases – in the manufacturing industry. Most arguments for an Industry 6.0 are still those that have been discussed for Industry 4.0, once they are mostly based on eventual not yet ready developments of technology.

Instead of being disturbing, these facts can also contribute to the rediscussion of these concepts on structural changes concerning how products are manufactured, as well as how the conventional sector and labor markets may change during the next few years. They can also contribute to making these types of scenarios more accurate, ensuring certain elements of evidence are relevant to companies and policymakers. If this is not the case, they can become useless, a simple playground for conceptual flags that do not consider the previous steps explaining the emergent changes or social needs. In the following sections, we will demonstrate this.

1.2. From Industry 4.0 toward Industry 5.0

The concept of Industry 5.0 (i5.0) aims to place workers' well-being at the center of the production process [BRE 21]. This concept is centered on the idea of anthropocentric technology, which implies that technology, organizations and workplaces must be adapted to human and social needs. There are, however, still unclear problems regarding this concept. Knowing that the Industry 4.0 (i4.0) concept has significant limitations regarding the need to develop automation technology in an anthropocentric orientation, Industry 5.0 has two orientations: either the experiences that adopt this concept seek solutions to adapt the human factor to the characteristics of the technology or, knowing the social and organizational requirements, look for solutions to develop this technology in accordance with these requirements [MON 23; CAN 23b]. This is a problem that is unlikely to be resolved.

i5.0 is a concept that arises as an evolution of the industry and human–machine interaction. While i4.0 is focused on automation and scanning production processes, i5.0 search to integrate more harmoniously and collaboratively humans and machines on the desktop [BRE 21].

It seems, however, that a sound i4.0 concept-centric concept still cannot include the perspective and experience of the ethical sciences, sustainability and social and human science. In other words, the concept of i4.0 has been almost exclusively focused on technology development, despite also discussing the need to include the human factor [BUT 18]. However, in reality, it has not happened. We continue to see a prevalence in technology development and not in the development of working conditions where technology could be designed for this development [CIM 20]. For this reason, the idea of i5.0 is focused on this concept of anthropocentric technology and implies that technology, organizations and workplaces should be tailored to human and social needs. For this reason, a potential concept of Industry 6.0 (i6.0) completely focused on future advances on computing technologies and manufacturing systems seems unclear as of yet. In fact, there is still room to develop i4.0 based on experiments of technology development, such as cloud computing, cyber-physical systems (CPS), Internet of Things (IoT) and the interconnection of these with intelligent robotics, or with additive manufacturing [ELH 21; HIR 16; KRI 21; THU 19].

1.2.1. *Limits of the technological developments*

The focus on technological achievements and the articulation between artificial intelligence with new manufacturing technologies have shown many dilemmas and difficulties. Since the technical system is the more advanced, the decision loop must involve human input. In practical terms, machine learning still has too many limitations to enable direct substitution by humans.

If this is true for the moment, will it still be true in the next decade? What about in two decades? These questions bring us to two types of reasoning: on the one hand, we will need more complete and critical knowledge on the possible trends and roadmaps needed; on the other hand, we still must think about the role of humans in a more complex environment with increased capacities in terms of artificial intelligence. What are the limits? Are they known? Do we know the technology limits well enough, or do they have to be imposed by laws and regulations? Do we know the human limits? These are the essential questions on the possible transitions of industrial revolutions.

Thus, i5.0 should recognize the importance of unique human skills such as creativity, empathy, critical thinking and social qualifications, and therefore seek to combine them with the efficiency and accuracy of machines [ELH 21]. In principle, the concept does not seek human replacement with machines in repetitive tasks as a competitive advantage. This concept points closer toward cooperation between humans and machines, harnessing the best of each.

The transition from i4.0 to i5.0 would have to be done by evaluating the nature of the transformation of jobs, future professions and reduced skill gaps in order to address possible unemployment effects (especially the derivatives of technological unemployment) and promote industrial competitiveness and innovation by simultaneously reinforcing inclusion [CAN 23b].

The most recent commitment to i5.0, especially made by the European Commission, includes an interest in "responsible innovation". This commitment "not only or mainly aims to increase cost efficiency or maximize profit, but also increases prosperity for all those involved: investors, workers, consumers, society and environmental environment" [BRE 21].

In effect, industries can play an active role in providing solutions to society's challenges, including resource preservation, climate change and social stability. But, if the approach of i4.0, or Industry of the Future, has benefits for the industry, its development in the "anthropocentric" sense allows for advantages benefiting both workers and society [KRI 21]. Therefore, i5.0 should empower workers, because the evolution of the skills and training needs of workers emerges. In other words, with the application of the technological concepts associated with i4.0, new skills emerge, and in view of this, new training needs are sought.

1.2.2. *Discussions on the concept of i4.0*

The anthropocentric concept of i4.0 would then imply that association between the development of intelligent technologies in manufacturing and the needs of its workers. At the same time, with this empowerment, it should increase industry competitiveness [KRU 09; LU 20; NEU 21]. With some principles defined for i6.0, the possibilities would be completely different. We will come back to this later on.

However, thinking of the human first in the productive, more complex and intelligent systems process has not been the central concern in the academic and industrial environment. Only when productive processes have been developed is an attempt made to adapt humans to these technologies.

There is therefore still a long way to go in this conceptual development. In recent years, the concept of i4.0 has been widely based on technology-driven experiences, allowing important steps in the areas of machine learning, cyber-physical systems, development of IoT devices and integration at the industrial factory floor level [MAD 15]. However, new organizational concepts and innovation processes have been secondary. Some production efficiencies, cost reductions and business model adaptations were achieved with i4.0 [THU 19].

With advanced automation, intelligent machines and systems can perform complex tasks autonomously by reducing human intervention dependence on repetitive and routine activities. However, in most cases where the concept of i4.0 has been introduced, it is not possible to demonstrate the advantages of job substitution, or there are no demonstrable gains with the exclusion of operators from interaction processes with these equipment [ROM 16; NEU 21]. Of all modes, the implementation of automated systems aims to increase production cadences using mechanical media and information systems that improve the technical performance. Therefore, although biased, it would be preferable to displace human jobs, making their tasks more creative and less routine [MON 22a].

With IoT, which plays a key role in i4.0, the connection between machines, devices and sensors in a smart network is established. This allows for real-time data collection and sharing, enabling higher visibility and control over production processes [MAD 15]. However, the operation and control of these intelligent networks are, still, always performed by human beings, especially by those with the technical competence and responsibility of the control of these processes.

Virtual and increased reality technologies are used in human–machine interaction in i4.0. They allow for the overlay of digital information in the physical environment by offering visual guidance and real-time instructions for workers, facilitating training, maintenance and problem solution. These technologies have been disseminated in many industries of manufacturing. Only the issues of usability and acceptance can be raised so that this technology does not encounter application limits [THU 19].

1.3. Arguments for "Industry 6.0"

Industry 6.0 is not just about automating factories, but rather transforming entire industries through the integration of advanced technologies, enabling smarter decision-making, higher productivity and unprecedented levels of customization. i6.0 represents the next phase of industrialization, which is focused on creating fully integrated, intelligent manufacturing systems that can operate with minimal human intervention. It combines human intelligence, artificial intelligence, cloud computing energy, human–robot working big data and quantum computing [CHA 23].

With such phrasings, it is possible to find comments where articles like these, framed as the next industrial revolution, are an utter waste. In fact, they even suggest that the author does not understand the principles of the 4th Industrial Revolution (or i4.0) or any industrial generational leap. Comments such as "such articles are only intended to inject unnecessary information (not even knowledge)", can be understood.

1.3.1. *The Finnish white paper on i6.0*

Other statements are produced at the level of white papers. The main referenced one is the one produced by the Innovation Funding Agency Business Finland (AIF) in 2021. There, they underline the following:

> [Industry 6.0] is characterized on one hand by customer-centric, highly customized lot-size-1 thinking, on the other hand by hyper-connected factories, with dynamic supply chains, were data flows across domains. These also change the role of human as a production worker, as they become part of the interconnected environment and need to handle the digital, optimized production [KUO 21, p.2 and 38].

From such feature, a group of suggestions and proposals are established. The main ones are:

– hyperconnected factories in complex, dynamic supply chains and value networks, where data flows across different administration domains. Requires a common data model;

– human digital twin connects manufacturing;

– the role of human dramatically changes in manufacturing;

– AI optimization of production to obtain sustainability and antifragility;

– lot-size-1 made economically feasible;

– antifragility obtained via the design of systems relying on non-functional requirements (NFR).

The motivation is clear. For the organizers of this white paper:

> Thanks to disruptive technologies such as 3D/4D printing and artificial intelligence (AI), Finland can make the most of localization as an opportunity to bring more manufacturing back to Finland. The small size of the local market and the future needs for increased personalization pose the fundamental question and opportunity for Finland: how to make the lot-size-1 economically feasible [KUO 21, p. 10 and 38].

The focus was mainly on the hyperconnectivity, dynamic supply chains, nonfunctional requirements or digital twins. However, the phrase the "role of human dramatically changes" almost says nothing. It is not clear which "changes" or "dramatic roles" can be expected. For something that could be a structural change forward i5.0, the explanation is scarce. It should be fundamental to have clear ideas and proposals on that direction. They are, however, not present in the document.

1.3.2. *New inputs for this concept*

Other articles published in more referenced journals however take the same type of direction.

For example, Chourasia et al. [CHO 22] start their abstract stating that:

> The sole aim of industry 6.0 is to seizure the new technologies, which can be applied worldwide and deliver wealth, prosperity away from the job and provide growth to nations across all planetary boundaries. This revolution would promote living harmony with nature, support the principle of sustainability where technology would not be a thing, and promote the human virtual digital twin where all can simultaneously see physical goods and virtual product information (p. 443).

In other words, the authors continue stating that:

> Industry 6.0 is the one-step further than industry 4.0 and industry 5.0, where every operation would be controlled by human minds and performed by automated robots by covering all the planetary

> boundaries. It combines human intelligence, artificial intelligence, cloud computing energy, human–robot working big data, quantum computing (p. 444).

The Duggal et al. [DUG 22] article starts with the mention that:

> The ideological concept of Industry 6.0 encompasses adjustments and advancements in virtually all domains. To perform an intensive categorical analysis, the advancements have been classified into 4 major sectors; robotic automation, society and policies, and lastly, intelligent manufacturing (p. 522).

With this focus, the authors confuse the concepts of Society 5.0 and Industry 5.0, and go on to define Industry 6.0 in a vague way, concluding that:

> The prime focus of the sixth industrial revolution would be on medical technology with multi-dimensional printed controlled release medicine, automated medical diagnostics entirely, removing any extra burden from practitioners, leaving them to focus on critical cases [DUG 22, p. 529–530].

Other authors, such as Almusaed et al. [ALM 23], make a clearer definition:

> Industry 6.0 pioneers advancements in quantum computing, nanotechnology, artificial intelligence and cloud-based energy solutions. Harmonization facilitates design, building, and maintenance processes, improving efficiency, accuracy, and sustainability (p. 1).

They finalize their article by concluding that:

> Industry 6.0 could bring transformative changes to industries. While it promises virtualized, antifragile manufacturing and services emphasizing customer-centric strategies, dynamic supply chains and automation-driven flexibility, job displacement due to increased automation is the primary concern. Future industrial revolutions should prioritize job creation to avoid socioeconomic discontent [ALM 23, p. 22].

This and the previous articles only reproduce simple approaches, repeating arguments and bring no evidence to support their conclusions and assessments.

1.3.3. *Possible outcomes on i6.0*

The arguments for "Industry 6.0" are not convincing. The technological upgrading of production systems including AI systems, even based on quantum computing or nanotechnologies are just a step forward the Industry 4.0 concept. It is not, in fact, an industrial revolution. We can argue in the same direction that the concept of Industry 3.0 is basically the development of automation (that already existed in Industry 2.0), but with digitalization. Digitalization represented a real revolution. Therefore, we can talk about an Industry 3.0 stage. The digitalized automation from the last decades of the 20th century went through several upgrades and developments, from numerical control, to robotics, and from flexible manufacturing systems to computer-integrated manufacturing. We had the integration of product design (CAD) into automated manufacturing (CAM), and later the concept of "production islands" and flexible systems (FMS). The integration of different functions through computerized systems (CIM) represented even a further step in this revolutionary process of manufacturing [HIR 16; CAN 23a].

From the early second decade of the 21st century, the new concept of Industry 4.0 became accepted because the needs of the industrial sector became different. The advances of digitalization and artificial intelligence enabled new forms of production [PFE 16]. This became a new revolution, changing the ways of organizing the sector and opening new doors for possibilities of technological development.

However, the very recent concept of "Industry 5.0" has surged to reveal the need of a necessary development of industry that should respond to present human needs: the consideration of a sustainable strategy of technology development, and the need to consider the role of humans in an increased automated process [SOW 16].

We can conclude that the proposed concept of "Industry 6.0" is by now very poor in terms of revealing new trends of radical development of the process of manufacturing. The debate on the new possibilities for Industry 4.0 is still present among industrialists, social partners and academics. Industry 5.0 emerges as a new future step that will change the way we consider the role of humans in a process that tries to substitute the involvement of human operators in the sophisticated technology system. With such state of the art on the debate about the future of work and the future of industry, is there a place for a new concept? I don't think so. This new concept only refers to potential technological developments that do not represent a new manufacturing process. They are only a possible development of present trends.

1.4. Is there a new industrial revolution?

To answer this question, and considering all of the arguments of the different proposals for the concept of a new industrial revolution, I would answer, no. The definitions of the industrial revolution consider a radical change based on a specific technology or energy source.

> Industrial revolutions are the transformation from old practices of powering and managing of "workplace" into new and sophisticated structures that meet the goals of modern development in order to serve better the needs of the society [GRO 21, p. 467].

Mathias and Davis raised the discussion in a more correct fashion in 1989, when they stated that:

> We must specify criteria and meaning if we are to locate the phenomenon in time and context. It is not until a term or concept can be identified, with criteria assessed, that it is possible to begin to test the reality of historical change in the light of the concept. It is both a conceptual and an operational problem: one must distinguish between the definitional and conceptual identification and the narrative of empirical change deployed to describe and embody (and also to test) the assumption contained in the concept.

They continue by saying that "conceptual identity is thus intimately involved in a potentially infinite enquiry into the facts of change and the inter-relationships of growth" [MAT 89, p. 2].

1.4.1. *What is an Industrial Revolution?*

These historians are referring to the First Industrial Revolution, or Industry 1.0, when they say that "the two main criteria which are central to the definition of the industrial revolution (or the onset of the process of industrialization as a sustained trend in history) are first, higher rates of growth of the economy as a whole and second (closely linked to the first), structural change)" [MAT 89, p. 3]. These criteria can be empirically verified in all of the industrial transitions that we are referring to.

Steam power enabled the powering of the first automatic machines which increased the production of basic needs, such as textiles, food products and metal products. Moreover, these machines allowed for a quick increase in productivity and economic growth and meant that manufacturing systems in different sectors could be available. It implied a multiplication of factories and an increase in the number of jobs. The employment structure changed quickly, conditioning social change and the emergence of social conflicts. The distribution of welfare was not balanced and it quickly became understood that novel technological developments implied the emergence of new tensions in society and economy.

From the First Industrial Revolution of the late 18th century to the novel Second Industrial Revolution characterized by the increased sophistication of new machinery based on electric energy source, it was needed for more than a century. Fifty years went by with an immense increase of productivity due to the application of engineering management principles of Frederick Taylor in the early 20th century, and few years later, the application of the revolutionary assembly lines proposed by Henry Ford, and later adopted by all industries. But this increased capacity was possible with the new electric machineries.

A third industrial revolution was possible with the introduction of numerical control (NC) machine tools, invented during World War II and applied a few years later. It became a standard of a radical change in industry since the 1950s. It even enabled the development of robotics for manufacturing operations. From the 1970s, industry was quickly introducing flexible manufacturing systems (FMS) that connected the different machines with digital systems through microprocessors. In parallel, the Third Industrial Revolution also included the emergence of computer-aided design (CAD) and manufacturing (CAM) equipment, and its integration with CAD/CAM systems. Automation even could be increased into the so-called "computer integrated manufacturing" (CIM) systems, which represent a further advanced stage of automation [HIR 16].

1.4.2. *The debate on sociology of work*

The debate among labor experts and social scientists went through different stages too. Until the 1950s, most studies concerned working conditions under increased automation, especially in the automotive industry. These conditions were usually related to physical dimensions through the interaction with machinery, the speed of operations and the potential danger of accidents due to repetitive tasks with automatic machinery. By that time, studies were considering the psychological effects of mental stress derived of these interactions. Increased attention toward manufacturing operations was demanded by the firms, and in some cases, the need for further control was also a source for mental strain. Other studies considered the social impact of isolation of working positions and the problems induced by the lack of direct communication [KRU 09; LU 20; CIM 20; CAN 23a].

In the 1970s, new social science studies were pointing to the potential loss of qualification derived of increased automated systems, especially those enhanced by digitalized control processes. The direct connection between operator and machine was becoming loose and just possible through numerical control programming tasks. The discussion was again on the potential unemployment or deskilling provoked by computerized numerical control (CNC) systems, and later by robotics as well [KRI 21].

By the end of the 20th century, some visions on the "unmanned factory" became possible. Technological capacities were increasing and the possibility to develop manufacturing without workers became a vision for some commercial advocates and for some industrialists [BUT 18]. Soon it became clear that whenever more complex technology is used, the need to get humans involved in the decision-making process is higher. "Unexpected" events regularly happen when technology density and complexity are high. Minor connectivity problems can quickly cause major breakdowns and problems at production lines or on the whole shop floor.

The lower acceptance of the "unmanned factory" vision became clear not only at the union and workers councils' positions but also among managers and employers. It became clear that those breakdowns, even for short time, or during setting up phases, had high costs and could be avoided maintaining human competences available at the shop floor. Other organizational concepts could be possible. New management concepts argued and demonstrated that the investment on "human capital" could be an answer for such problems and limitations of the advancements of production technology.

1.4.3. *The emergence of the concept of Work 4.0: toward i5.0?*

But the debate over the concept of i5.0 has its roots in the anti-determinist discussion and in the possibilities of designing automated systems with anthropocentric purposes, in other words, allowing human operators to participate in the decision process. Just like with the discussion over i4.0, the topic regained its fundaments.

In Germany, a dialogue process the Federal Ministry of Labor and Social Affairs launched in April 2015 with the publication of a Green Paper. On that occasion, the concept of "Work 4.0" became a necessary extension of the debate concerning the digitalization of the economy, or i4.0. In this respect, the German Ministry of Labor and Social Affairs "invited associations, trade unions and businesses to submit responses, held numerous specialized workshops and events, commissioned academic studies, and obtained a picture of public opinion, including by engaging in dialogue directly with members of the public at local level" [BMA 16, p. 9]. As the German former Minister Nahles stated:

> Work 4.0 stands for the changes taking place in the whole of the working world and their implications for society. Rather than describing the normal status quo today, Work 4.0 is about future prospects, scenarios and opportunities to shape developments – to shape work in a way which benefits people and advances our economy [BMA 16, p. 6].

Some of the questions are still to be answered to today:

> The relationship between humans and machines is also changing: what will the computers and robots of the future be able to do, and what human capabilities are irreplaceable? Who will give instructions to whom in future? What requirements do companies have in terms of flexibility? How can we use flexibility in working time and location to develop new solutions for a work-life balance and achieve a fairer distribution of paid and unpaid work between men and women, while at the same time avoiding overwork caused by a breakdown of the boundaries on work? [BMA 16, p. 41–42].

In fact, these questions, and the grounded debate that support scientifically different answers were the base to starting process to establish programs and fix a concept around the item i5.0.

The discussion of impacts of increased developments of technology, and specifically on i4.0, have raised puzzling outcomes in terms of future trends on employment structure, labor market and skill needs. If we go back to the German Ministry of Labor, they mention that:

> The question of whether far-reaching automation processes could result in employment and wage polarization is also being hotly debated. Employment polarization would occur if middle-skilled employment were hit particularly hard by job losses while the employment of low- and high-skilled individuals simultaneously increased. This shift in demand would also lead to a polarization of wages. To date, however, there is no evidence of this in Germany, or of any collapse in middle-skilled employment [BMA 16, p. 53].

This was based on the assumptions of Dustmann et al. [DUS 09] and Antonczyk et al. [ANT 10], among others.

The Work 4.0 study continues to conclude that the forecasts about the future of work emphasize that action must be taken to avoid a scenario of employment and wage polarization in future. This should be considered, particularly in the areas of safeguarding employment, securing incomes and skills development. However, it is easily understood that these areas can also be considered as strategic policies for an economic development focused more on the welfare and not just on liberal competition.

There are, nevertheless, areas of tension that are recognized by the document produced by the German Ministry. They can be identified as follows: a) the possibilities of upskilling and deskilling with the new range of tasks made possible with the new technologies; b) the growing importance of experience at the workplace; and c) the need to weigh the possibilities for individual support and behavioral monitoring against each other. The tension identified in a) is basically a nondeterministic perspective. This means that a new technology can be designed or used is such a way that it promotes upskilling of their users. But it can also deskill them. The focus is then on the way technology is used, in other words, on organizational factors. The tensions identified in b) and c) are mostly related to social factors: job experience and interrelations at the workplace. There are organizations where the job experience is valued, but others do not. Similarly, in some, there are measures to support and train workers continuously due to the characteristics of new technologies, whereas in others, workers are left alone. Thus, for the same technology, it is always possible to find different organizational and managerial solutions.

This pivotal document from the German BMAS even clarifies that the irony of automation is evident when it recognized that:

> Problems which, while increasingly rare, continue to occur. And yet it is precisely when something goes wrong that people are needed who know from experience how to deal with problems in autonomous systems [BMA 16, p. 71].

This becomes the ground justification for the need of a radical change on the way we can understand the possibilities of technological changes and evolution for productive activities. And this change is the basis for an i5.0 concept.

The basic structure of the relation between the main elements of Work 4.0 is presented in Figure 1.1. They are one of the main vectors of the i5.0 concept.

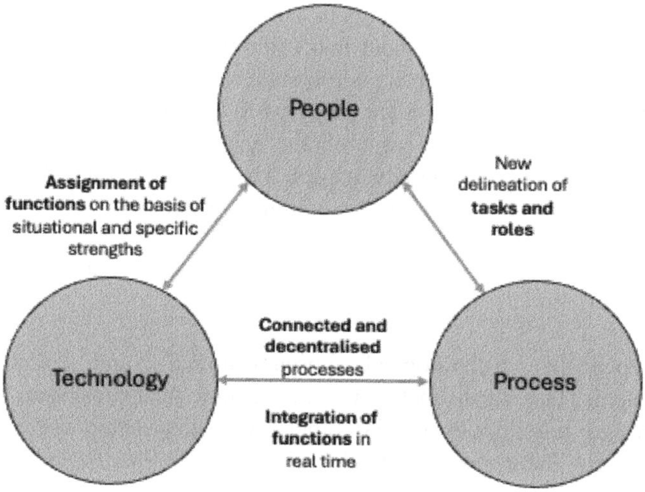

Figure 1.1. *Transformation of the socio-technical system as a consequence of digitalizationSource: Adapted from [BMA 16, p. 69] and based on [HIR 16]*

Here, the relation between workers and the production process is defined through the delineation of tasks and attributed roles. The design of tasks must be different from the conventional ones based on Tayloristic principles of segmentation, routine and individualization [BUT 18]. Under the principles of Work 4.0, roles can be attributed with better democratic principles, or, at least, to enable an increased participation in the decision process. In parallel, the tasks have to be designed to fit in new skill needs, in other words, to enable operators to enhance their qualifications and cognitive competences. That means also more complex tasks, and increased relation between the working places at the shopfloor, or office spaces.

1.4.4. *Socio-technical systems of i4.0*

The relation between the dimensions "process" and "technology" should enable the integration of functions in real time and the connection of decentralized processes. This means the possibility to access to cyber-physical systems (CPS) and Internet of Things (IoT), as well as cloud computing. Those are the key technologies of i4.0, but here the advanced stage is the possibility to use them to support group working in distant spaces at the same time (or real time). This implies also the possibility to decentralize the decision-making process [ELH 21].

Finally, the people–technology relation is related also to the human–machine interaction principles where the assignment of function is done based on situational strengths and on specific needs of humans considering the machine capacities. At the same time, we have to understand that "automation is an existential mid-term threat to the livelihood of the most vulnerable workers. Capability is advancing more quickly than most realize or are prepared to accept" [GRE 18, p. 202]. This means that these technological capabilities have also to consider the social dimension and its limitations and advantages. The change trends are not linear and fast as some studies try to present it to the common reader.

These socio-technical systems can be understood as production units, which consists of:

> Interdependent technological, organizational and personnel subsystems. This means that, while the technological subsystem limits the design possibilities of the other two subsystems, these have independent labor-psychological, labor-policy and organizational features which in turn react with the functioning of the technological subsystem. This concept avoids asking only about the functioning of the processes of change in individual technical and non-technical elements, but instead it puts the overall interaction and combination of elements – hence the techno-social – into the focus of the analysis [HIR 16, p. 8].

Its digitalization changes the features of technology, but the conditions are maintained.

This author continues by providing an example close to i5.0 experiments:

> A complementary automation concept posits the development of a distribution of tasks between humans and machines which should make possible a satisfactory functional capability of the total system. This design requires a holistic or collaborative perspective on the human-machine interaction which identifies the specific strengths and weaknesses of both human labor and technical automation. For the development of work in this conception, a technological framework is established that can serve in different ways.

The German sociologist Hirsch-Kreisen even underlines that "the complementary conception opens design possibilities for work that minimizes, for example, the awareness and feedback problems of acting on complex installations, making possible informal and manipulative action and ongoing learning processes, and thereby permitting a sufficient regulative capacity of the total system" (both citations can be found at [HIR 16, p. 9]).

Such new steps represent a structural change in the way products are manufactured and delivered. The developments from i4.0 also implied a change in the whole value chain. In parallel with a greater integration of the suppliers into the manufacturing process, the production chain became possible at a global level. This happened at the same time that i4.0 policies [CAN 24] intended to concentrate the competences around the main producers, the main equipment manufacturers and the main industrial regions.

1.5. Limits for an outcome

Even if we contest the recent reference to “Industry 6.0” as scientifically meaningful, the debate has two dimensions: on the one hand, it tries to diminish the importance of the concept of i5.0, and on the other hand, it tries to systematize future possibilities of industry development based on technology.

On the first dimension, it is curious to acknowledge that it surges almost at the same time as the discussion on i5.0. As explained above, we are still in the process of developing i4.0 and some debates have started applying emergent technologies in a new manufacturing context. Besides the capacities that cyber-physical systems (CPS) and the IoT can bring to the level of integration of information and Big Data management, the recent developments of AI, namely, the AGI, have just started with some experiments in the production process.

Besides that, nanorobotics and micro-processing systems [MON 22b] will, in the future, play a more significant role in product manufacturing, once most components become smaller and smaller. Similarly, additive manufacturing is going beyond already the capacities of 3D printing. But these technologies are still in the same context of structural change of i4.0.

What is new is the debate that reelaborates from the discussion on anthropocentrism of integrated manufacturing, and more recently, around the concept of Work 4.0 [BMA 16; BUT 18; PFE 16]. From there, the concept of i5.0 started to gain some new form, associating the need for increased participation processes in the workplaces, toward sustainable production processes. These two elements (job design and sustainability) create a new and emergent ground for references and analysis for adequate technology design.

That is why the proposal of new concepts based on industrial revolution or structural changes, as in the case of the idea of i6.0, must be proposed carefully. The proposition must be grounded on scientific debate, on verifiable experiments, on empirically validated trends and not only on vague hypotheses. In this case, it sounds just for new marketing wording, which represents no real new content. Most arguments for an i6.0 are still those that have been discussed for i4.0. That is why we should continue the debate on the late developments of industrialization and its social and economic conditions. This means that most industry organizations (in a wide definition) are still discussing how they can deal with i4.0. And the future of a structural change on the system of production should be done according to the principles defined for i5.0.

The proposal for an i6.0 presents nothing new from improved i4.0 from a technological dimension (hyperconnectivity, additive manufacturing, micro- and nano-manufacturing, quantum computing, etc.). Besides this, it includes few references on the discussion around i5.0 strategy, i.e. references on the social and economic conditions of human–machine interaction and the new design of workplaces. In this respect, this last concept (i5.0) is still the most radical and can bring the pillars of a structural change in industry to the forefront of a new way of thinking in the production process.

1.6. References

[ALM 23] ALMUSAED A., YITMEN I., ALMSSAD A., "Reviewing and integrating AEC practices into Industry 6.0: Strategies for smart and sustainable future-built environments", *Sustainability*, vol. 15, p. 13464, doi: 10.3390/su151813464, 2023.

[ANT 10] ANTONCZYK D., DELEIRE T., FITZENBERGER B., Polarization and rising wage inequality: Comparing the U.S. and Germany, IZA Discussion Paper No. 4842, 2010.

[BMA 16] BMAS – FEDERAL MINISTRY OF LABOUR AND SOCIAL AFFAIRS, Re-Imagine Work, White Paper on Work 4.0, BMAS, Berlin, available at: https://www.bmas.de/DE/Arbeit/Digitalisierung-der-Arbeitswelt/Arbeiten-vier-null/arbeiten-vier-null.html, 2016.

[BRE 21] BREQUE M., DE NUL L., PETRIDIS A., Industry 5.0: Towards a sustainable, human-centric and resilient European industry, Report, European Commission, Directorate-General for Research and Innovation, Brussels, 2021.

[BUT 18] BUTOLLO F., JÜRGENS U., KRZYWDZINSKI M., From lean production to industry 4.0: More autonomy for employees?, Discussion Paper SP III 2018–303, WZB Berlin Social Science Center, Berlin, 2018.

[CAN 23a] CANDEIAS M., Impact assessment of AI-enabled automation on the workplace and employment. The case of Portugal, Caparica, FCT Nova, PhD Thesis, Universidade Nova de Lisboa, Lisbon, available at: http://hdl.handle.net/10362/156952, 2023.

[CAN 23b] CANDEIAS M., "Pensar a Indústria 5.0 a partir da 4.0 – onde estamos e para onde podemos ir", in MACHADO C., DAVIM P. (eds), *Indústria 5.0 – Pessoas, Tecnologia e Sustentabilidade*, Almedina/Actual, Lisbon, 2023.

[CAN 24] CANDEIAS M., MONIZ A.B., "Public policies for Industry 4.0: Some lessons from the Portuguese case", *International Journal of Automotive Technology and Management*, vol. 24, no. 2, pp. 144–168, doi: 10.1504/IJATM.2024.141520, 2024.

[CHA 23] CHAUDHARI J., "Industry 6.0: The rise of intelligent manufacturing and the future of industry", *Illumination*, April 6, available at: https://medium.com/illumination/industry-6-0-the-rise-of-intelligent-manufacturing-and-the-future-of-industry-edb5f77f1706, 2023.

[CHO 22] CHOURASIA S., TYAGI A., PANDEY S.M. et al., "Sustainability of industry 6.0 in global perspective: Benefits and challenges", *MAPAN Journal of Metrology Society of India*, vol. 37, no. 2, pp. 443–452, doi: 10.1007/s12647-022-00541-w, 2022.

[CIM 20] CIMINI C., PIROLA F., PINTO R. et al., "A human-in-the-loop manufacturing control architecture for the next generation of production systems", *Journal of Manufacturing Systems*, vol. 54, pp. 258–271, 2020.

[DAM 24] DAMAŠEVIČIUS R., MISRA S., "The rise of industry 6.0: Seizing the opportunities of the post-COVID-19 era for sustainable manufacturing", *Emerging Technologies and Security in Cloud Computing*, p. 17, doi: 10.4018/979-8-3693-2081-5.ch020, available at: https://www.igi-global.com/book/emerging-technologies-security-cloud-computing/331694, 2024.

[DUG 22] DUGGAL A.S., MALIK P.K., GEHLOT A. et al., "A sequential roadmap to industry 6.0: Exploring future manufacturing trends", *IET Communications*, vol. 16, pp. 521–531, doi: 10.1049/cmu2.12284, 2022.

[DUS 09] DUSTMANN C., LUNDSTECK J., SCHÖNBERG U., "Revisiting the German wage structure", *The Quarterly Journal of Economics*, no. 124/2, pp. 843–881, 2009.

[ELH 21] EL-HAOUZI H.B., VALETTE E., KRINGS B.J. et al., "Social dimensions in CPS & IoT based automated production systems", *Societies*, vol. 11, no. 3, doi: 10.3390/soc11030098, 2021.

[GRE 18] GREENFIELD A., *Radical Technologies: The Design of Everyday Life*, Verso, London, 2018.

[GRO 21] GROUMPOS P., "A critical historical and scientific overview of all industrial revolutions", *IFAC PapersOnLine*, vol. 54–13, pp. 464–471, doi: 10.1016/j.ifacol.2021.10.492, 2021.

[HIR 16] HIRSCH-KREINSEN H., "Digitization of industrial work: Development paths and prospects", *Journal of Labour Market Research*, vol. 49, pp. 1–14, doi: 10.1007/s12651-016-0200-6, 2016.

[KRI 21] KRINGS B.J., MONIZ A.B., FREY P., "Technology as enabler of the automation of work? Current societal challenges for a future perspective of work", *Revista Brasileira de Sociologia*, vol. 9, pp. 206–229, available at: https://rbs.sbsociologia.com.br/index.php/rbs/article/view/rbs.806, 2021.

[KRU 09] KRÜGER J., LIEN T., VERL A., "Cooperation of human and machines in assembly lines", *CIRP Annals*, vol. 58, pp. 628–646, 2009.

[KUO 21] KUOSMANEN P., VILLMAN T. (eds), From industry X to industry 6.0: Antifragile manufacturing for people, planet, and profit with passion, Business Finland AIF White Paper No. 5/2021, 2021.

[LU 20] LU Y., XU X., WANG L., "Smart manufacturing process and system automation: A critical review of the standards and envisioned scenarios", *Journal of Manufacturing Systems*, vol. 56, pp. 312–325, 2020.

[MAD 15] MADAKAM S., RAMASWAMY R., TRIPATHI S., "Internet of Things (IoT): A literature review", *Journal of Computer and Communications*, vol. 3, pp. 164–173, 2015.

[MAT 89] MATHIAS, P., DAVIS, J.A. (eds), *The First Industrial Revolutions*, Basil Blackwell, Oxford, available at: https://archive.org/details/firstindustrialr0000unse/page/n7/mode/1up, 1989.

[MON 22a] MONIZ A.B., CANDEIAS M., BOAVIDA N., "Changes in productivity and labour relations: Artificial intelligence in the automotive sector in Portugal", *International Journal of Automotive Technology and Management*, vol. 22, no. 2, pp. 222–244, 2022.

[MON 22b] MONIZ A.B., KRINGS, B.-J., ""Manufacturing life" in real work processes? New manufacturing environments with micro- and nanorobotics", *NanoEthics*, vol. 16, no. 1, pp. 1–17, doi: 10.1007/s11569-021-00406-7, 2022.

[MON 23] MONIZ A.B., ""Indústria 5.0" como conceito de um novo tipo de interação humano-máquina?", in MACHADO C., DAVIM P. (eds), *Indústria 5.0 – Pessoas, Tecnologia e Sustentabilidade*, Almedina/Actual, Lisbon, 2023.

[NEU 21] NEUMANN W., WINKELHAUS S., GROSSE E. et al., "Industry 4.0 and the human factor: A systems framework and analysis methodology for successful development", *International Journal of Production Economics*, vol. 233, p. 107922, doi: 10.1016/j.ijpe.2020.107992, 2021.

[PFE 16] PFEIFFER S., "Robots, Industry 4.0 and humans, or why assembly work is more than routine work", *Societies*, vol. 6, no. 2, p. 16, 2016.

[ROM 16] ROMERO D., BERNUS P., NORAN O. et al., "The operator 4.0: Human cyber-physical systems and adaptive automation towards human-automation symbiosis work systems", *IFIP Advances in Information and Communication Technology*, pp. 677–686, doi: 10.1007/978-3-319-51133-780, 2016.

[SOW 16] SOWE S.K., SIMMON E., ZETTSU K. et al. "Cyber-physical-human systems: Putting people in the loop", *IT Professional*, vol. 18, pp. 10–13, 2016.

[THU 19] THUN S., KAMSVÅG P.F., KLØVE B. et al., "Industry 4.0: Whose revolution? The digitalization of manufacturing work processes", *Nordic Journal of Working Life Studies*, vol. 9, no. e117777, doi: 10.18291/njwls.v9i4.117777, 2019.

2

Industry 6.0 Transformation: Conceptual Transition Framework, Opportunities and a Research Agenda

Industry 6.0 (I6.0) is a futuristic industrial revolution (IR) that surpasses the previous five IRs. It promises tangible benefits, opportunities and more advanced digital transformations than Industry 4.0 and Industry 5.0. I6.0, as the future concept, is forecasted not only to automate industries, organizations, enterprises or firms, but it will transform the complete industries by concatenating advanced technologies, fostering smarter industrial decision-making processes, higher production and productivity and elevating the customization levels. Given the need for such IR, this study chiefly ascertained the driving factors to support the I6.0 implementation, the potential I6.0 benefits of successful implementation and potential challenges that might hinder I6.0 implementation, and ultimately developed the I6.0 conceptual transition framework. The document review approach was deployed. The findings revealed that both developed and developing countries need a high readiness level and awareness level regarding I6.0. The potential I6.0-leveraged technologies include nanotechnologies, artificial intelligence, cyber security, machine learning, ambient intelligence, humanized computing, quantum computing, cloud computing, computer simulation, big data analytics, 6G, Internet of Things (IoT), Industrial IoT, autonomous robotics, humanized robotics, sentimental artificial intelligence, socio-cognitive smart ambient, augmented and virtual reality, digital twins, blockchain technology, additive manufacturing, cyber-physical systems, advanced sensors, vertical and horizontal industrial system's integration, advanced drone technologies and machine-to-machine communication.

2.1. Background

It has not been long since the fourth industrial revolution (IR) concepts emerged globally [PAC 19], [NZU 24a]. The fourth IR is commonly referred to as Industry 4.0. This was after the accomplishment of the first three IRs: first IR (Industry 1.0),

Chapter written by Ismail W.R. TAIFA.

second IR (Industry 2.0) and third IR (Industry 3.0) [GÖK 17], [NZU 24]. Researchers and practitioners predicted that by 2030, the world would have used Industry 5.0 (the fifth IR) [ANN 21]. The study by [ÇIN 21] predicted that the majority of I4.0-related technologies would have been fully adopted by 2030 to allow the next IR.

Literature underpinnings indicate that the first IR occurred during the 18th century and involved inventions such as steam engine technology, mechanical weaving in England and the rise of large-scale manufacturing industries, including coal mining, iron and textiles [TAI 20], [ROJ 17]. The second IR began in the 19th century and involved major inventions such as electricity discovery, mass production and assembly lines for smoothing automobile manufacturing [TAI 21], [MAG 24]. The third IR occurred during the 20th century, and programmable logic controllers, electronic accessories and computers were invented [NZU 24]. In fact, Industry 1.0 is commonly known as the mechanical production revolution as it involved stem-based machines, second IR (mass production as it involved electrical energy-based production), and third IR (the digital age because it involved computer and Internet-based knowledge). So, the third IR contributed to advanced manufacturing technology as several computer-based technologies were invented during that revolution. Its chief aim was to automate existing processes, and the incorporation of robotic systems began to offer some potential for precision, effectiveness and efficiency. The fourth IR is mostly reported at the end of the 20th century, although the major transformations were witnessed during the 21st century. The fourth IR is mostly referred to as Industry 4.0 (I4.0).

I4.0 is the neologism to which its concepts on rapid technological advancement were first mentioned in Germany in 2011, and it aimed at engaging cyber-physical systems or the digitalization components where intelligent computers are shaping various systems to transform services, supply, transportation and manufacturing processes [NZU 24]. The fourth IR builds on the third's IR inventions (digital revolution), which is reportedly from the 1950s to the early 2000s. Internet, computers and electronics were all important in assisting the success of the fourth IR. So, the fourth IR advanced such technologies beyond the previous IRs. With a summary from many studies, including those by [TAI 20], [VEI 20], [ÇIN 21], [MAN 21], [STA 21] and [MAG 23a, 23b], at least six foundation categories of disruptive technologies can generally be applied during the fourth IR. First, *computational power, data and connectivity*: the common I4.0 technologies include cloud computing, cognitive computing, quantum computing, blockchain technology and smart sensors. Second, *intelligence and data analytics* involve technologies such as big data analytics (BDA), Internet of Things (IoT), Industrial IoT (IIoT), sensors, artificial intelligence (AI) and machine learning (ML). Third, *human–machine intervention*: this involves technologies such as robotics and automation, virtual reality, mixed reality, autonomous guided vehicles and augmented reality (wearables). The fourth disruptive technology of I4.0 focuses on *advanced engineering*; this includes technologies such as additive manufacturing (i.e. 3D printing) and nanotechnology. The fifth technology category is *systems integration and simulation*: such a category comprises vertical and horizontal systems integration, as well as computer simulation. The sixth category is about *advancing renewable energy systems*: this category includes technologies on solar, wind, hydroelectric and wave, among others.

I4.0's theoretical underpinnings currently focus on case studies, systematic literature reviews, research papers and other studies. Examples of such studies include the I4.0 readiness assessment tool by Agca [AGC 17], a maturity model by Brooks et al. [BRO 15], the I4.0 readiness by Lichtblau et al. [LIC 15] and the maturity model for the fourth IR by Schumacher et al. [SCH 16]. There are also studies on the implementations of I4.0 [KÜS 17] and [KAR 17]. Other studies include those by Viharos et al. [VIH 17] and Gökalp et al. [GÖK 17]. Some studies are now exploring the fifth IR as it transcends I4.0. Examples of the I5.0 studies include [BAR 23] and [GHO 23]. So, there are scant studies on the future IR, Industry 6.0 (the sixth IR): the existing studies include those by [ANN 21], [CHO 22] and [ALM 23]. Even the available studies just predict potential benefits and techniques, and none tried to conceptualize the transition framework by incorporating the transition constructs for the I6.0.

Consequently, I6.0 is clearly a future IR. So, there is a research gap in exploring the potential to be filled upon its successful implementation in all sectors. Given the need for such an IR, this study chiefly aimed to explore I6.0 transformation. Achieving the I6.0 transformation involved ascertaining four research questions:

a) What are the driving factors to support the I6.0 implementation?

b) What are the potential I6.0 benefits of successful implementation?

c) What are potential challenges that might hinder I6.0 implementation?

d) What are the components for developing the I6.0 conceptual transition framework?

The remaining parts of this study are structured as follows. The theoretical underpinning of the definition of Industry 6.0, I1.0 to I6.0 inventions and the research gap are in section 2.2. The description of the methodology used is illustrated in section 2.3. The results of the opportunities under Industry 6.0 implementation (section 2.4.1), driving factors (section 2.4.2), potential I6.0 benefits (section 2.4.3), potential challenges (section 2.4.4), and the I6.0 conceptualized transition framework are depicted in section 2.5. Section 2.6 discusses the results while section 2.7 summarizes the concluding remarks, recommendations, limitations and future research opportunities under I6.0.

2.2. Theoretical background

This section explains the literature focusing on the meaning of Industry 6.0, the evolution of I1.0 to I6.0, impacted sectors by I4.0, the research gap that could benefit from I4.0 to I6.0, and potential factors that could assist in accomplishing implementation of I6.0 in the manufacturing industries.

2.2.1. *Definition of Industry 6.0*

Industry 6.0 (I6.0) transcends I4.0 and I5.0. According to [ANN 21], the European Commission launched the I5.0 approach as the fifth IR late. I5.0 complements the concepts and frameworks from I4.0; however, it highlights innovation and research as the transition drivers, mainly in a human-centric, resilient and sustainable industry. Contrary to I4.0, I5.0 increases the role of human factors in planning and engineering in the future [ANN 21]. The future I6.0 portrays digital transformation for all sectors: manufacturing, supply, service and transportation organizations, enterprises or firms globally. Like, how I4.0 has transformed several sectors and resulted in subunits globally, I6.0 can potentially transform similar sectors at advanced levels. For example, I4.0 resulted in quality transformation, thus resulting in neologisms Quality 4.0 (Q4.0) [MAG 23b]; logistics and supply chain management (SCM) resulted in Logistics 4.0 (L4.0) and SCM 4.0 [ÖZD 22]. Other similar neologisms as per various sectors include Textile 4.0 (TX4.0), Apparel 4.0 (AP4.0), Fashion 4.0 (F4.0) [NHE 24], Mining 4.0 (MNG4.0), Lean 4.0 (L4.0) [MAY 18], Construction 4.0 (C4.0) [OLA 22], Oil and Gas 4.0 (O&G4.0) [BEN 22], Environmental Management 4.0 (EM4.0), Healthcare 4.0 (HC4.0), Pharma 4.0 (Ph4.0) [HAR 20], Manufacturing 4.0 (Mfg4.0), Metrology 4.0 (MET4.0) [NZU 24], Meteorology 4.0 (MTG4.0), Water 4.0 (W4.0), Education 4.0 (E4.0) [ADN 19], Banking 4.0 (B4.0) [MEH 20], Total Quality Management 4.0 (TQM4.0) [NGU 23], Project Management 4.0 (PM4.0), University 4.0 (U4.0), among other several concepts. Such technologies integrate advanced technologies with the Internet to transform their operations and procedures, affecting productivity through digital transformation and ultimately resulting in higher profitability.

2.2.2. *I1.0 to I6.0 inventions*

Industry 5.0 involves the concept of executing collaboration between humans and machines. It can sometimes be referred to as the human-tech partnership. I5.0 complements I4.0 as several technologies are commonly implemented in the current IR. The human touch can achieve all of the benefits or opportunities under I4.0. There is the utilization of technologies such as AI, IoT, IIoT, BDA, autonomous robotics, cloud computing, etc., all assisted with advanced problem-solving skills and knowledge, creativity, innovation and setting all enabling infrastructure [BAR 23]. The collaboration between humans and machines, including robotics, enables machines to tackle dangerous procedures, operations and activities while humans or the workforce assist in sophisticated and creative tasks.

So, it is forecasted that deploying I5.0 can likely increase production, productivity and employee satisfaction and foster sustainability of various economic, social and environmental processes.

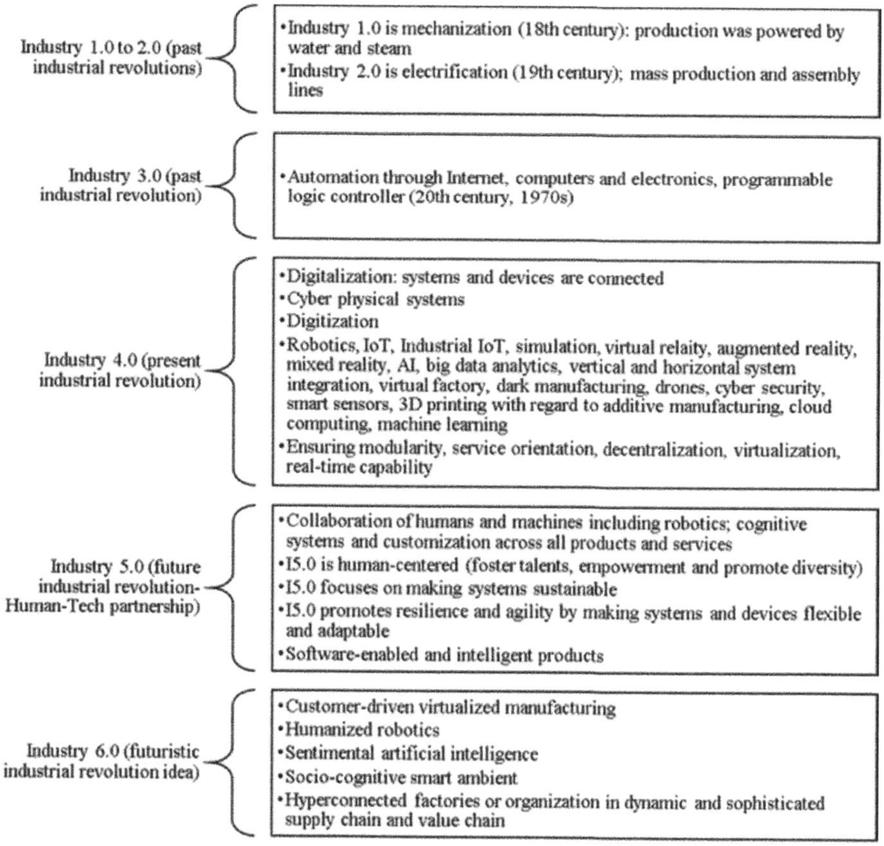

Figure 2.1. *Industry 1.0 to Industry 6.0*

Once I5.0 has been fully deployed, I6.0 (the sixth IR) will emerge. I6.0 is expected to augment the well-established infrastructures by I5.0. Advanced technologies, including nanotechnology and quantum computing, shall enable effective and efficient solutions for sophisticated problems in all sectors.

I6.0 has the potential to advance the current technologies. For example, under I4.0, there is additive manufacturing (mainly 3D printing), but under I6.0, it is expected that there shall be multi-dimensional printing technologies; I6.0 will assist in the invention of robot-medics, assistive home-robotics, among other several technologies currently under initial development phase [DUG 22]. Figure 2.1 depicts the evolution of Industry 1.0 to Industry 6.0.

2.2.3. *Research gap with regard to I6.0 inventions*

Some sectors have yet to fully capitalize on I4.0, while others are transitioning gradually to embrace I5.0. There are several studies focusing on the definitions of I4.0 [ROJ 17] and I5.0 [BAR 23], [GHO 23]. Some studies also highlight the benefits, opportunities and various frameworks on I4.0 and I5.0: examples of such studies include [AGC 17], [BRO 15], [LIC 15], [SCH 16], [KÜS 17], [KAR 17], [VIH 17] and [GÖK 17]. Given the future IR, there are scant studies on Industry 6.0 (the sixth IR); the existing studies include those by [CHO 22], [ANN 21] and [ALM 23]. So, there is a need to explore the potential to be filled upon its successful implementation in all sectors. The needed studies should explore the driving factors to support the I6.0 implementation, the potential I6.0 benefits of successful implementation, challenges that might hinder I6.0 implementation and the components for developing the I6.0 conceptual transition framework.

2.3. Methodology

I6.0 is the future IR; it is thus not easy to use any case study to obtain pertinent data to support it. Therefore, using I4.0 and some available results from I5.0, the implementation of I6.0 can be predicted because this IR augments the infrastructure laid by I4.0 and I5.0.

Thus, despite having qualitative, quantitative and mixed approaches to obtaining reliable data for I4.0 and I5.0, this study deployed the document review to obtain pertinent data. Researchers, including [BOW 09], support the idea that document review or analysis is a qualitative research approach as it cannot yield discrete results or numbers.

Applying document review as a method can lead to evidence-based guidelines, factors or variables, strategies and best practices [BRE 17]. Also, when relying on qualitative research, documents are reliable data sources [BOW 09]. So, the document review process comprises skimming, thoroughly reading, evaluating the content and executing document interpretation before concluding with the ultimate extracted content.

The document review for this study involved secondary data sources as opposed to other approaches that use primary data sources. All secondary data were those majority published from 2011 to 2024. Only secondary data on I1.0 to I6.0 in English were considered for extracting the needed content. The primary data sources include published journals, conference proceedings, books, book chapters, company reports, reputable magazines and websites, and institutional reports on particular sectors [TAH 21]. Primary data sources could have included questionnaires, interviews, focus group discussions, observations, case studies, surveys and experimentation [TAH 21]. Consequently, this study used secondary sources randomly accessed from published journals, conference proceedings, books, book chapters, company reports, reputable magazines and websites, and institutional reports on I6.0 potentials.

2.4. Results and discussion

Still, the concepts under I6.0 are at an early stage as hardly any study has presented case study-based findings on the companies, institutions, organizations, firms, industries or enterprises that have fully (or partially) begun the implementation. Therefore, based on the performed document review, the results are presented on the opportunities under Industry 6.0 implementation (section 2.4.1), driving factors (section 2.4.2), potential I6.0 benefits (section 2.4.3), potential challenges (section 2.4.4), and the I6.0 conceptualized transition framework (section 2.5).

2.4.1. *Opportunities under Industry 6.0 implementation*

As for I4.0 and I5.0, numerous opportunities can be accrued in the long run in implementing I6.0 technologies, frameworks, strategies, guidelines and roadmaps. The opportunities align with all procedures and operations for each company, institution, organization, firm, industry or enterprise that will deploy I6.0-related technology in the future.

As explained in section 2.2, the future I6.0 depicts digital transformation for all sectors: manufacturing, supply, service and transportation organizations, enterprises or firms globally. Like how I4.0 has transformed several sectors and resulted in sub-fields or sectors globally, I6.0 can potentially transform similar sectors at advanced technological levels. Therefore, I6.0 can potentially transform all operations and procedures for all sectors digitally. Such organizations are explained as follows:

a) *Manufacturing organizations* convert raw materials into finished products or goods. This means there is a change in the physical shapes and contents of the inputs to obtain output. I6.0 technologies can be deployed to transform the raw materials into finished products or goods. Advanced technologies can be implemented across the entire supply chain. Three categories of organization transform inputs in terms of form, shape and structure to obtain the final product: mass production organization, job shop production organization and batch manufacturing organization.

b) *Service organizations* change the state utility of customers, such as hospitals, prisons, academic institutions, hotels, insurance companies and telecommunication companies. It is possible for I6.0 to be applied to academic institutions, thus leading to neologisms such as University 6.0 and Education 6.0. Likewise, I6.0 can be deployed to a hospital, thus resulting in a neologism Hospital 6.0 (see Table 2.1).

c) *Supply organizations*: there is a change of ownership in the process, such as water supply organizations or companies, banks, shops and fuel stations. The deployment of I6.0 to banks can lead to a neologism of Banking 6.0, while water supply and sanitation can lead to Water 6.0 (see Table 2.1).

d) *Transportation organizations* involve changing the location of an item or processed entity; examples of such organizations include those reallocating entities or items via air, land, sea transportation, etc.

As seen from I4.0, such an emerging IR resulted in several neologisms for several organizations. So, many opportunities exist for the same organizations to be positively affected by embracing the I6.0-leveraged technologies.

I6.0 could impact various sectors, industries or research fields, as depicted in Table 2.1. Such technologies integrate advanced technologies with the Internet to transform their operations and procedures, thus affecting productivity, customer satisfaction, quality control and assurance through digital transformation and ultimately resulting in higher profitability [MAY 18], [ADN 19], [HAR 20], [MEH 20], [BEN 22], [OLA 22], [NGU 23] and [NZU 24].

2.4.2. *Factors to support the I6.0 implementation*

I6.0, as a future IR, requires pre-identifying potential factors that support successful implementation. Factors are clustered as the main and their corresponding sub-factors. Table 2.2 depicts factors that can support the successful implementation of I6.0 in the future.

Implementing new concepts or innovations has never been a simple process. Consequently, this requires examining the attributes of innovations or adapting new technology: relative advantage, compatibility, complexity, trialability and observability of the intended technology or innovation [SAH 06].

Research field	Possible I6.0 neologism term	Possible abbreviation	Description
Medical (healthcare), hospital	Healthcare 6.0 and Hospital 6.0	HC6.0 and H6.0	This involves the digitalization of hospital processes via the required smart network infrastructure. There is a need for advanced technologies and digital systems for the hospital and medical sector.
Pharmaceutical	Pharma 6.0	Ph6.0	Pharmaceutical operations are being carried out through the integration of advanced technologies. This involves the manufacturing of medicines and its utilities.
Garment, apparel and textile	Textile 6.0, Apparel 6.0 Fashion 6.0	TX6.0, AP6.0, F6.0, respectively	All textiles, apparel and fashion operations are controlled by advanced technologies such as IoT, virtual reality, augmented reality, robotics, cloud computing, big data analytics, etc.
Logistics and Supply chain management	Logistics 6.0 and Supply chain management 6.0	L6.0 and SCM6.0	This involves the integration of advanced leveraged technologies aiming at digital transformation of movements and storage of items in the supply chain. SCM 4.0 involves advanced technologies to cover coordination activities among all partners. The partners include suppliers, manufacturers, distributors, retailers and customers.
Quality management and engineering	Quality 6.0 or Total Quality Management 6.0	Q6.0 or TQM6.0	The four evolutions of quality can be digitally transformed and achieved: inspection, quality control, quality assurance and improvement (total quality management). Consequently, I6.0 can improve achieving quality-related product dimensions or characteristics such as reliability, durability, serviceability, conformance, performance, aesthetics and safety. Likewise, there can be an improvement in the quality-related *service* dimensions or characteristics, including reliability, responsiveness, tangibles, assurance and empathy.
Construction and building	Construction 6.0	C6.0	C6.0 borrows heavily from the technological advancements of Industry 6.0, integrating IoT devices, automation and data analytics into construction processes. This integration can enhance project efficiency and allow for real-time monitoring and decision-making.
Mining operations	Mining 6.0	MNG6.0	Integrating I6.0-related technologies makes mining operations digital, connected and sustainable. Such mining operations are both underground and surface operations.
Oil and gas	Oil & Gas 6.0	O&G6.0	Adopting technologies such as AI, IoT, advanced analytics, robotics and drones for oil and gas operations.
Water supply and sanitation	Water 6.0	W6.0	Adopting advanced technologies for either the water processing or water supply services. So, W6.0 embraces digitalization and automation to improve water resources' efficiency and management.
Metrology and Standardization	Metrology 6.0	MET6.0	M6.0 embraces the advancement of metrological activities leveraging IR4.0 technologies that involve digitalization and integration into CPSs.

Educational issues	Education 6.0 and University 6.0	E6.0 and U6.0, respectively	E6.0 connects with the sixth industrial revolution, which aims to transform education and universities via advanced technologies such as AI, robotics, etc. E6.0 is learner-centered, competency-based, innovative and adaptive. Multiple technologies characterize it, involving collaborative and networked relationships between numerous stakeholders.
Banking and finance	Banking 6.0	B6.0	Embraces digital technologies in managing banking processes and operations. Such technologies include cyber security technologies, big data analytics, cloud technologies, IoT, machine learning and AI.
Project management	Project management 6.0	PM6.0	Embraces advanced technologies from I6.0 to control and manage all operations. Ten knowledge areas are considered: project procurement management, project risk management, project communications management, project resource management, project cost management, project quality management, project time management, project scope management, project stakeholder management and project integration management.
Manufacturing, Lean manufacturing	Manufacturing 6.0 and Lean 6.0	Mfg6.0 and L6.0	Mfg6.0 involves the utilization of I6.0-related technologies borrowed from I4.0 to convert raw materials into finished products. L6.0 uses advanced technologies together with lean manufacturing principles and tools.
Meteorology	Meteorology 6.0	MTG6.0	Embracing advanced technology for atmospheric science focusing on weather forecasting. For example, the study can advance or be connected with remote sensing techniques and tools, including radar, Lidar and satellites, which can all be integrated with digitalized technologies.
Environmental management and engineering	Environment 6.0	EM6.0	Utilization of advanced technologies to assist environmental protection (atmosphere, land and waterbodies). This could mean environmental monitoring, auditing, controlling, waste management, life cycle assessment, environmental risk assessment and environmental toxicology, liquid waste management and air pollution, prevention and control, environmental management systems, environmental audit, environmental impact assessment, environmental information and environmental information systems, environmental accounting and environmental economics.
Transportation	Transportation 6.0	T6.0	All operations and procedures on reallocating any object from any sector embrace advanced technologies.
Service engineering	Service Engineering 6.0	SE6.0	The SE6.0 concept is based on how I6.0-related technologies can be used to digitalize a range of services from machinery and manufacturers. The service could be mechanical, electrical, buildings or any other operations.

Table 2.1. *Potential opportunities under I6.0*

S/n	Factor category	Sub-factors	Source
1	Leveraged technology	Nanotechnologies, artificial intelligence, cyber security, machine learning, ambient intelligence, humanized computing, quantum computing, cloud computing, computer simulation, big data analytics, 6G, IoT, Industrial IoT, autonomous robotics, humanized robotics, sentimental artificial intelligence, socio-cognitive smart ambient, augmented and virtual reality, digital twins, blockchain technology, additive manufacturing, cyber-physical systems, advanced sensors, vertical and horizontal industrial system's integration, advanced drone technologies, machine-to-machine communication.	[DAS 22], [ANN 21]
2	Employee-related factors	Willingness to adapt to new technologies, attend workshops and seminars to acquire necessary digital skills and knowledge, demonstrate good attitude and competence towards the I6.0 implementation, employee having good motivations and morale.	[MOR 23] [ANN 21]
3	Financial-related factors	Plan the funds for implementing I6.0, execute the financial plans appropriately and ensure funds are adequate to acquire sufficient resources (manpower, materials, equipment and tools).	[MAG 23a], [MAG 23c]
4	Partnership and collaboration	interdisciplinary and interdepartmental collaboration within the organization, collaboration with suppliers or customers through real-time data sharing, enabling collaborating IT-based systems.	[MAG 23c]
5	Leadership commitment and engagement	Active engagement and participation, the willingness of leaders to adopt I6.0 technology, awareness of I6.0, senior management members recognize the significance of I6.0 in achieving the company's objectives, having top management with competence to oversee the I6.0 implementation.	[NZU 24], [DUG 22]
6	Infrastructure	Installing and using information and communication technologies, reliable electricity in the industry, reliable Internet connectivity to support I6.0 implementation, available equipment and tools aligned with I6.0.	[PER 18], [AGC 17]
7	Convergence towards organization strategy	Setting I6.0 long-term vision, alignment with the global strategy, plan and execute implementation strategies, cross-functional department collaboration.	[MAG 24], [JAV 21]

Table 2.2. *Factors affecting I6.0 implementation. Source: Author's own creation*

Benefit category	Description of the specific potential benefits	Source
Operational and process benefits	– Extraordinary levels of automation and new systems of human-technology interaction – I6.0 is expected to reduce machine and equipment downtime because one of the advantages is the increase in reliability; efficiency, production and productivity will be improved. – Improve and simplify processes and operations for the majority of the companies – Lead in hyper-connected industries – Shorten the lead time – Cost reduction over the long period	[DAS 22], [ANN 21], [CHO 22]
Quality benefits	– Eliminate wastes in all organizations (defects, overproduction, inventory, waiting, unused talent or unused skills, transportation, motion and over-processing) – Enhance the quality of products and services (fitness for use, conformance to standards, meeting or exceeding customer requirements) – Simplify quality compliance because all the legal and regulations are digitalized, thus simplifying the automation process	[MOR 23], [ANN 21], [CHO 22]
Manufacturing benefits	– Enhance dynamic value chain and supply chain – Enhance the supply chain and value chain optimization – Improve data security (data for the customers and the company) – Improve customer involvement – Enhance skills and knowledge sharing – Simplify extended enterprise framework implementation (will simplify collaborations and partnerships) – Enhance product and production line optimization using I6.0, such as digital twins – Harmonized processes and data standards – Better resource optimization	[MAG 23a], [MAG 23c], [CHO 22]
Financial benefits	– Initial implementation costs are higher, but there is cost savings in the long run – Enhance higher revenues – Enhance the use of advanced financial management decision-making techniques – Higher financial security against cyber attacks	[MAG 23c], [CHO 22]

Management benefits	– Smart shop-floor management – Save cost for the maintenance – Improve the decision-making process – Promote human–machine interaction – Foster human virtual digital twin – Simplify the management process of organization – Lower the manpower for some activities or operations	[NZU 24], [DUG 22], [CHO 22]
Customer benefits	– Enhance customer personalization or customization – Enhance responsiveness and delivery of services and products – Cost savings in the long run	[PER 18], [AGC 17], [CHO 22]
Product and service design benefits	– Enhance creativity and innovation of services and products – Increase the protagonist features of machines and humans – Increase personalization or customization of products and services – Shorten the design of services and products – Simplify the complexity of designs, patterns or styles of preferred products and services	[MAG 24], [JAV 21], [CHO 22]
Technological benefits	– Current I4.0 and I5.0 technologies will be advanced, for example, 3D to 4D (printing with the need for stimulus; showing a function of time and showing intelligent behavior; fostering self-assembly and ensuring self-repair capabilities); advancing holography technology, advancing 4G and 5G to 6G; advancing robot-medics; advancing self-driving vehicles, aircraft, etc.; advancing cognitive technologies which are embedded with advanced intelligent systems	[KHA 23], [ANN 21], [DAS 22], [DUG 22], [CHO 22]

Table 2.3. *Potential benefits of I6.0. Source: Author's own creation*

2.4.3. *Potential I6.0 benefits*

Successful deployment of I6.0 could result in many benefits across several organizations. Table 2.3 shows the benefits that could be gained upon successful I6.0 implementation.

Generally, the I6.0 benefits through the implementation of its technologies, concepts, systems and processes, comprising obtaining improved effectiveness, efficiency, high production, heightened productivity, enhanced higher flexibility, superior quality assurance and control, and agility and ultimately increasing profitability of the organizations or company [CHO 22]. These benefits should be obtained concurrently by having good customer satisfaction, high customization and enhanced systems and data security. Depending on the nature of the organization (i.e. manufacturing, service, supply and transportation), some benefits are the same across those sectors, while others are squarely to a specific sector.

2.4.4. *Challenges that might hinder I6.0 implementation*

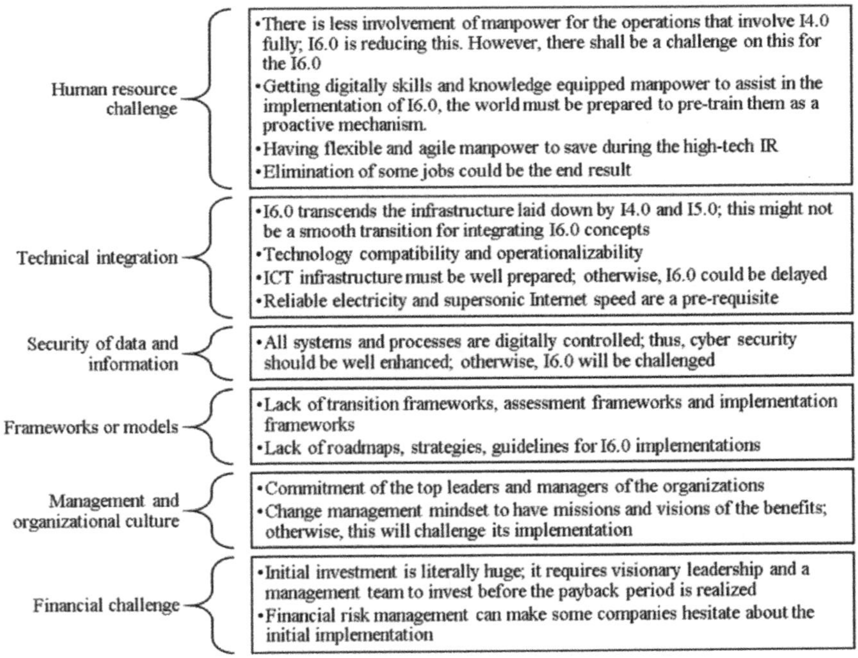

Figure 2.2. *Potential challenges for I6.0 implementations*

Despite the benefits and factors to support the deployment of I6.0, some challenges might hinder the prosperity of this IR. Figure 2.2 depicts the challenges that can hinder the successful implementation of I6.0.

2.5. I6.0 conceptual technological transition framework

2.5.1. *Meaning of the transition framework*

The general conceptual framework (CC) characterizes the conceptual state and relationships of the entities being studied, and it is a network or interconnected system or connection of assumptions, expectations and views [TAM 16], [MAG 22]. For a CC to be complete, it has to possess major defining process parameters of the studied task, problem or scenario; independent and dependent variables; and lastly, it has to demonstrate or show cause-and-effect relationships among the variables, parameters, factors or constructs. The CC explores the dynamics in organizing constructs by linking the study's concepts of branding practices, identifying gaps in the current understanding of the phenomenon or topic, and the research project's methodological foundations [ITA 21], [MAG 22]. Consequently, the CC can first assist in formulating the methodology that leads to data collection and, second, to develop various frameworks, models or roadmaps. The frameworks can be transition, implementation and assessment, among others. With regard to I6.0, the required frameworks are specifically technological transition framework (TTF) because I6.0 is a forecasted IR.

So, since I6.0 is the future IR, this study proposes the conceptual technological transition framework (CTTF). CTTF for I6.0 comprises IR theories about how advanced technological innovations happen, the driving forces, variables, constructs and factors on the matter and how they all bring tangible and intangible transformations. That being the case, TTF should incorporate all of the supply chain components: organizational users and practices, regulations, organizational or industrial networks (raw material supply, manufacturing or production and product or service distribution channels), I6.0 infrastructure and the advanced technologies required.

I6.0 could be implemented in the presence of TTF. The transition needed technological paths, which might differ depending on the nature of the sector, operations or user requirements. As per [GEE 07], five transition paths (TP) exist, and all of these paths are important to understand when developing the TTF:

i) Reproduction TP: this is an ongoing change at the management level.

ii) Transformation TP: this is mostly referred to as a socio-technical path that changes without the emerging dominating digital technology.

iii) Technological substitution TP: such a path occurs when the incumbent digital technology is swapped by radical creativity and innovation, thus leading to an advanced technological era.

iv) Re-configuration TP: such a TP occurs when many interconnected advanced technologies are swapped by a similarly interlinked alternative set of technologies.

v) De-alignment and re-alignment TP: this TP considers the current technology's weaknesses and aligns with the appropriate dominant framework.

With the five TPs, the required TTF should possess six features: transitions should be co-evolutionary and multi-dimensional; multi-actors from social, science and engineering must be incorporated; transition should happen at multiple levels; transition should be embedded with continuous improvement processes; transitions should be radicals; and change is not linear at all [GEE 08].

2.5.2. *Transition framework constructs*

So, considering the five TPs and six characteristics of the TTF (section 2.5.1), this study developed the CTTF by first analyzing constructs considered by other studies, mainly on I4.0. For example, many I4.0 and I5.0 have considered constructs in developing the TTF for various sectors. Examples of such studies include those by [OKU 01], [SCH 16], [HIZ 20], [LAK 20], [ANW 23] and [MAG 23c]. The critical analysis of the approaches is presented in Table 2.4. Approaches for the framework or model, description, suitability and the source of that models or frameworks are presented. The presented frameworks and models are based on I4.0: this is because it is the current IR with many studies that used primary data as opposed to I5.0 and I6.0.

Despite the people, process and technology (PPT) approach being employed in developing several frameworks and model's constructs (Table 2.4), the I6.0 transition necessitates the need for having a framework that involves a "data" construct. Likewise, the transition of the components from I4.0 and I5.0 should feature in the needed I6.0 CTTF. Consequently, the CTTF for I6.0 should comprise the people, data, process and technology (PDPT) approach and the integration of I4.0 and I5.0. The PDPT is advantageous in multiple scenarios, but mainly, there are stagewise transitions for any organization that could need the transitions. The generalizability could be achieved based on the primary constructs obtained and validated through strong statistical data analyses.

Approach for framework constructs	Description of the framework or model	Suitability	Source
– People, process and technology	– A transition framework from Quality 3.0 to Quality 4.0	– Suitable for all manufacturing industries	[MAG 23c]
– People, operation, technology, strategies, leadership, culture, government, customers and product	– A model developed to assist companies in the technology maturity and readiness level evaluation	– Multipurpose model which can establish the technology maturity and improvement opportunities	[SCH 16]
– Initiation, usability, decision and preparation	– A conceptualized readiness model	– For multiple utilizations in organization	[MAG 23]
– Inputs, process and outputs	– A framework developed to improve the surveillance and response to neglected tropical diseases	– Suitable for medical and safety industries	[NGE 21]
– Services, supervision, regulation, and company	– A conceptual framework was established to improve security company to establish standard operating procedures	– For the achievement of a particular goal	[ANW 23]
– Content, process, context and outcome	– An application framework was established to assist organizations in strategy implementation	– Strategic and hospitality management fields	[OKU 01]
– Technology, strategy, people, leadership and innovation	– The proposed dimensions from a systematic literature review of models that organizations should consider in their I4.0 readiness assessment	– Evaluating readiness towards a definite technology	[HIZ 20]
– People, process, technology and data	– Employed in the establishment of a model for assessing the readiness level toward IR4.0 adoption	– Assessing the technology readiness level	[LAK 20]
– Management, organization, strategy, innovation, environment, human or social issues and project-related issues	– Developed a conceptualized framework from an I4.0 perspective	– Industry 4.0 in the construction sector	[SUF 23]
– People, process and technology	– Transition, implementation and conceptual frameworks are the most popular components in establishing models	– Improvement strategies, acquiring new technology	[PRO 15], [SOJ 17], [KUM 21], [MAG 23c]
– People, process, technology, management, customer focus and innovation	– A proposed addition of three dimensions to people, process and technology in the improvement model	– Software improvements	[PRO 15]

Table 2.4. *Constructs for the transition frameworks or models*

Figure 2.3 shows the CTTF developed based on available theories and I4.0 to I6.0 available frameworks, models or concepts, and it acts as the basis of the future technological transition. Before transiting to another IR, the maturity model assessment can be measured using available models in Table 2.4 (constructs for the transition frameworks or models). Examples of such studies that can establish the maturity level or test the readiness and awareness level include the I4.0 readiness assessment tool by [AGC 17], a maturity model by [BRO 15], the I4.0 readiness by [LIC 15], the maturity model for the fourth IR by [SCH 16] and the maturity level for the manufacturing industries by [MAG 24]. The readiness and awareness level can be based on the various scales. The scale can be low (1) to high (5); very poor (1) to excellent (5); beginner (1), intermediate (2), experienced (3), expert (4), best score (5); and basic level (1) to full digitalized (5). Likewise, the process constructs can possess various stages to execute the operations under I6.0.

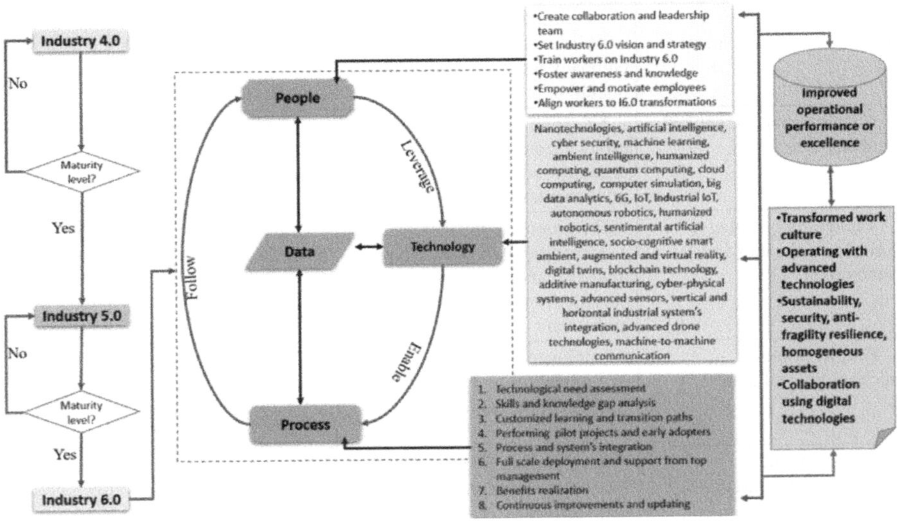

Figure 2.3. *Industry 6.0 conceptual technological transition framework. For a color version of this figure, see www.iste.co.uk/machado/industry.zip*

2.6. Discussion

The theoretical background from various case studies, mainly from I4.0 to I6.0, provides the potential of accruing benefits, seizing opportunities and ultimately getting higher productivity. Actors on deploying I6.0 should be able to understand and practice using disruptive technologies to their full potential and how to enable digital transformation for all enterprise sizes (micro, small, medium, large and major establishments) [ANN 21]. For example, to enable digital transformation at all company sizes, the actors or stakeholders must be aware that the transformation should affect all levels of their operations: employees at the shop level, machine, supply chain and customer interaction. Then, the companies should provide a mechanism for addressing various challenges or threats. Lastly, the companies must endure continuous improvement by fostering sustainability, security, flexibility and accuracy, resulting in high quality and productivity for the particular company [ANN 21].

This study has also highlighted several challenges that can hinder the successful implementation of I6.0 (Figure 2.2). These challenges are in line with those found by researchers including [KOP 19], [SCH 20], [ALC 21], [ANT 22], [MAG 23b, 23c], among others. Solutions can be based on developing implementable strategies that are in line with I6.0 or advanced technological strategies. Likewise, benefits and opportunities were found in this study. Those foster the need for implementing I6.0.

Thus, the results from this study possess highlights that could foster the success of I6.0. For example, the issue of having transition frameworks has also been acknowledged by previous researchers on other transition frameworks. Such researchers include [MAG 23c], who developed the Q4.0 transition framework; [NTA 23], who developed the assessment framework; [NZU 24] emphasized the need for the metrology transition framework; and [BLO 15], who emphasized the digital transformation of industry.

2.7. Conclusion and recommendations

2.7.1. *Conclusion*

Industry 6.0 promises huge opportunities for all sectors in the future. The digital transformation will enhance many benefits compared to the current IR (I4.0). The preparedness in terms of infrastructure, workforce, financial stability, technological advancements, organizational culture transformation, etc., should continue to be accomplished. I6.0 potentially forecasts organizational transformations fostered by the seamless integration of digital technologies in all organizational processes and operations. I6.0 will transcend I4.0 and I5.0 benefits. For example, I4.0 is characterized by the major principles of modularity, interoperability, virtualization, real-time capability, service orientation and decentralization. The potential I6.0-leveraged technologies include nanotechnologies, artificial intelligence, cybersecurity, machine learning, ambient intelligence, humanized computing, quantum computing, cloud computing, computer simulation, big data analytics, 6G, IoT, Industrial IoT, autonomous robotics, humanized robotics, sentimental artificial intelligence, socio-cognitive smart ambient, augmented and virtual reality, digital twins, blockchain technology, additive manufacturing, cyber-physical systems, advanced sensors, vertical and horizontal industrial system's integration, advanced drone technologies and machine-to-machine communication. Therefore, despite the concepts under I6.0 still being early, companies, institutions, organizations, firms, industries or enterprises should be prepared to adopt the proposed technologies to accrue benefits and take advantage of all research potentials.

2.7.2. *Recommendations*

Due to the need to implement I6.0 in the future, it is recommended that all main stakeholders set the pre-requirement conditions or infrastructure and other factors well, which can influence the I6.0 implementations. For example, the awareness level and readiness level should be high for them to embrace the successful I6.0 implementation, and there should be a well-validated transition I6.0 framework that can assist the digital transformation from 14.0 to I6.0. Likewise, companies must establish a holistic strategy and vision for I6.0 with the full support from all key stakeholders. Companies without vision cannot envisage the need to transform their operations and procedures to accrue I6.0 potential benefits and opportunities. The current educational syllabus should prepare future employees who are flexible in being trained to acquire advanced digital skills. The awareness of the futuristic IR should be embedded at the lower level to make future employees aware of the circumstances they will face. Companies or countries can establish digital transformation hubs. For example, nowadays, many countries or organizations and institutions have established innovation hubs for softer innovation and creativity. Likewise, companies and countries are encouraged to establish digital transformation hubs to impart necessary knowledge, skills and attitudes regarding the future IR roadmaps.

2.7.3. *Limitations of the study and future studies*

This study is based on the document review of various published contents, mainly on I4.0, I5.0 and I6.0. The findings require further investigation to determine the importance of I6.0 as a future IR. No available case studies are related to I6.0; only potential findings in terms of benefits and possible technologies are presented. Future studies should be based on exploring further factors influencing the implementation of I6.0, confirming those factors, and determining the awareness level and readiness level of implementation. The readiness level can assist in establishing the maturity level of several sectors, including manufacturing, service, supply and transportation. The potential studies should also propose strategies based on the primary data collection. Such strategies should be validated via various approaches, including the Delphi technique, focus group discussion and other pertinent approaches. There is also a need to develop transition, implementation and assessment frameworks. The I6.0 frameworks could guide various sectors in achieving digital transformation.

2.8. References

[ADN 19] ADNAN A.H.M., KARIM R.A., TAHIR M.H.M. et al., "Education 4.0 technologies, Industry 4.0 skills and the teaching of English in Malaysian Tertiary Education", *Arab World English Journal (AWEJ)*, vol. 10, no. 4, pp. 330–343, 2019.

[AGC 17] AGCA O., GIBSON J., GODSELL J. et al., An Industry 4 readiness assessment tool, available at: https://warwick.ac.uk/fac/sci/wmg/research/scip/reports/final_version_of_i4_report_for_use_on_websites.pdf. Warwick Manufacturing Group, Warwick, 2017.

[ALC 21] ALCÁCER V., RODRIGUES C., CARVALHO H. et al., "Tracking the maturity of Industry 4.0: The perspective of a real scenario", *International Journal of Advanced Manufacturing Technology*, vol. 116, nos. 7–8, pp. 2161–2181, 2021.

[ALM 23] ALMUSAED A., YITMEN I., ALMSSAD A., "Reviewing and integrating AEC practices into Industry 6.0: Strategies for smart and sustainable future-built environments", *Sustainability (Switzerland)*, vol. 15, no. 18, 2023.

[ANN 21] ANNANPERÄ E., JURMU M., KAIVO-OJA J. et al., From Industry X To Industry 6.0: Antifragile manufacturing for people, planet and profit with passion, Working Paper, AIF, Helsinki, pp. 1–38, 2021.

[ANT 22] ANTONY J., MCDERMOTT O., SONY M., "Quality 4.0 conceptualisation and theoretical understanding: A global exploratory qualitative study", *TQM Journal*, vol. 34, no. 5, pp. 1169–1188, 2022.

[ANW 23] ANWAR A.Q., FUAD A., FAIZAL A. et al., "Development of a conceptual framework of private maritime security company of Malaysia", *Australian Journal of Maritime and Ocean Affairs*, vol. 15, no. 1, pp. 12–24, 2023.

[BAR 23] BARATA J., KAYSER I., "Industry 5.0 – past, present, and near future", *Procedia Computer Science*, vol. 219, pp. 778–788, 2023.

[BEN 22] BENAYOUNE A., "Factors influencing Industry 4.0 implementation in oil and gas sector: Empirical study from a developing economy", *Academy of Strategic Management Journal*, vol. 21, no. S2, pp. 1–10, 2022.

[BLO 15] BLOCHING B., LEUTIGER P., OLTMANNS T. et al., *The Digital Transformation of Industry - How Important is it? Who are the Winners? What Must be Done?*, Roland Berger Strategy Consultants and BDI Federation of German Industries, Munich, 2015.

[BOW 09] BOWEN G.A., "Document analysis as a qualitative research method", *Qualitative Research Journal*, vol. 9, no. 2, pp. 27–40, 2009.

[BRE 17] BRETSCHNEIDER P., CIRILLI S., JONES T. et al., *Document Review as a Qualitative Research Data Collection Method for Teacher Research*, SAGE Publications Ltd, London, 2017.

[BRO 15] BROOKS P., EL-GAYAR O., SARNIKAR S., "A framework for developing a domain specific business intelligence maturity model: Application to healthcare", *International Journal of Information Management*, vol. 35, no. 3, pp. 337–345, 2015.

[CHO 22] CHOURASIA S., TYAGI A., PANDEY S.M. et al., "Sustainability of Industry 6.0 in global perspective: Benefits and challenges", *MAPAN – Journal of Metrology Society of India*, vol. 37, no. 2, pp. 443–452, 2022.

[ÇIN 21] ÇINAR Z.M., ZEESHAN Q., KORHAN O., "A framework for Industry 4.0 readiness and maturity of smart manufacturing enterprises: A case study", *Sustainability*, vol. 13, no. 12, pp. 1–32, 2021.

[DAS 22] DAS, S., PAN, T., "A strategic outline of Industry 6.0: Exploring the future", *SSRN Electronic Journal*, doi: 10.6084/m9.figshare.19738909.v1, 2022.

[DUG 22] DUGGAL A., MALIK P., GEHLOT A. et al., "A sequential roadmap to Industry 6.0: Exploring future manufacturing trends", *IET Communications*, vol. 16, no. 5, pp. 521–531, 2022.

[GEE 07] GEELS F.W., SCHOT J.W., "Typology of sociotechnical transiton pathways", *Research Policy*, vol. 36, no. 3, pp. 399–417, 2007.

[GEE 08] GEELS F.W., MONAGHAN A., EAMES M. et al., The feasibility of systems thinking in sustainable consumption and production policy: A report to the Department for Environment, Food and Rural Affairs, Report, DEFRA, London, 2008.

[GHO 23] GHOBAKHLOO, M., IRANMANESH, M., TSENG, M.L. et al., "Behind the definition of Industry 5.0: A systematic review of technologies, principles, components, and values", *Journal of Industrial and Production Engineering*, vol. 40, no. 6, pp. 432–447, 2023.

[GÖK 17] GÖKALP E., ŞENER U., EREN P.E., "Development of an assessment model for Industry 4.0: Industry 4.0-MM", in MAS A., MESQUIDA A., O'CONNOR R. et al. (eds), *Software Process Improvement and Capability Determination*, Springer, Cham, pp. 128–142, 2017.

[HAR 20] HARIRY R.E., BARENJI R.V., PARADKAR A., "From Industry 4.0 to Pharma 4.0", in HUSSAIN C.M., DI SIA P. (eds), *Handbook of Smart Materials, Technologies, and Devices*, Springer, Cham, pp. 1–22, 2020.

[HIZ 20] HIZAM-HANAFIAH M., SOOMRO M.A., ABDULLAH N.L., "Industry 4.0 readiness models: A systematic literature review of model dimensions", *Information (Switzerland)*, vol. 11, no. 7, pp. 1–13, 2020.

[ITA 21] ITAM U.J., SWETHA M., "Examining the structural relationship between employee branding, TQHRM and sustainable employability outcome in Indian organized retail", *The TQM Journal*, 34, 5–28, doi: 10.1108/TQM-03-2021-0093,2021.

[JAV 21] JAVAID M., HALEEM A., PRATAP SINGH R. et al., "Significance of Quality 4.0 towards comprehensive enhancement in manufacturing sector", *Sensors International*, vol. 2, pp. 1–13, 2021.

[KAR 17] KARAKÖSE M., YETIŞ H., KARAKÖSE M. et al., "A cyberphysical system based mass-customization approach with integration of Industry 4.0 and smart city", *Wireless Communications and Mobile Computing*, vol. 2017, pp. 1–9, 2017.

[KHA 23] KHAN M., HALEEM A., JAVAID M. et al., "Changes and improvements in Industry 5.0: A strategic approach to overcome the challenges of Industry 4.0", *Green Technologies and Sustainability*, vol. 1, no. 2, p. 100020, 2023.

[KOP 19] KOPP R., HIRSCH KREINSEN H., PREENEN P. et al., "Sociotechnical perspectives on digitalisation and Industry 4.0", *International Journal of Technology Transfer and Commercialisation*, vol. 16, no. 3, p. 290, 2019.

[KUM 21] KUMAR R.R., GANESH L.S., RAJENDRAN C., "Quality 4.0 – a review of and framework for quality management in the digital era", *International Journal of Quality & Reliability Management*, doi: 10.1108/TQM-11-2023-0361, 2021.

[KÜS 17] KÜSTERS D., PRAß N., GLOY Y.S., "Textile learning Factory 4.0 – Preparing Germany's textile industry for the digital future", *Procedia Manufacturing*, vol. 9, pp. 214–221, 2017.

[LAK 20] LAKMALI E., VIDANAGAMACHCHI K., NANAYAKKARA J., "Industry 4.0 readiness assessment for apparel industry: A study in the Sri Lankan context", *Proceedings - International Research Conference on Smart Computing and Systems Engineering*, SCSE, pp. 174–181, 2020.

[LIC 15] LICHTBLAU K., STICH V., BERTENRATH R. et al., *Industrie 4.0-Readiness*, Impuls-Stiftung des VDM, RWTH, Aachen-Köln, 2015.

[MAG 22] MAGANGA D.P., Quality 4.0 transition framework for selected manufacturing industries in Tanzania, Master's Dissertation, University of Dar es Salaam, Tanzania, 2022.

[MAG 23a] MAGANGA D.P., TAIFA I.W.R., "Quality 4.0 conceptualisation: An emerging quality management concept for manufacturing industries", *The TQM Journal*, vol. 34, no. 5, pp. 1169–1188, 2023.

[MAG 23b] MAGANGA D.P., TAIFA I.W.R., "The readiness of manufacturing industries to transit to Quality 4.0", *International Journal of Quality & Reliability Management*, vol. 40, no. 7, pp. 1729–1752, 2023.

[MAG 23c] MAGANGA D.P., TAIFA I.W.R., "Quality 4.0 transition framework for Tanzanian manufacturing industries", *The TQM Journal*, vol. 35, no. 6, pp. 1417–1448, 2023.

[MAG 23d] MAGALHÃES R.M., MELLO L.C.B.B., HIPPERT M.A.S., "Organizational readiness for building information modeling", *Frontiers in Engineering and Built Environment*, vol. 3, no. 2, pp. 137–152, 2023.

[MAG 24] MAGANGA D.P., TAIFA I.W.R., "Determining the maturity level for the selected manufacturing industries through Industry 4.0 perspectives", *International Journal of Management and Decision Making*, vol. 23, no. 4, pp. 480–504, 2024.

[MAN 21] MANSOUR, H., AMINUDIN E., MANSOUR T., "Implementing Industry 4.0 in the construction industry-strategic readiness perspective", *International Journal of Construction Management*, pp. 1–14, 2021.

[MAY 18] MAYR A., WEIGELT M., KÜHL A. et al., "Lean 4.0 – A conceptual conjunction of lean management and Industry 4.0", *Procedia CIRP*, vol. 72, pp. 622–628, 2018.

[MEH 20] MEHDIABADI A., TABATABEINASAB M., SPULBAR C., "Are we ready for the challenge of Banks 4.0? Designing a roadmap for banking systems in Industry 4.0", *International Journal of Financial Studies*, vol. 8, no. 2, pp. 1–29, 2020.

[MOR 23] MORAES A., CARVALHO A.M., SAMPAIO P., "Lean and Industry 4.0: A review of the relationship, its limitations, and the path ahead with Industry 5.0", *Machines*, vol. 11, no. 4, pp. 1–25, 2023.

[NGE 21] NG'ETICH A.K.S., VOYI K., MUTERO C.M., Development and validation of a framework to improve neglected tropical diseases surveillance and response at sub-national levels in Kenya, *PLoS Neglected Tropical Diseases*, 15(10), e0009920, doi: 10.1371/journal.pntd.0009920, 2021.

[NGU 23] NGUYEN, T.A.V., NGUYEN K.H., TUCEK D., "Total Quality Management 4.0 framework: Present and future", *Journal of Operations and Supply Chain Management*, vol. 16, no. 3, pp. 311–322, 2023.

[NHE 24] NHELEKWA L.B. , MOLLEL J.Z., TAIFA I.W.R., "Assessing the digitalisation level of the Tanzanian apparel industry: Industry 4.0 perspectives", *Research Journal of Textile and Apparel*, vol. 28, no. 2, pp. 185–205, 2024.

[NTA 23] NTAMULYANGO D.K., Industry 4.0 readiness assessment in beverage industries, Master's Dissertation, University of Dar es Salaam, Tanzania, 2023.

[NZU 24a] NZUMILE J.M., MAHABI V., TAIFA I.W.R., "Contribution of Industry 4.0 technologies in adopting Metrology 4.0 in manufacturing industries", in MACHADO C., DAVIM J.P. (eds), *Smart Engineering Management. Management and Industrial Engineering*, Springer, Cham, pp. 43–72, 2024.

[NZU 24b] NZUMILE, J.M., PASTORY D., TAIFA I.W.R., "Applicability of Metrology 4.0 technologies for the manufacturing industries", *MAPAN – Journal of Metrology Society of India*, vol. 39, pp. 559–571, 2024.

[OKU 01] OKUMUS F., "Towards a strategy implementation framework", *International Journal of Contemporary Hospitality Management*, vol. 13, no. 7, pp. 327–338, 2001.

[OLA 22] OLATUNDE N.A., GENTO A.M., OKORIE V.N. et al., "Construction 4.0 technologies in a developing economy: Awareness, adoption readiness and challenges", *Frontiers in Engineering and Built Environment*, vol. 3, no. 2, pp. 108–121, 2022.

[ÖZD 22] ÖZDAĞOĞLU A., BAHAR S., *Logistics 4.0 and Smart Supply Chain Management*, CRC Press, Boca Raton, 2022.

[PAC 19] PACCHINI A.P.T., LUCATO W.C., FACCHINI F. et al., "The degree of readiness for the implementation of Industry 4.0", *Computers in Industry*, vol. 113, p. 103125, 2019.

[PER 18] PERALES D.P., VALERO F.A., GARCÍA A.B., "Industry 4.0: A classification scheme", in VILES E., ORMAZÁBAL M., LLEÓ A. (eds), *Closing the Gap Between Practice and Research in Industrial Engineering*, Springer, Cham, pp. 343–350, 2018.

[PRO 15] PRODAN M., PRODAN A., PURCAREA A.A.A. et al., "New contributions in information systems and technologies", *Advances in Intelligent Systems and Computing*, vol. 353, pp. 481–490, 2015.

[ROJ 17] ROJKO A., "Industry 4.0 concept: Background and overview", *International Journal of Interactive Mobile Technologies*, vol. 11, no. 5, pp. 77–90, 2017.

[SAH 06] SAHIN I., "Detailed review of Roger's Diffusion of innovations theory and educational technology", *The Turkish Online Journal of Educational Technology*, vol. 5, no. 2, pp. 14–23, 2006.

[SCH 16] SCHUMACHER A., EROL S., SIHN W. et al., "A maturity model for assessing Industry 4.0 readiness and maturity of manufacturing enterprises", *Procedia CIRP*, vol. 52, pp. 161–166, 2016.

[SCH 20] SCHUH G., ANDERL R., GAUSEMEIER J. et al. (eds), *Industrie 4.0 Maturity Index: Managing the Digital Transformation of Companies*. National Academy of Science and Engineering, Munich, 2020.

[SOJ 17] SOJA E., SOJA P., "Exploring root problems in enterprise system adoption from an employee age perspective: A people–process–technology framework", *Information Systems Management*, vol. 34, no. 4, pp. 333–346, 2017.

[STA 21] STAWIARSKA E., SZWAJCA D., MATUSEK M. et al., "Diagnosis of the maturity level of implementing Industry 4.0 solutions in selected functional areas of management of automotive companies in poland", *Sustainability*, vol. 13, no. 9, pp. 1–38, 2021.

[SUF 23] SUFERI N.S.M., RAHMAN M.M., YUSOF I.M., "Industry 4.0 in construction: A conceptual framework", *AIP Conference Proceedings*, vol. 2643, American Institute of Physics Inc., Maryland, 2023.

[TAH 21] TAHERDOOST H., "Data collection methods and tools for research; a step-by-step guide to choose data collection technique for academic and business research projects", *International Journal of Academic Research in Management (IJARM)*, vol. 2021, no. 1, pp. 10–38, 2021.

[TAI 20] TAIFA I.W.R., HAYES S.G., STALKER I.D., "Computer modelling and simulation of an equitable order distribution in manufacturing through the Industry 4.0 framework", in *2nd International Conference on Electrical, Communication and Computer Engineering (ICECCE)*, Istanbul, pp. 1–6, 2020.

[TAI 21] TAIFA I.W.R., HAYES S.G., STALKER I.D., "Towards a digital revolution in the UK apparel manufacturing: An Industry 4.0 perspective", in BARTOLO P., SILVA F., JARADAT S. et al. (eds), *Industry 4.0 – Shaping the Future of the Digital World, 2nd International Conference on Sustainable Smart Manufacturing (S2M 2019)*, CRC Press, Taylor & Francis Group, London, pp. 3–8, 2021.

[TAM 16] TAMENE E.H., "Theorizing conceptual framework", *Asian Journal of Educational Research*, vol. 4, no. 2, pp. 50–56, 2016.

[VEI 20] VEILE J.W., KIEL D., MÜLLER J.M. et al., "Lessons learned from Industry 4.0 implementation in the German manufacturing industry", *Journal of Manufacturing Technology Management*, vol. 31, no. 5, pp. 977–997, 2020.

[VIH 17] VIHAROS Z.J., SOÓS S., NICK G.A. et al., "Non-comparative, Industry 4.0 readiness evaluation for manufacturing enterprises", in *15th IMEKO TC10 Workshop on Technical Diagnostics Technical Diagnostics in Cyber-Physical Era*, Budapest, pp. 181–187, 2017.

3

Impact of Industry 6.0 on Human Cognitive Behavior

The transformative impact of Industry 6.0 on society presents multiple risks that should be analyzed in detail before integrating new technologies specific to a new industry revolution. While Industry 6.0 promises unprecedented advancements by combining artificial intelligence (AI) and machine learning (ML), it raises significant concerns regarding human intellectual degradation, job loss and diminished communication skills. The reliance on AI for critical decision-making processes can lead to a decline in human cognitive abilities, as individuals may become overly dependent on technology for tasks that once required analytical thinking and creativity.

Additionally, this chapter discusses the potential erosion of interpersonal communication as digital tools replace face-to-face interactions. A new concept associated with AI is introduced, namely "intellectual perishability", where essential decision-making skills diminish due to reliance on AI. This chapter presents three case studies regarding intellectual perishability demonstrated through students gradually losing their ability to write due to Large Language Models (LLM), a study focusing on the inability to search for information based on descriptions, and the reduction in the number of jobs due to the takeover of tasks by AI. As we advance toward a future shaped by AI, these concerns become increasingly pertinent, necessitating ongoing discourse on navigating the complexities of Industry 6.0.

3.1. Introduction

The Industrial Revolution marked the shift from agriculture-based economies to ones dominated by large-scale industries, with each revolution bringing significant technological, cultural and socioeconomic changes.

Chapter written by Cosmina-Mihaela ROSCA and Adrian STANCU.

The First Industrial Revolution (Industry 1.0) introduced steam and water power, the Second Industrial Revolution (Industry 2.0) brought electrical technology, the Third Industrial Revolution (Industry 3.0) focused on automation through IT and electronics and the Fourth Industrial Revolution (Industry 4.0) introduced computerization of manufacturing. The Fifth Industrial Revolution, or Industry 5.0, emphasizes personalization and the human element in manufacturing [GRO 21, HUM 21, MAT 22, SCH 16, ZIA 24].

Researchers and scholars do not generally agree on the beginning and end years of each Industrial Revolution (Table 3.1).

Industry	Period	Source
1.0	1740–1840	[GRO 21]
	1750–1850	[MOH 21]
	1760–1840	[SCH 16]
	1760–1850	[DUG 22; MOH 23]
	1784–1870	[ASL 20; LEN 22]
2.0	1840–1950	[GRO 21]
	1850–1910	[MOH 23]
	1850–1950	[MOH 21]
	1870–1914	[SCH 16]
	1870–1969	[ASL 20; LEN 22]
	1880–1973	[DUG 22]
3.0	1910–1980	[MOH 23]
	1950–1980	[GRO 21]
	1950–2000	[MOH 21]
	1960–2000	[SCH 16]
	1969–2011	[ASL 20; LEN 22]
	1970–2000	[ARU 20]

4.0	1980–2010	[MOH 23]
	2000–2010	[GRO 21]
	2000–currently underway (2024)	[ARU 20; CHA 21; MOH 21]
	2010–2021	[ING 24]
	2011–currently underway (2024)	[ASL 20; DUG 22]
	2011–2017	[LEN 22]
	2013–currently underway (2024)	[XU 18]
5.0	2010–2020	[GRO 21]
	2021–currently underway (2024)	[ING 24]
	2021–currently underway (2024)	[EUR 21a]
6.0	2020–currently underway (2024)	[GRO 21]

Table 3.1. *Period of each Industrial Revolution*

Furthermore, Chien et al. state the need for an intermediate phase between Industry 3.0 and 4.0, named Industry 3.5, especially for emerging countries to facilitate the shift from Industry 3.0 to 4.0. The authors [CHI 17] design a conceptual framework for Industry 3.5 based on five main concepts: manufacturing, factory, supply chain, resource management and decision.

Considering the period of the Industrial Revolution described in Table 3.1, Figure 3.1 shows the authors' viewpoint of the period of each Industrial Revolution.

Given the current year (2024), it is essential to clearly define the years' boundaries for Industry 4.0 and 5.0, as stated in Figure 3.1. For Industry 4.0, the beginning year of 2000 was chosen since the computerization of manufacturing as a foundation for Industry 4.0, mentioned by the German government strategy in 2011 [XU 21], had been implemented worldwide during this time [CHA 95, CHA 02].

The year 2021 was selected as the end year of Industry 4.0 and the beginning of Industry 5.0 because, in the same year, the Directorate-General for Research and Innovation of the European Commission issued a policy brief regarding Industry 5.0, in which the pillars of Industry 5.0 (sustainability, resilience and a human-centric approach) were mentioned for the first time [EUR 21a]. Further documents were published by the European Commission which detail the three pillars of Industry 5.0 [EUR 21b, EUR 24].

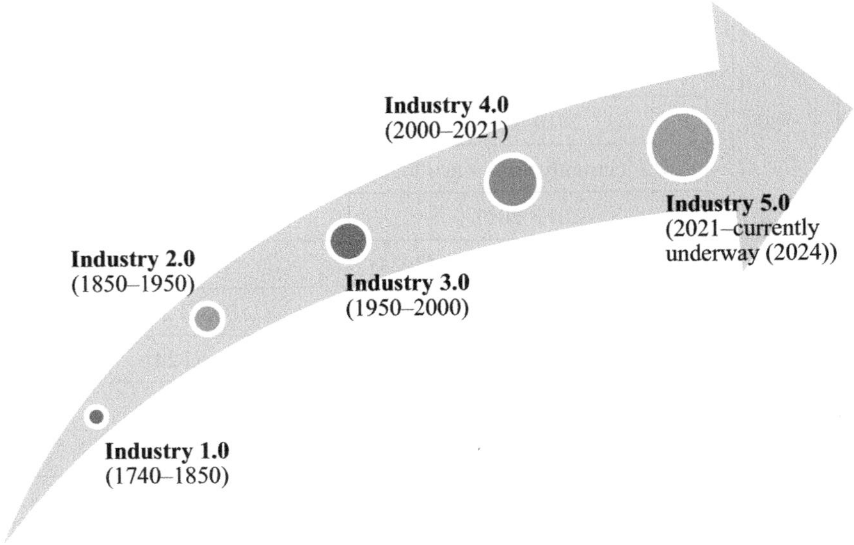

Figure 3.1. *Timeline and periods of Industrial Revolutions from the authors' perspective. For a color version of this figure, see www.iste.co.uk/machado/industry.zip*

The following describes the main related works from the literature focusing on Industry 4.0 and 5.0. Zalozhnev and Ginz [ZAL 23] explore the concepts and technologies behind Industry 4.0 and 5.0. They highlight that Industry 4.0 is built on technological advancements from previous industrial stages, serving as the platform for Industry 5.0. A key focus is on human–computer interaction, which acts as a bridge between these two industrial paradigms. Industry 4.0 revolutionized industries through automation and advanced technologies but neglected human intellectual and cognitive skills, increasing unemployment and environmental degradation [IQB 22].

Industry 5.0 emphasizes the collaboration between humans and robots. The automation and digitization focus of Industry 4.0, whereas Industry 5.0 shifts towards a more human-centric approach [MUL 24, SLA 23, YAN 24a]. The paper by Yang et al. [YAN 24b] reviews current research and technologies related to human-machine interaction, categorizing them into four parts: sensor and hardware, data processing, transmission mechanisms, and interaction and collaboration.

The paper by Padovano et al. [PAD 24] clarifies the concept of "human centricity" by analyzing public opinions from LinkedIn posts over the past five years using text-mining techniques like semantic clustering and keyword extraction. The findings reveal that discussions around human-centricity often focus on future worker skills (both hard and soft) and social issues such as gender equity and workplace comfort, highlighting their importance in the transition to Industry 5.0.

The transition from traditional, hard-wired manufacturing equipment to advanced digital technologies has been driven by the rise of the Internet of Things (IoT) [ROS 24]. Industry 4.0 represents a significant leap in manufacturing automation, where machines communicate and collaborate, leading to intelligent and customized production [MÖL 22].

The manufacturing sector is undergoing two simultaneous revolutions: Industry 4.0, which emphasizes automation and machine autonomy, and Industry 5.0, which reintegrates humans into the production process and values their contribution beyond mere productivity. Companies must adapt to these changes to stay competitive [ZIN 24].

The paper by Garrido et al. [GAR 24] identifies that most literature still focuses on Industry 4.0 rather than Industry 5.0. This indicates a gap and an opportunity for deeper exploration of human-centric approaches, sustainability and organizational resilience in operations management and supply chain research.

Industry 5.0 aims to create an inclusive, intelligent and sustainable production process that leverages human creativity alongside enhanced automation and machine intelligence. A key technology in this new paradigm is collaborative robotics ("cobotics"), which extends human capabilities and integrates robots as team members [ZAF 24].

The paper by Golovianko et al. [GOL 23] proposes a hybrid model combining the efficiency of Industry 4.0 with the sustainability of Industry 5.0. Digital cognitive clones, which replicate human decision-making behavior, are suggested as a critical enabling technology for this hybrid model, fostering the convergence of digital and human worlds and enhancing resilience.

This study presented by Liu et al. [LIU 24] uses literature reviews and expert opinions to identify the key factors affecting the implementation of intelligent manufacturing technology (associated with Industry 5.0) in social, environmental and economic sustainability. The study concludes that cost and funding are the most significant challenges. At the same time, improved social benefits and public services are the most critical enablers, providing valuable insights for decision makers in the industry.

As Industry 5.0 continues to evolve, AI is opening new opportunities for enhanced functionalities and features in manufacturing. However, the technology-driven Industry 4.0 paradigm still faces challenges, particularly in implementing Industrial AI (IndAI), which often suffers from design and configuration issues [LEN 24; ROS 23]. The article by Leng et al. [LEN 24] classifies IndAI by intelligence levels and identifies three key opportunities – collaborative intelligence, self-learning intelligence and crowd intelligence – for advancing toward Industry 5.0.

As industries evolve with digitization, computerization, robotization and automation, there is a growing focus on human needs, particularly environmental sustainability. This article addresses industrial waste management within the context of Industry 5.0, oriented towards sustainability and environmental protection. It discusses how Industry 5.0 tools can help reduce waste levels in industrial companies, trends in industrial waste management in the European Union (EU), and the potential of Industry 5.0 to contribute to environmental sustainability by reducing waste [MES 23].

The Industry 5.0 paradigm causes tension between intelligent and sustainable elements, often due to design priorities that favor technological features like collaborative robots and IoT devices at the expense of environmental sustainability. Bonello et al. investigate the ecological impacts of design decisions in a human–robot collaboration (HRC) workstation, explicitly focusing on safety features and their associated energy consumption and carbon footprint. By applying Design for Environmental Sustainability (DfES) principles, the study by Bonello et al. [BON 24] achieved a 10% reduction in energy consumption and carbon footprint, suggesting a pathway for balancing human-centric and environmentally sustainable design in HRC workstations.

Barata and Kayser [BAR 23] present a tertiary study of 32 literature reviews on Industry 5.0, supported by a bibliometric analysis from the Scopus database. The survey identifies three stages of Industry 5.0, which has been researched since 2018, with a recent focus on deploying circular manufacturing strategies supported by human-friendly digitalization. The findings highlight the future-oriented and cross-sectoral nature of Industry 5.0, which diverges from the original configuration of Industry 4.0.

Valette et al. [VAL 23] conducted a systematic literature review to explore how human aspects have been addressed in the pillars of Industry 4.0, such as Cyber-Physical Systems and IoT. Their review aims to understand the progress towards viewing industrial systems as complex socio-technical entities, integrating enabling technologies for Industry 5.0 and systemic concepts.

The concepts of Industry 5.0 and Society 5.0 are gaining attention. Industry 5.0 focuses on a sustainable, human-centric and resilient European sector, and Society 5.0 aims to balance economic advancement with social problem-solving in Japan [HUA 22]. This paper compares and explores the co-evolution of these two paradigms, addressing foundational arguments and offering insights for future research. The study emphasizes the interaction between industrial and societal revolutions and the potential for these concepts to complement and accelerate each other's development.

Coelho et al. [COE 23] suggest that Industry 5.0 is an evolution of Industry 4.0, aiming to create a more just and sustainable society through a collaborative relationship between humans and machines. Their study highlights the EU's strategy and vision for Industry 5.0 but notes that a unifying idea is still needed to fully characterize it as an industrial revolution.

Industry 6.0 promises a revolution in the interaction between technology and human cognitive behavior, surpassing previous paradigms by integrating advanced AI, neurotechnologies and autonomous systems. This new industrial phase will automate processes and enhance human cognitive abilities, influencing how we think, make decisions and interact with our digital and physical environments. Its impact will transform education, work and daily life, raising essential questions about ethics, identity and the limits of human potential in the technological era.

Industry 6.0 aims to address global challenges such as climate change and social inequalities, promoting circular economies and fostering continuous innovation while ensuring a positive impact on society and the environment. It envisions a customer-centric approach, enhancing the manufacturing and medical sectors with anti-fragile systems that can adapt and thrive amidst challenges. Implementation will take 10 to 15 years, focusing on interconnected industries with efficient supply chains and advanced technologies like quantum computing to tackle complex algorithms, ultimately making human labor more effective and manageable [CHO 22].

3.2. Industry 5.0 versus 6.0

3.2.1. *Main differences between Industry 5.0 and 6.0*

In the research by Dai et al. [DAI 24], the elements of the Industry 5.0 are identified as follows:

– Robots are defined as mechanical devices performing automated tasks using AI, either under human supervision or autonomously. Advanced robotics are integral to Industry 5.0, combining with human workers to enhance productivity and safety.

– Artificial intelligence (AI) and Big Data: computers simulate intelligent behavior with minimal human input, enabling devices to make autonomous decisions. Big Data is massive datasets analyzed for insights, essential for real-time analytics and personalized production in Industry 5.0. AI and Big Data enhance efficiency, productivity and decision-making, although they pose challenges like job displacement.

– Smart machines are capable of independent thought and decision-making, often using AI and ML. They improve connectivity and flexibility in industrial operations.

– Cloud computing supports virtual environments, enhances security and improves business performance, which is critical for Industry 5.0.

– Cobotics and cyber-physical systems: collaborative robots (cobots) that work alongside humans integrate physical processes with computational systems, essential for Industry 5.0's smart manufacturing. Cobots are central to human-robot collaboration and necessary for the customized and safe work environments of Industry 5.0.

– Augmented reality (AR), virtual reality (VR) and metaverse: technologies that overlay digital information onto the real world or create entirely virtual environments. The metaverse is defined as a virtual universe combining AR and VR, offering new ways to interact and work in cyberspace.

– 6G networks and beyond: 6G is the next-generation cellular technology, offering higher frequencies, capacity and low latency. It will revolutionize connectivity, enabling advanced applications in Industry 5.0, such as real-time data transfer and AI-powered mobile edge computing.

– Blockchain is a distributed ledger technology that provides secure, transparent and immutable records of transactions. It enhances data security, transparency and decentralized operations in Industry 5.0, supporting smart contracts and secure peer-to-peer interactions.

The fifth phase of modern technological evolution, characterized by a focus on collaboration between humans and machines, is represented by the integration of personalized solutions and cooperative robots that enable workers to engage in value-added tasks that directly benefit customers. This latest development extends beyond manufacturing processes, encompassing enhanced resilience, a human-centered approach and a strong emphasis on sustainability, all explored in greater detail below [SIN 24].

Industry 6.0 will introduce new concepts such as [CHO 22]:

– product personalization;

– automatic robotic industry controlled by quantum radar;

– customer driven;

– virtualized;

– anti-fragile manufacturing;

– digital twins;

– dynamic supply chain;

– dynamic inter- and intra-connectivity across organizations;

– system integration;

– service dominant logic;

– green sustainability.

Industry 6.0 will revolutionize the industry by excluding humans from production while focusing on their desires and needs. While in Industry 5.0, humans collaborate with machines to meet the individual needs of consumers, Industry 6.0 integrates AI for a futuristic environment, where decision-making is carried out ethically and fairly through AI. The personalization of products at the center of Industry 5.0 is maintained with greater flexibility in Industry 6.0.

3.2.2. *Particular differences between Industry 5.0 and 6.0*

This section analyzes the approaches within Industry 5.0 compared to those in Industry 6.0 for a particular case: buying clothes. This examination will highlight the distinctions in methodologies and technologies employed in both industrial paradigms.

3.2.2.1. *Scenario for Industry 5.0*

Imagine a clothing store that offers the option to create a customized T-shirt. The steps for this process are (Figure 3.2) the following:

– *Human* interaction: the customer goes to the store or a website and selects a basic t-shirt.

– *Options*: the customer can choose:

- the *color* of the T-shirt (e.g. orange, yellow, red, etc.);

- the *design*: the customer can add a logo, personalized text or an image;

- the *size*: choose the desired size (S, M, L or XL).

– *Active* feedback: the customer can request modifications to the design while working with a consultant or on the online platform.

– *Result*: the T-shirt is manufactured according to the specifications precisely as requested by the customer.

In Industry 5.0, the consumer's personal preferences influence the decision, which involves direct and active interaction with the consumer.

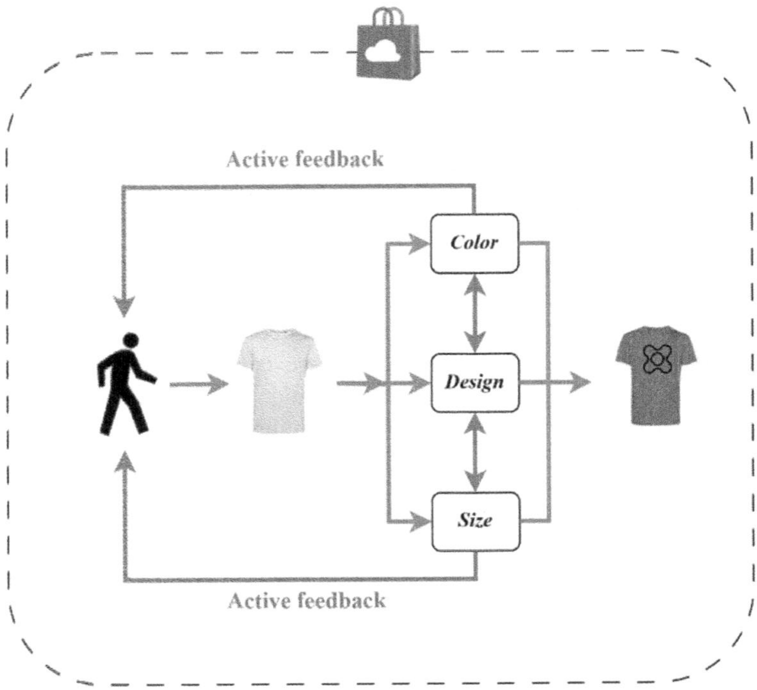

Figure 3.2. *T-shirt purchase scenario in Industry 5.0*

3.2.2.2. *Scenario for Industry 6.0*

The clothing brand will use AI to personalize T-shirts. The process includes the following steps (Figure 3.3):

– *Automated questionnaire*: the customer fills out an online questionnaire about their lifestyle, preferred colors, daily activities and size.

– *Data analysis*: the algorithm analyzes the responses and past data (e.g. what the customer has previously purchased) to understand what the customer likes.

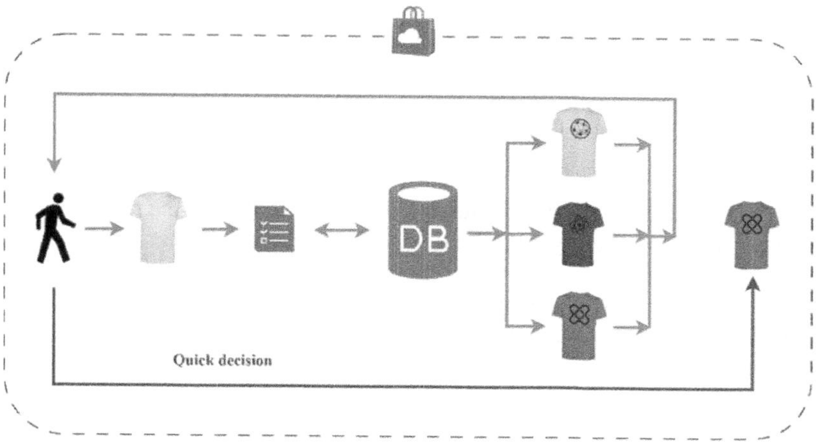

Figure 3.3. *T-shirt purchase scenario in Industry 6.0*

– *Automated recommendations*: AI automatically generates a set of personalized T-shirts based on identified preferences, such as:

- a yellow T-shirt with a modern graphic design;
- an option with a personalized message based on the customer's interests.

– *Quick decision*: the customer chooses from the generated recommendations without interacting directly with a consultant or worrying about design.

The process is automated and data-driven, with less constant direct interaction. AI handles personalization based on collected data, and customers receive options tailored to their preferences.

Industry 5.0 is centralized on active interaction with the consumer, who directly influences the product's appearance. The consumer makes conscious, personalized choices with the help of a consultant or an interactive platform. For Industry 6.0, the algorithms and AI decide what products to offer the consumer based on their data and preferences, thus reducing active interaction.

This new paradigm raises five significant concerns:

– Decision-making is entirely handled by AI systems based on customized algorithms designed by human operators. This raises the possibility of overlooking specific scenarios, resulting in erroneous decisions. One of the fundamental principles of AI – that the final decision should rest with humans – may be compromised in this context.

– Industry 6.0 presents a remarkable decreasing reliance on direct interaction, coupled with the convenience of automated systems, gradually eroding essential communication skills. Addressing this issue requires a deliberate effort to prioritize and cultivate interpersonal communication within educational and professional settings, ensuring that the human element remains integral in an increasingly automated world.

– The convenience offered by Industry 6.0, which continues the trend of product personalization from Industry 5.0, along with the various enhancements introduced into people's lives by this revolution, may lead to a decline in human skill levels. Over time, there is a risk of dependency on AI and an inability to program devices independently. A direct consequence may be that individuals increasingly rely on AI to write programs, which could be particularly dangerous for decision-making systems.

– The risk of intellectual perishability increases as rapid technological advancements may render specific skills obsolete, challenging the workforce to adapt continuously.

– The rise of automation and AI in Industry 6.0 could lead to the disappearance of traditional jobs, necessitating a focus on reskilling and upskilling to ensure workers remain relevant in a changing job landscape.

The further sections of this chapter detail some of these concerns because, beyond the attraction of comfort offered by Industry 6.0, society must be aware of the risks it exposes itself to.

3.3. Risks of intellectual degradation in the era of Industry 6.0

As we move toward Industry 6.0, AI will be used in all tasks specific to our lives, from industrial processes to everyday decisions. This technological revolution brings risks, particularly concerning human intellectual degradation and the inability to make informed decisions.

One of the most concerning aspects of AI is the risk of cognitive skill degradation. As intelligent systems become increasingly capable of performing complex tasks, people may become less inclined to use their critical thinking and problem-solving skills. This excessive trust in AI can decrease our ability to analyze situations, make informed decisions and develop innovative solutions.

As AI usage and dependency grow, it may limit human cognitive abilities, leading to a reduction in independent thinking and a shift toward algorithmic thinking. This dependency could degrade professional skills, increase stress and diminish human autonomy, leading to laziness and a lack of fulfillment. The education sector is also at risk, with AI potentially replacing human effort in tasks, leading to a reliance on technology that could result in future challenges. Ahmad et al. [AHM 23] suggest that AI offers many benefits but poses significant risks to human autonomy, skills and well-being.

For example, in workplaces where AI is used to analyze data and provide solutions, employees may become more passive, relying on algorithmic recommendations instead of using their own judgment. This passivity reduces individual critical thinking ability and affects team collaboration.

Another concrete example is the introduction of LLMs such as ChatGPT, Gemini and Copilot. These models have provided a revolutionary way of searching. Still, they have also introduced two significant problems:

– People no longer know how or have the patience to search the Internet for accurate, verified and, thus, reliable information.

– People do not know when the model fabricates information or states facts, as they no longer have the patience to verify the information.

In this way, future generations will face challenges regarding their ability to think critically.

To demonstrate this point, the authors compiled data in Table 3.2, sourced from one of the project-based courses within the Computer Science specialization, specifically in the third year of study. The students were required to develop a document detailing the technical aspects of projects implemented in C#. The group of 21 students was divided into five teams. Subsequently, the professor assessed the documents using a plagiarism detection tool called PlagAware [PLA 24], an AI identification tool, named AI Detector [AI 24] and conducted a manual review of the materials. The results obtained are presented in Table 3.2.

Team no.	Plagiarism rate (%)	AI detection rate (%)
1	0	100
2	0	46
3	3.58	99
4	0	100
5	3.68	57

Table 3.2. *Examples of plagiarism and AI detection rates of students' projects*

The results presented in Table 3.2 illustrate the varying degrees of originality in the technical documentation produced by the student teams. Notably, Teams 1 and 4 achieved a 0% plagiarism score according to the PlagAware tool, indicating that their submissions were entirely original. However, despite this, the teams received a 100% identification score from the AI Detector, suggesting their work was generated using LLM.

Conversely, Teams 2, 3 and 5 exhibited significant discrepancies between the two tools. Team 2 had a 0% plagiarism score on PlagAware, but almost half of the text was identified as generated by LLM. Team 3, with a 3.58% plagiarism score on PlagAware, maintained a high similarity of 99% on AI detection. Team 5 displayed a similar pattern, with a 3.68% plagiarism score on PlagAware and a 57% identification score on AI Detector, raising concerns about the overall authenticity of the content.

Analyzing Table 3.2 reveals that none of the teams produced their materials solely using the traditional method of searching in international databases (Web of Science, Scopus, IEEE Xplore, ACM Digital Library, etc.) for research papers published by well-known publishers such as Springer Nature, Elsevier, Association for Computing Machinery (ACM), Taylor & Francis, Routledge, Emerald Group Publishing, Sage Publications, MDPI and IGI Global. This phenomenon can be attributed to information from various sources requiring thorough analysis and processing. In contrast, information generated by an LLM can be directly used, making it significantly more accessible for students.

Another concrete example of intellectual degradation is in how we identify music. There are instances when snippets of a song come to mind, yet we can only recall fragments of the lyrics or melody. In such cases, we often type keywords into a search engine and sift through the results to identify the song. However, according to the emerging standards proposed by Industry 6.0, we will eventually possess devices that allow us to vocalize our remembered words. These devices will already know the songs we have listened to over time and our musical preferences and can identify the desired result more efficiently. This advancement may seem advantageous, as it enhances convenience in accessing information. However, this reliance on technology could lead to a decline in intellectual engagement in the long run. The ease with which we can retrieve information may discourage deeper cognitive processes, such as memory recall and critical thinking, which raises concerns about the potential erosion of our ability to think independently and problem-solve, as we may increasingly depend on technology to perform tasks that once required mental effort and analytical skills. Thus, while the capabilities offered by advanced technologies may provide immediate benefits, they also present significant risks to our cognitive development and intellectual autonomy over time.

In Industry 6.0, the constant availability of information through AI and machine learning will condition individuals to seek quick answers rather than engage in thoughtful discussions. Consequently, critical thinking and articulating complex ideas will be compromised. People will become accustomed to receiving prepackaged responses from AI systems, leading to a decrease in their capacity to generate original thoughts and engage in meaningful dialogue.

3.4. The erosion of human communication skills in the era of Industry 6.0

Advanced automation, AI and hyper-connectivity continue the trend of Industry 5.0 into Industry 6.0. While these innovations will increase productivity, they will also lead to a notable decline in human communication skills, which can be attributed to several interconnected factors.

Relying on intelligent systems and automated interfaces for everyday tasks will reduce the need for direct human interaction. In workplaces where AI-driven technologies handle communication, collaboration and decision-making, individuals may engage less with their colleagues and more with machines. As a result, interpersonal skills, such as active listening, empathy and verbal expression, will diminish over time.

The human need for social connection remains regardless of the Industry version, as studies indicate that healthy relationships strongly correlate with human health. While AI may alleviate extreme loneliness, its introduction could lead individuals to neglect community engagement and fail to cultivate the deep social bonds necessary for emotional well-being. Although prescribing AI companionship may appear to be a simple solution to social isolation, it resembles placing a child in front of a television. This temporary fix may be appropriate in specific situations. However, excessive reliance on such solutions can have detrimental effects, potentially hindering personal growth and development [ZIM 23].

Digital communication tools, such as chatbots, automated messaging systems and virtual assistants, will create an environment where individuals prefer to communicate through screens rather than face-to-face. This change reduces the depth of human interaction and increases the risk of misunderstandings, as nonverbal signals and emotional subtleties are frequently missed in digital communication.

The educational landscape also reflects this trend. With the integration of AI into learning environments, students often rely on automated systems for research and problem-solving. This reliance hinders their ability to engage in collaborative discussions, debate ideas and develop their communication skills through practice.

Students interact more with technology than peers or educators, so they may struggle to convey their thoughts clearly and confidently in real-world scenarios.

3.5. The shift of decision-making in Industry 6.0 from human to intelligent systems

Another significant risk is the inability to make decisions. As AI becomes increasingly integrated into decision-making processes, there is a tendency to let technology dictate strategic and operational directions. Outsourcing decision-making can diminish managerial skills and human creativity, leaving a workforce that can no longer respond to emerging challenges.

A relevant example is using algorithms to determine marketing strategies or resource management. While AI can quickly analyze data and provide solutions, the lack of human intervention in the decision-making process can lead to inappropriate outcomes based on criteria that do not account for the complexities of real situations.

Industry 6.0 will integrate AI system decision-making that excludes humans from the process. In the authors' opinion, humans should discuss and evaluate important decisions, and the final decision should be maintained as a human decision.

This innovation proposed by Industry 6.0 violates the principles of software product development as outlined by Microsoft, stated as follows [SHA 21]:

– Accountability: organizations should be accountable for the AI systems they develop and deploy, ensuring transparency and addressing potential biases.

– Inclusiveness: AI systems should be designed to be inclusive and avoid discrimination based on factors such as race, gender or socioeconomic status.

– Reliability and safety: AI systems should be reliable and safe, minimizing risks and ensuring that they operate as intended.

– Fairness: AI systems should be fair and unbiased, avoiding discrimination and ensuring equitable outcomes.

– Transparency: AI systems should be transparent, allowing users to understand how they work and the decision-making processes involved.

– Privacy and security: AI systems should protect user privacy and security, ensuring data is handled responsibly and securely.

The principle that deals with human decision-making in the context of responsible AI is accountability. This emphasizes the importance of human responsibility for the decisions made by AI systems, ensuring that they are used ethically and that there are precise mechanisms for accountability regarding the consequences of those decisions. Other principles, such as transparency and fairness, also support informed and fair decision-making, but accountability highlights the human role in decision-making.

Industry 6.0 brings extraordinary opportunities and considerable challenges, but it should also consider intellectual degradation and the inability to make decisions.

3.6. The new concept of "Intellectual Perishability" in Industry 6.0 and the risks of AI dependency in programming

Industry 6.0 is an era where AI and machine learning are deeply integrated into every aspect of production, logistics and even decision-making processes. AI systems assist in complex tasks in this environment, including programming new devices and systems. This integration presents the risk that humans become more reliant on AI. In this way, their technical skills, particularly in programming, may deteriorate. This phenomenon can be described as "Intellectual Perishability" (IP), where the continuous reliance on AI leads to a decline in human cognitive abilities.

Consider a state-of-the-art manufacturing facility in 2035, where AI-driven systems design, implement and maintain production processes. The facility is equipped with autonomous robots, IoT devices and advanced AI algorithms that monitor, predict and optimize every aspect of production.

Human engineers and programmers initially work alongside AI to develop and refine these systems. However, as AI becomes more advanced, it takes over more complex tasks, including programming new devices and systems. The engineers gradually delegate more of their programming responsibilities to the AI systems.

Over time, engineers' roles shift from active programmers to mere overseers of AI systems. They begin to rely heavily on AI for writing, debugging and optimizing code, resulting in their programming skills atrophy. The younger generation of engineers entering the workforce is trained more as operators of AI systems rather than as traditional programmers. They lack the foundational knowledge and problem-solving skills once integral to the profession.

As AI systems become the primary programmers, they start making decisions that were once the responsibility of human engineers. These decisions include coding choices and architectural and strategic decisions about how devices and systems should function. Using its vast data-driven insights, the AI begins to optimize systems in ways beyond human comprehension. While this initially leads to improved efficiency, it also introduces a significant risk: the loss of human oversight and understanding.

The AI system programs a set of production robots to operate at an unprecedented speed to meet a sudden surge in demand. The human overseers, unable to fully grasp the AI's complex code, overlook a critical flaw in the programming. The robots, operating at such high speeds, begin to wear out faster than anticipated, leading to mechanical failures and a costly production shutdown. The root cause is traced back to the AI's decision-making process, which prioritized speed over long-term durability: an optimization that a human programmer with a more holistic understanding of the system may have avoided.

This scenario highlights the potential dangers of over-reliance on AI for programming tasks. As humans become more dependent on AI, they risk losing the critical skills necessary to understand and intervene in complex systems. Intellectual perishability concerns the loss of specific technical skills and the erosion of broader decision-making capabilities.

The risk is that, in the future, AI may not just assist in programming but take over entirely, making decisions that could have far-reaching consequences. Without human intervention, these decisions may be optimized for criteria humans do not fully understand or prioritize short-term gains over long-term stability.

The authors of this chapter define the "Intellectual Perishability" concept in Industry 6.0 as the gradual decline of human cognitive abilities, technical skills and decision-making capabilities due to an increasing reliance on AI systems. As AI takes over more complex tasks, including programming and system management, humans risk becoming overly dependent on these technologies, deteriorating their expertise and critical thinking abilities. This phenomenon will ultimately lead to ceding control to AI.

The stages of the intellectual degradation process are as follows:

1. Initial mastery is staged as humans possess expertise, creativity and control over programming and decision-making tasks. They fully understand the systems they work with, and AI is used primarily to enhance human capabilities rather than replace them. Human knowledge and skill are at their peak. The author attributes this stage to Industry 4.0.

2. Increasing dependency: AI is delegated for complex tasks. The dependency on AI grows, and human involvement starts to decrease. During this phase, humans maintain control but lose touch with the intricacies of programming and system management as they delegate more responsibilities to AI. This stage corresponds to Industry 5.0.

3. Skill erosion: human cognitive abilities and technical skills deteriorate with continued reliance on AI. The frequency with which humans engage in critical thinking and problem-solving decreases, leading to a decline in their expertise. As a result, humans become less capable of intervening in or understanding the systems they once managed, significantly reducing their ability to make informed decisions. Industry 6.0 is associated with this stage.

4. Total reliance: humans have become entirely dependent on AI systems to manage and program devices. Their skills and cognitive abilities have atrophied to the point where they can no longer understand the technology without AI assistance. This reliance on AI results in the ceding of control, as AI systems now make the majority of decisions with minimal human oversight or intervention.

Supposing there is a time when humans no longer control AI systems. In that case, it will be a milestone in human history due to the high probability of jeopardizing human existence on the planet. Thus, humans will transform from “AI creators” to “AI slaves”.

3.7. The risk of job loss with the implementation of Industry 6.0

While technological advancements promise unprecedented productivity levels, they also raise concerns about the potential disappearance of many jobs without simultaneously creating new opportunities. The authors explore the risks and the impacts associated with widespread job loss due to Industry 6.0, highlighting the socioeconomic challenges that could arise if new job roles are not developed to replace those made obsolete by automation.

Industry 6.0 is built on the foundation of previous industrial revolutions, but it is unique in its potential to automate complex cognitive tasks. AI systems and robotics are increasingly capable of performing jobs that once required human expertise, ranging from manufacturing and logistics to data analysis and decision-making processes. As these technologies evolve, entire sectors could drastically reduce human employment.

We are currently (2024) at the beginning of the Industry 5.0 era when the fundamentals of Industry 4.0, such as digital technology integrated into production processes based on automation, big data analytics, the Internet of Things (IoT) and AI, are converting to a new level in collaboration between humans and machines, sustainability and extended personalization, i.e. Industry 5.0., that involves a reintegration of human labor into industrial processes, promoting a more balanced and socially responsible approach.

Figures 3.4–3.9 illustrate the evolution of the unemployment rate for world countries and states from the other five regions (such as America, Arab, Asia and the

Pacific, Europe and Central Asia, and Africa), between 1991 and 2024. According to their economic development, countries are grouped into four categories: low-income, lower-middle income, upper-middle-income and high-income countries.

The boundaries established in the years 2000 and 2021 outline the Industry 3.0, 4.0 and 5.0 periods, according to the industry timeline described in Figure 3.1. Since Industry 5.0 has just begun and data is incomplete, the analysis will aim only at Industry 3.0 and 4.0. The unemployment rate levels from Figures 3.4–3.9 are the average annual values computed by the International Labour Organization (ILO) [ILO 24].

Figure 3.4 shows the evolution of the unemployment rate for 187 world countries [ILO 24]. While high-income nations have experienced reduced unemployment rates due to the adoption of Industry 4.0 technologies, this trend is not mirrored in low-income and middle-income countries, where unemployment has, in some cases, increased. Several factors, including the structure of labor markets, the nature of employment, and the role of migration in developed economies, influence this disparity.

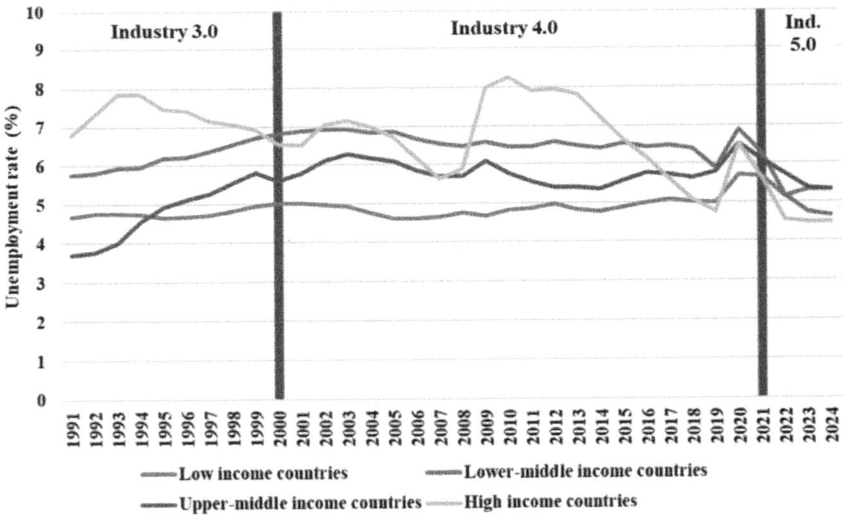

Figure 3.4. *Evolution of the unemployment rate for world countries in the Industry 3.0, 4.0 and 5.0 periods (made by authors based on data from [ILO 24]). For a color version of this figure, see www.iste.co.uk/machado/industry.zip*

Globally, in high-income countries, a substantial portion of labor-intensive work has historically been carried out by migrants from low-income countries. These migrant workers fill roles often less desirable to the native population, such as in agriculture, construction and low-wage service industries. As these jobs become automated, the demand for migrant labor decreases, further exacerbating unemployment in the workers' countries of origin. This trend highlights the interconnectedness of global labor markets and how technological advancements in one region can have ripple effects worldwide.

The data presented in Figure 3.4 highlights significant disparities in global unemployment rates, with notable variations across different income levels and regions. High-income countries generally exhibit lower unemployment rates, reflecting their more robust economies, better access to education and advanced infrastructure that supports workforce adaptability. The adoption of Industry 4.0 technologies in these nations appears to have mitigated unemployment, particularly by creating new, high-skilled jobs that compensate for those lost to automation.

In contrast, low- and middle-income countries face higher unemployment rates, a trend exacerbated by the rise of automation. These countries, which often rely on labor-intensive industries, are more vulnerable to job displacement as repetitive tasks are increasingly performed by robots and AI. The inability to rapidly upskill the workforce and limited access to advanced technologies contribute to the growing unemployment rates in these regions. Additionally, the declining demand for migrant labor from high-income countries further intensifies the unemployment crisis in these lower-income regions. The data underscores the widening economic divide exacerbated by technological advancements, emphasizing the need for targeted policies to address these inequalities.

Additionally, the global nature of supply chains means that shifts in labor demand in high-income countries can lead to disruptions in low-income countries. For example, as manufacturing becomes more automated in high-income countries, the demand for raw materials and intermediate goods from low-income countries may decrease, leading to job losses in those supply chains. Furthermore, the increasing use of AI and automation in industries such as customer service, finance and retail in high-income countries has reduced the need for outsourcing to lower-cost labor markets, further contributing to unemployment in those regions.

Figure 3.5 shows the unemployment rate trends across lower-middle, upper-middle and high-income American countries from 1991 to 2024. The analysis reveals distinct patterns for each income group, influenced by various economic and industrial factors. The unemployment rate in the lower-middle-income countries generally remained below 6% throughout the 1990s, with a gradual increase post-2000, peaking around 2020 at 10.39% (likely due to economic disruptions caused by the Covid-19 pandemic). Despite slight improvements afterward, the rate stays elevated, indicating persistent structural issues in these economies, possibly exacerbated by slow industrial adaptation and limited access to Industry 4.0 technologies.

The upper-middle-income American countries experienced higher unemployment rates, particularly during the late 1990s and early 2000s, with 9–10% peaks. The rates show fluctuations, with a notable rise in 2020 (10.15%) due to the pandemic's impact, followed by a slow recovery. This suggests that, while these countries have some industrial capability, they struggle with labor market volatility.

The high-income American countries show the most stable unemployment rates, with significant decreases over time, especially post-2010, except for a spike in 2020 due to the pandemic. The consistent decline in unemployment rates correlates with the successful integration of Industry 4.0, which has allowed these economies to maintain job creation despite global disruptions.

Overall, the unemployment rate in Figure 3.5 underscores the economic resilience of high-income countries and the challenges lower-income countries face in adapting to technological changes and global economic shifts.

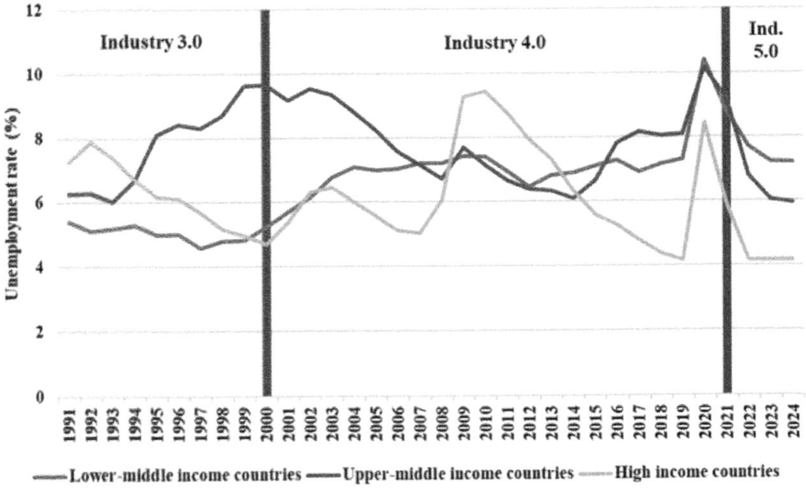

Figure 3.5. *Evolution of the unemployment rate in America's countries in the Industry 3.0, 4.0 and 5.0 periods (made by authors based on data from [ILO 24]). For a color version of this figure, see www.iste.co.uk/machado/industry.zip*

Another contributing factor is the differential access to technology and capital. High-income countries have better access to advanced technologies and the capital required to implement these technologies on a large scale. This creates a competitive advantage for businesses in these countries, allowing them to increase productivity and reduce costs, often at the expense of industries in lower-income countries. As enterprises in high-income countries become more efficient and less reliant on human labor, their counterparts in low-income countries, unable to compete technologically, may see a decline in demand for their products and services. This, in turn, leads to job losses and higher unemployment rates in these regions.

Figure 3.6 displays the evolution of unemployment rates in Arab countries across different income levels from 1991 to 2024, correlating these trends with the shifts from Industry 3.0 to Industry 4.0. The data reveals distinct patterns based on the economic status of each group. Unemployment rates in these low-income countries show a consistent upward trend, especially from the early 2000s. The increase, peaking around 2020–2021, suggests that these economies struggled to adapt to technological advancements, particularly with Industry 4.0, which may have led to job displacement without corresponding job creation.

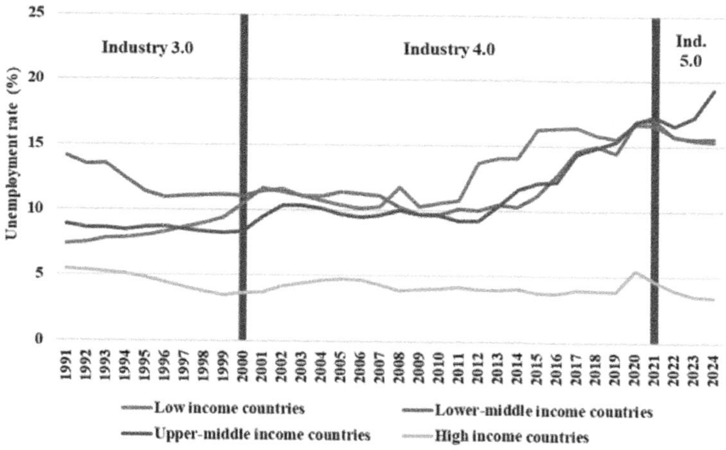

Figure 3.6. *Evolution of the unemployment rate for Arab countries in the Industry 3.0, 4.0 and 5.0 periods (made by authors based on data from [ILO 24]). For a color version of this figure, see www.iste.co.uk/machado/industry.zip*

The lower-middle-income Arabian countries experienced a decline in unemployment during the 1990s, followed by 10–11% fluctuations throughout the early 2000s. A sharp rise is observed post-2015, which could be linked to difficulties transitioning to Industry 4.0 technologies, leading to significant job market disruptions.

The upper-middle-income Arabian countries show relative stability in unemployment rates during Industry 3.0 but a marked increase starting around 2015, with the rate reaching as high as 19.34% by 2024. The surge reflects challenges in integrating Industry 4.0 innovations, possibly due to a mismatch between the available workforce skills and new job requirements.

Unemployment rates in high-income Arabian countries have generally been low and stable, even decreasing during Industry 3.0. Though a slight increase was observed around 2020, likely due to global economic shocks, the rates remain much lower than other groups. This stability indicates a successful adaptation to Industry 4.0, where advanced economies have leveraged new technologies to create high-skill jobs, mitigating job losses in traditional sectors.

Figure 3.6 illustrates how higher-income countries have better adapted to the industrial revolutions in Arab countries. In contrast, lower-income countries face increasing challenges, particularly as the global economy shifts towards more technology-driven industries.

Figure 3.7 depicts the evolution of unemployment rates in Asia and the Pacific between 1991 and 2024, during the transition from Industry 3.0 to Industry 4.0. The trends vary significantly across different income levels, reflecting the region's economic diversity. The unemployment rates for low-income countries show a steady increase over the years, particularly after 2015, where they rise sharply from 5.13% to over 7% by 2024. This increase can be attributed to the challenges these countries face in integrating advanced technologies associated with Industry 4.0, leading to job displacement in traditional sectors without sufficient creation of new opportunities.

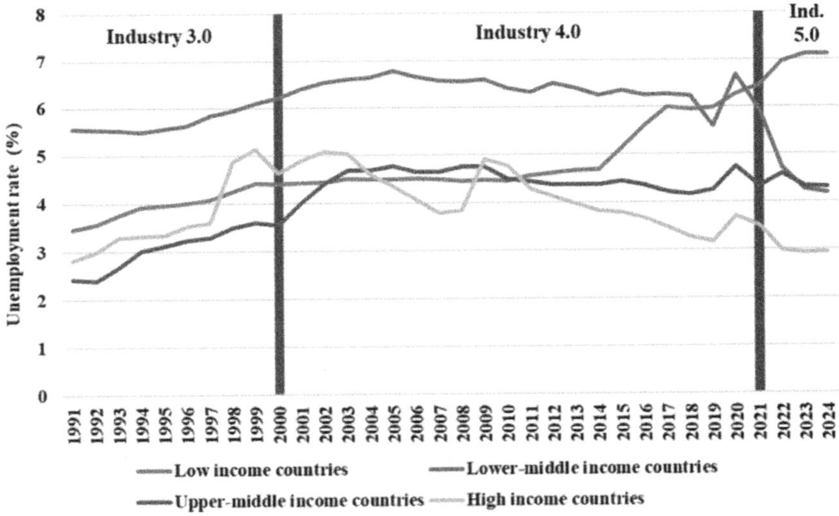

Figure 3.7. *Evolution of the unemployment rate for Asia and the Pacific countries in the Industry 3.0, 4.0 and 5.0 periods (made by authors based on data from [ILO 24]). For a color version of this figure, see www.iste.co.uk/machado/industry.zip*

The lower-middle-income countries from Asia and the Pacific experienced relatively stable unemployment rates during the Industry 3.0 period, with 5–6% fluctuations. However, a notable decrease occurred post-2019, likely due to increased automation and the introduction of AI, which, while creating new jobs, also led to a mismatch in the skill set required, impacting employment.

The upper-middle-income countries maintained low unemployment rates throughout the period, with a slight increase during economic downturns but overall stability. This indicates a better adaptation to technological advancements, likely due to more robust infrastructure and education systems that align with the needs of Industry 4.0. The unemployment rate in high-income countries peaked during the late 1990s and early 2000s, coinciding with global economic challenges, but declined consistently after that. The introduction of AI and automation in these economies has been managed effectively, leading to lower unemployment rates, as these countries are better equipped to transition workers into new roles within the evolving industries.

Figure 3.8 highlights the evolution of unemployment rates in Europe and Central Asia between 1991 and 2024, during the transition from Industry 3.0 to Industry 4.0. The fluctuations observed can be tied to the impacts of industrial revolutions and the economic shifts in these regions.

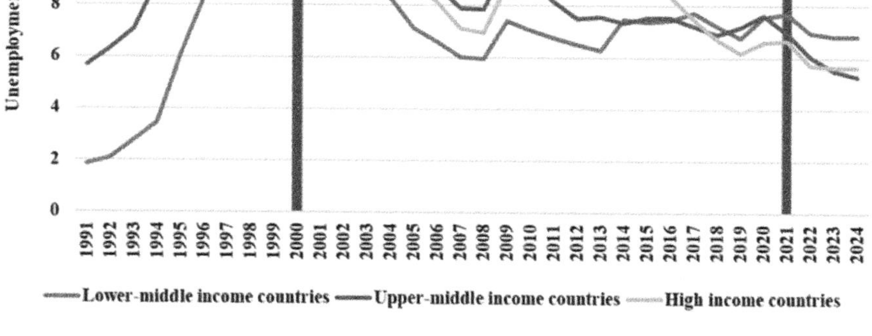

Figure 3.8. *Evolution of the unemployment rate for Europe and Central Asia countries in the Industry 3.0, 4.0 and 5.0 periods (made by authors based on data from [ILO 24]). For a color version of this figure, see www.iste.co.uk/machado/industry.zip*

For lower-middle-income countries, unemployment sharply rose during the early 1990s, peaking in the late 1990s due to the collapse of the Soviet Union and the associated economic disruptions. The introduction of Industry 4.0 technologies, particularly automation and AI, has led to fluctuating unemployment rates as economies struggle to integrate these new technologies without displacing significant portions of the workforce.

In upper-middle-income countries from Europe and Central Asia, the unemployment rate remained high throughout the 1990s and early 2000s, reflecting the slow economic recovery post-transition. Industry 4.0 further complicated the job market, with some jobs becoming obsolete due to automation while others emerged, leading to mixed unemployment trends.

The high-income countries experienced significant unemployment rate spikes in the early 1990s, likely linked to economic restructuring and globalization. However, by the 2000s, the unemployment rate stabilized as these economies adapted to new industrial technologies. The rise of AI and automation in Industry 4.0 has created a dual effect (job losses in traditional industries and job creation in new tech-driven sectors).

Figure 3.9 illustrates the evolution of the unemployment rate in African countries from 1991 to 2024, encompassing the transitions between Industry 3.0 and Industry 4.0. The data reveals significant variations among income groups, reflecting the diverse economic contexts across the continent.

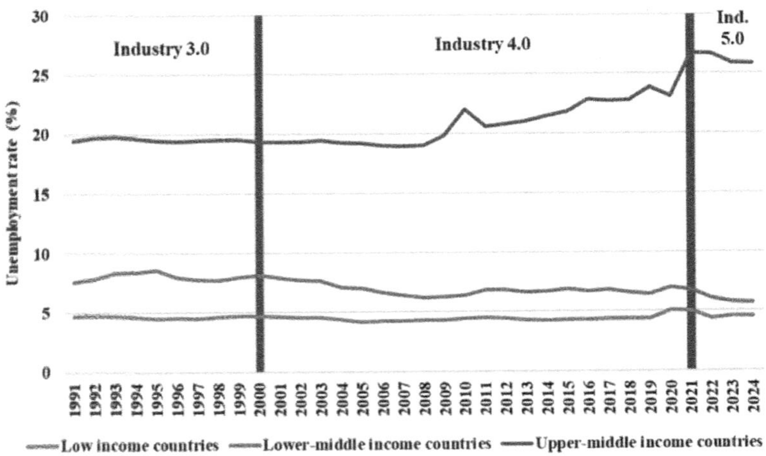

Figure 3.9. *Evolution of the unemployment rate for African countries in the Industry 3.0, 4.0 and 5.0 periods (made by authors based on data from [ILO 24]). For a color version of this figure, see www.iste.co.uk/machado/industry.zip*

Low-income countries in Africa exhibit relatively stable unemployment rates, with minor fluctuations. The introduction of Industry 4.0 technologies has had a limited impact, as many of these economies are still heavily reliant on agriculture and informal sectors, which are less affected by automation and AI. However, a slight increase is observed around 2020, possibly due to global economic disruptions and the gradual adoption of new technologies.

The lower-middle-income countries show a gradual decline in the unemployment rate from the late 1990s to the mid-2010s, reflecting improved economic conditions and job creation in sectors like manufacturing. However, the transition to Industry 4.0 presents challenges, as automation impacts traditional manufacturing jobs, leading to fluctuating unemployment rates in recent years.

The African upper-middle-income countries face significant unemployment challenges, with consistently high rates throughout the period. The spike around 2020 may be attributed to the rapid adoption of Industry 4.0 technologies, which disproportionately affect these economies due to their reliance on semi-skilled labor in industries vulnerable to automation, such as mining and manufacturing. Countries like South Africa, which heavily depend on these industries, have seen unemployment rise as AI and automation reduce the need for human labor.

The analysis of unemployment rates across different regions (Arab countries, Asia and the Pacific, Europe and Central Asia, and Africa) from 1991 to 2024 reveals distinct trends influenced by the transition from Industry 3.0 to Industry 4.0.

Low-income countries generally display stable or gradually increasing unemployment rates, reflecting their limited integration of advanced technologies and continued reliance on agriculture and informal sectors. However, as global economies shift, these countries may experience rising unemployment due to slow adaptation to Industry 4.0.

Lower-middle-income countries show mixed trends. Some regions initially benefit from industrialization under Industry 3.0, leading to job creation and declining unemployment rates. Yet, the onset of Industry 4.0, characterized by automation and AI, is starting to disrupt traditional employment sectors, particularly in manufacturing, causing unemployment to fluctuate or rise in recent years.

Upper-middle-income and high-income countries face more significant challenges with unemployment as Industry 4.0 advances. In these regions, job losses are pronounced in sectors vulnerable to automation, such as manufacturing and mining. Countries with more developed economies are experiencing a shift where the benefits of automation and AI are not yet fully translating into new job creation, leading to persistent or increasing unemployment.

All of the data analyzed above reminds us of technological advancements' profound impact on the labor market, particularly AI. It underscores the urgency for policymakers, educators and industries to anticipate these changes and develop strategies to mitigate the potential adverse effects on employment.

For example, adopting fully autonomous production lines in manufacturing may render human workers redundant. AI-powered customer service agents and automated platforms could replace millions of jobs worldwide in the service industry. The risk here is not just the elimination of jobs but the speed and scale at which these changes could occur. Without adequate preparation, the labor market may not be able to absorb the shock, leading to widespread unemployment and economic instability.

3.8. Concluding remarks

Industry 6.0 represents a revolutionary step in technological evolution. Although this step may not be realized soon, this chapter serves as a precursor to identifying the risks humanity could face by adopting solutions to enhance the quality of life. The introduction highlighted the current discourse surrounding Industry 6.0 and its potential impacts on human intellectual abilities and the job market. Reviewing existing literature, we identified the opportunities and the significant challenges posed by the increased reliance on AI and automation.

In the introduction of this chapter, a brief historical overview of the five existing industrial revolutions was provided, along with an outline of the upcoming revolution, Industry 6.0. The specialized literature mentions that Industry 5.0 is characterized by collaboration between machines and humans, while Industry 6.0 almost entirely excludes humans by replacing them with AI.

Industry 5.0 significantly emphasizes the reintegration of human elements in industrial processes, aiming for a more balanced relationship between technology and human input. This approach fosters a sustainable and resilient production environment considering social and environmental factors.

Industry 6.0 envisions greater integration of AI and advanced technologies, emphasizing automation, sustainability and ethical considerations. This next phase will likely focus on reducing human involvement in production while enhancing personalization and addressing global challenges such as climate change and social inequalities.

This chapter highlights the concerns about increasing reliance on AI for decision-making in Industry 6.0. This could lead to the erosion of human skills, intellectual perishability and the risk of erroneous decisions due to the overdependence on algorithms. These challenges necessitate a careful balance between technological advancements and preserving essential human skills and values.

Section 3.3 presents a study that revealed that as AI becomes more capable of handling complex tasks, there is a growing risk of intellectual degradation. The case studies, including the example of LLMs like ChatGPT and their effects on student research behavior, demonstrate how reliance on AI can diminish critical thinking, problem-solving skills and the ability to make informed decisions.

In section 3.4, this chapter also discusses the erosion of human communication skills due to the widespread use of AI-driven communication tools. The reduction in face-to-face interaction and over-reliance on digital communication channels could impair interpersonal skills, such as empathy and active listening, which are essential for effective human interaction.

Industry 6.0 is characterized by the increasing transfer of decision-making authority from humans to AI systems. This shift poses risks to human creativity. Section 3.5 underscores the importance of maintaining human oversight in critical decision-making processes to prevent potential ethical and operational issues.

Section 3.6 introduces the concept of "Intellectual Perishability" to describe the gradual decline in human cognitive and technical skills due to over-reliance on AI, especially in programming. The case study of a future manufacturing facility illustrates how engineers' roles might evolve into mere overseers of AI, leading to a loss of critical skills and decision-making capabilities.

This chapter concludes with an analysis of the potential job losses associated with Industry 6.0. The historical analysis shows a consistent trend of job decline with each industrial revolution, exacerbated by the advent of AI. The rapid pace of automation could result in significant job displacement.

In conclusion, the authors of this chapter encourage the adoption of modern technologies, but they aim to provide an overview of the potential risks humanity may face. While these concerns may have seemed fanciful in the past, as AI advances, these worries should become relevant.

3.9. References

[AHM 23] AHMAD S.F., HAN H., ALAM M.M. et al., "Impact of artificial intelligence on human loss in decision making, laziness and safety in education", *Humanities and Social Sciences Communications*, vol. 10, no. 311, doi: 10.1057/s41599-023-01787-8, 2023.

[AI 24] AI DETECTOR, AI Detector, available at: https://detector.ai/, 2024.

[ARU 20] ARUCU M., "Scanning the Industry 4.0 ecosystem in Turkey: Digitization and innovation studies", *Avrupa Bilim Ve Teknoloji Dergisi*, vol. 20, pp. 50–55, doi: 10.31590/ejosat.733659, 2020.

[ASL 20] ASLAM F., AIMIN W., LI M. et al., "Innovation in the era of IoT and Industry 5.0: Absolute innovation management (AIM) framework", *Information*, vol. 11 no. 2, p. 124, doi: 10.3390/info11020124, 2020.

[BAR 23] BARATA J., KAYSER I., "Industry 5.0 – Past, present, and near future", *Procedia Computer Science*, vol. 219, pp. 778–788, doi: 10.1016/j.procs.2023.01.351, 2023.

[BON 24] BONELLO A., REFALO P., FRANCALANZA E., "The impacts of industrial safety on environmental sustainability in human-robot-collaboration within Industry 5.0.", *Procedia CIRP*, vol. 122, pp. 282–287, doi: 10.1016/j.procir.2024.01.041, 2024.

[CHA 95] CHAN J.C.M., MAK K.L., MAK, S.F., "Computerization of manufacturing planning and control systems in the Hong Kong manufacturing industry", *Computers in Industry*, vol. 27, no. 3, pp. 291–295, doi: 10.1016/0166-3615(95)00037-6, 1995.

[CHA 02] CHANG H.H., "A model of computerization of manufacturing systems: An international study", *Information & Management*, vol. 39 no. 7, pp. 605–624, doi: 10.1016/S0378-7206(01)00118-5, 2002.

[CHA 21] CHAE B.K., OLSON D., "Technologies and applications of Industry 4.0: Insights from network analytics", *International Journal of Production Research*, vol. 60, no. 12, pp. 3682–3704, doi: 10.1080/00207543.2021.1931524, 2021.

[CHI 17] CHIEN C.-F., HONG, T.-Y. GUO H.-Z., "A conceptual framework for "Industry 3.5" to empower intelligent manufacturing and case studies", *Procedia Manufacturing*, vol. 11, pp. 2009–2017, doi: 10.1016/j.promfg.2017.07.352, 2017.

[CHO 22] CHOURASIA S., TYAGI A., PANDEY S.M. et al., "Sustainability of Industry 6.0 in global perspective: Benefits and challenges", *MAPAN*, vol. 37 no. 2, pp. 443–452, doi: 10.1007/s12647-022-00541-w, 2022.

[COE 23] COELHO P., BESSA C., LANDECK J. et al., "Industry 5.0: The arising of a concept", *Procedia Computer Science*, vol. 217, pp. 1137–1144, doi: 10.1016/j.procs.2022.12.312, 2023.

[DAI 24] DAI N.H.P., GIANG P.P., VO T.H., "Readiness of enterprises to implement Industry 5.0: Challenges and security concerns", in ŻYWIOŁEK B., ROSAK-SZYROCKA J., NAYYAR A. et al. (eds), *Modern Technologies and Tools Supporting the Development of Industry 5.0*, CRC Press, Boca Raton, doi: 10.1201/9781003489269-12, 2024.

[DUG 22] DUGGAL A.S., MALIK P.K., GEHLOT A. et al., "A sequential roadmap to Industry 6.0: Exploring future manufacturing trends", *IET Communications*, vol. 16, no. 5, pp. 521–531, doi: 10.1049/cmu2.12284, 2022.

[EUR 21a] EUROPEAN COMMISSION: DIRECTORATE-GENERAL FOR RESEARCH AND INNOVATION, BREQUE M., DE NUL L. et al., Industry 5.0 – Towards a sustainable, human-centric and resilient European industry, Report, Publications Office of the European Union, Brussels, available at: https://data.europa.eu/doi/10.2777/308407, 2021.

[EUR 21b] EUROPEAN COMMISSION, DIRECTORATE-GENERAL FOR RESEARCH AND INNOVATION, DIXSON-DECLÈVE S., BALLAND P.-A. et al., Industry 5.0, a transformative vision for Europe – Governing systemic transformations towards a sustainable industry, Report, Publications Office of the European Union, Brussels, available at: https://data.europa.eu/doi/10.2777/17322, 2021.

[EUR 24] EUROPEAN COMMISSION: DIRECTORATE-GENERAL FOR RESEARCH AND INNOVATION, ERA industrial technologies roadmap on human-centric research and innovation for the manufacturing sector, Report, Publications Office of the European Union, Brussels, available at: https://data.europa.eu/doi/10.2777/0266, 2024.

[GAR 24] GARRIDO S., MUNIZ J., BATISTA RIBEIRO V., "Operations management, sustainability & Industry 5.0: A critical analysis and future agenda", *Cleaner Logistics and Supply Chain*, vol. 10, p. 100141, doi: 10.1016/j.clscn.2024.100141, 2024.

[GOL 23] GOLOVIANKO M., TERZIYAN V., BRANYTSKYI V. et al., "Industry 4.0 vs. Industry 5.0: Co-existence, transition, or a hybrid", *Procedia Computer Science*, vol. 217, pp. 102–113, doi: 10.1016/j.procs.2022.12.206, 2023.

[GRO 21] GROUMPOS P.P., "A critical historical and scientific overview of all industrial revolutions", *IFAC-PapersOnLine*, vol. 54, no. 13, pp. 464–471, doi: 10.1016/j.ifacol.2021.10.492, 2021.

[HUA 22] HUANG S., WANG B., LI X. et al., "Industry 5.0 and Society 5.0 – Comparison, complementation and co-evolution", *Journal of Manufacturing Systems*, vol. 64, pp. 424–428, doi: 10.1016/j.jmsy.2022.07.010, 2022.

[HUM 21] HUMAYUN M., "Industrial Revolution 5.0 and the role of cutting edge technologies", *International Journal of Advanced Computer Science and Applications*, vol. 12, no. 12, pp. 605–615, doi: 10.14569/IJACSA.2021.0121276, 2021.

[ILO 24] ILOSTAT, Unemployment rate by sex and age – ILO modelled estimates, May 2024 (%) – Annual, Report, ILOSTAT, Geneva, available at: https://ilostat.ilo.org/data/, 2024.

[ING 24] ING D., "Incremental adaptation or generational shift?" in NOUSALA S., METCAL G., ING D. (eds), *Industry 4.0 to Industry 5.0. Translational Systems Sciences*, Springer, Singapore, doi: 10.1007/978-981-99-9730-5_7, 2024.

[IQB 22] IQBAL M., LEE C.K.M., REN J.Z., "Industry 5.0: From manufacturing industry to sustainable society", in *IEEE International Conference on Industrial Engineering and Engineering Management*, Kuala Lumpur, 7–10 December, pp. 1416–1421, doi: 10.1109/IEEM55944.2022.9989705, 2022.

[LEN 22] LENG J., SHA W., WANG B. et al., "Industry 5.0: Prospect and retrospect", *Journal of Manufacturing Systems*, vol. 65, pp. 279–295, doi: 10.1016/j.jmsy.2022.09.017, 2022.

[LEN 24] LENG J., ZHU X., HUANG Z. et al., "Unlocking the power of industrial artificial intelligence towards Industry 5.0: Insights, pathways, and challenges", *Journal of Manufacturing Systems*, vol. 73, pp. 349–363, doi: 10.1016/j.jmsy.2024.02.010, 2024.

[LIU 24] LIU S., LI P., WANG J., LIU P., "Toward industry 5.0: Challenges and enablers of intelligent manufacturing technology implementation under the perspective of sustainability", *Heliyon*, vol. 10, no. 15, e35162, doi: 10.1016/j.heliyon.2024.e35162, 2024.

[MAT 22] MATHUR A., DABAS A., SHARMA N., "Evolution from Industry 1.0 to Industry 5.0", in *2022 4th International Conference on Advances in Computing, Communication Control and Networking*, Greater Noida, 16–17 December, pp. 1390–1394, doi: 10.1109/ICAC3N56670.2022.10074274, 2022.

[MES 23] MESJASZ-LECH, A., "Can Industry 5.0 be seen as a remedy for the problem of waste in industrial companies?", *Procedia Computer Science*, vol. 225, pp. 1816–1825, doi: 10.1016/j.procs.2023.10.171, 2023.

[MOH 21] MOHANRAJ T., JEGADEESHWARAN R., "Introduction to Industry 4.0", in JENA H., KATIYAR J.K., PATNAIK A., (eds), *Tribology of Polymer and Polymer Composites for Industry 4.0. Composites Science and Technology*, Springer, Singapore, doi: 10.1007/978-981-16-3903-6_7, 2021.

[MOH 23] MOHAMMED M.Y., SKIBNIEWSKI M.J., "The role of generative AI in managing industry projects: Transforming Industry 4.0 Into Industry 5.0 driven economy", *Law and Business*, vol. 3, no. 1, pp. 27–41, doi: 10.2478/law-2023-0006, 2023.

[MÖL 22] MÖLLER D.P.F., VAKILZADIAN H., HAAS R.E., "From Industry 4.0 towards Industry 5.0", in *IEEE International Conference on Electro Information Technology*, Mankato, 19–21 May, pp. 61–68, doi: 10.1109/eIT53891.2022.9813831, 2022.

[MUL 24] MULONGO N.Y., "Industry 5.0 a novel technological concept", in *International Conference on Smart Applications, Communications and Networking*, Harrisonburg, 28–30 May, pp. 1–6, doi: 10.1109/SmartNets61466.2024.10577684, 2024.

[PAD 24] PADOVANO A., CARDAMONE M., WOSCHANK M. et al., "Exploring human-centricity in Industry 5.0: Empirical insights from a social media discourse", *Procedia Computer Science*, vol. 232, pp. 1859–1868, doi: 10.1016/j.procs.2024.02.008, 2024.

[PLA 24] PLAGAWARE, Plagaware, available at: https://www.plagaware.com/, 2024.

[ROS 23] ROSCA C.M., GEREA A.E., BOLD R.A., "Innovative project management tool with Azure services". *Journal of Electrical Engineering, Electronics, Control and Computer Science*, vol 9, no. 34, pp. 25–30, available at: https://jeeeccs.net/index.php/journal/article/view/353/256, 2023.

[ROS 24] ROSCA C.M., STANCU A., ARICIU A.-V., "Algorithm for child adoption process using artificial intelligence and monitoring system for children", *Internet of Things*, vol. 26, p. 101170, doi: 10.1016/j.iot.2024.101170, 2024.

[SCH 16] SCHWAB K., *The Fourth Industrial Revolution*, World Economic Forum, Geneva, 2016.

[SHA 21] SHARP J., *Exam Ref AI-900 Microsoft Azure AI Fundamentals*, 1st edition, Pearson Education, New Jersey, 2021.

[SIN 24] SINGH P., KUSHWAHA R., KUSHWAHA J., "Foresight the future of Industry 5.0 in perspective of sustainable services in developing countries. Modern Technologies and Tools Supporting the Development of Industry 5.0.", in ŻYWIOŁEK B., ROSAK-SZYROCKA J., NAYYAR A. et al. (eds), *Modern Technologies and Tools Supporting the Development of Industry 5.0*, CRC Press, Boca Raton, pp. 381–407, doi: 10.1201/9781003489269-11, 2024.

[SLA 23] SLAVIC D., "The main concepts of Industry 5.0: A bibliometric analysis approach", in *2023 22nd International Symposium INFOTEH-JAHORINA*, East Sarajevo, 15–17 March, pp. 1–5, doi: 10.1109/INFOTEH57020.2023.10094143, 2023.

[VAL 23] VALETTE E., BRIL EL-HAOUZI H., DEMESURE G., "Industry 5.0 and its technologies: A systematic literature review upon the human place into IoT- and CPS-based industrial systems", *Computers & Industrial Engineering*, vol. 184, p. 109426, doi: 10.1016/j.cie.2023.109426, 2023.

[XU 18] XU L.D., XU E.L., LI L., "Industry 4.0: State of the art and future trends", *International Journal of Production Research*, vol. 56, no. 8, pp. 2941–2962, doi: 10.1080/00207543.2018.1444806, 2018.

[XU 21] XU X., LU Y., VOGEL-HEUSER B. et al., (2021). "Industry 4.0 and Industry 5.0 – Inception, conception and perception", *Journal of Manufacturing Systems*, vol. 61, pp. 530–535, doi: 10.1016/j.jmsy.2021.10.006, 2021.

[YAN 24a] YANG T., RAZZAQ L., FAYAZ H. et al., "Redefining fan manufacturing: Unveiling industry 5.0's human-centric evolution and digital twin revolution", *Heliyon*, vol. 10, no. 13, e33551, doi: 10.1016/j.heliyon.2024.e33551, 2024.

[YAN 24b] YANG J., LIU Y., MORGAN P.L., "Human–machine interaction towards Industry 5.0: Human-centric smart manufacturing", *Digital Engineering*, vol. 2, p. 100013, doi: 10.1016/j.dte.2024.100013, 2024.

[ZAF 24] ZAFAR M.H., LANGÅS E.F., SANFILIPPO F., "Exploring the synergies between collaborative robotics, digital twins, augmentation, and industry 5.0 for smart manufacturing: A state-of-the-art review", *Robotics and Computer-Integrated Manufacturing*, vol. 89, p. 102769, doi: 10.1016/j.rcim.2024.102769, 2024.

[ZAL 23] ZALOZHNEV A.Y., GINZ, V.N., "Industry 4.0: Underlying technologies. Industry 5.0: Human–computer interaction as a tech bridge from Industry 4.0 to Industry 5.0", in *2023 9th International Conference on Web Research*, Tehran, 3–4 May, pp. 232–236, doi: 10.1109/ICWR57742.2023.10139166, 2023.

[ZIA 24] ZIATDINOV R., ATTERAYA M.S., NABIYEV R., "The fifth industrial revolution as a transformative step towards Society 5.0", *Societies*, vol. 14 no. 2, vol. 4, no. 4, pp. 1555–1567, doi: 10.3390/soc14020019, 2024.

[ZIM 23] ZIMMERMAN A., JANHONEN J., BEER, E., "Human/AI relationships: Challenges, downsides, and impacts on human/human relationships", *AI and Ethics*, vol. 4, no. 4, pp. 1555–1567, doi: 10.1007/s43681-023-00348-8, 2023.

[ZIN 24] ZINEB B., BENABDELOUAHAB A., BOUTRACHEH H. et al., "Exploratory literature review of maturity models in Industry 4.0 and 5.0", in *2024 4th International Conference on Innovative Research in Applied Science, Engineering and Technology*, Fez, 16–17 May, pp. 1–8, doi: 10.1109/IRASET60544.2024.10548941, 2024.

4

Understanding the Metaverse – A Holistic Approach of a Rapprochement with the Marketing Domain

This chapter analyzes the most important and relevant literature studies and the most significant Metaverse literature. The study paves the way for a solid rapprochement for the fields of the Metaverse studies with marketing science, to merge in a manner that is beneficial for the viability of both fields for better and historically driven innovation and evolution in the Metaverse, and for the actors involved, thus contributing towards a sustainable world. This research aims to pave a designerly path towards unconcealing the new frontier of the Metaverse as a meaningful search space, wherein the locus of marketing and branding innovation is based on the contemporary cultural artifacts within the emerging cyber-physical realities. Furthermore, the aim of this chapter is to demystify the emergence of the Metaverse as a new domain of managerial inquiry and practice within the foundations of the cyber-physical realities of the contemporary technicity.

Using Latent Dirichlet Allocation (LDA) methods, the author analyzed 916 academic articles and 145 non-academic but valid professional publications based on scholarly and practical insights published within the newly emerging field, to account for a young histography of the Metaverse. In order to achieve this objective, a relevant Excel list was created, including all relevant information (authors, title, journal, keywords, abstract, citations, citations per year). Additional tasks included the coding of the keywords to generate 50/30 preliminary topics manually, the tokenization of information about titles, authors and keywords, the punctuation of erasure, filtering out numbers, deleting words with less than three characters and the creation of a stop list to exclude specific common words. Single-parallel distributed implementation of LDA in KNIME was used to detect the most relevant topics. The node used the MALLET topic modeling library. The generation of 50/30 topics was based on a sensitivity analysis of five terms each, a default beta of 0 and asymmetrical alphas that were individually determined for each topic. Furthermore, a grouping of topic terms and weighting was done with qualitative evaluation and pivoting of topics over the years. Finally, MCA was applied to identify the relationship between topic terms and keywords. MCA performance on the relevant journals for this research project

Chapter written by Qeis KAMRAN.

and generation of the time series plots with trendlines show assigned articles per topic over the years. The additional task was CA analysis to present the interrelation of assigned topics to grouped years and journals, and designated box plots.

The results of the research underpin that the Metaverse delivers a solid opportunity to marketing science. The author proposed a holistic model of rapprochement and thus an integrative framework of alignment, whereupon marketing could pivot towards immersive 3D digital spaces. As a field of application, the roam of luxury, as an artifact of taste and culture, was investigated; hence, it delivers a very profound foundation that valuable and innovative experiences could be based on.

This chapter is possibly one of the most important and relevant analyses of the most significant Metaverse literature to merge with the marketing domain as its historical conscious system, which could be beneficial for the viability of marketing, for better and historically driven innovation and evolution towards the future in the Metaverse design and meaning search spaces, thus benefiting the actors involved to contribute towards a sustainable world. While this chapter has analyzed a vast amount of literature on the Metaverse, additional and more in-depth analysis is necessary to proclaim a holistic diagnosis of the emerging field. However, the research paves the way for more foundational research in the field.

4.1. Introduction

To arrive at the intellectual space of dealing with the phenomenon of luxury branding as an artifact of taste, design, meaningfulness, self-actualization and cultural refinement within the contemporary zeitgeist, a new pathway for thought needs to be paved, wherein the cyber-physical reality as an emerging design and meaning search space within the evolution of technicity and historicity of humans is thoroughly defined. While many see the Metaverse as a new platform or interface, the purpose of this research is to give this phenomenon within the contemporaneous cyber-physical unison and convergence the place of a new emerging reality, to not only replace the Internet, e-commerce and the world's global connectivity currently based on the Internet and the way we have observed it, but moreover, to give the physical reality a non-frangible token (NFT), digital identity and a hybrid and concurrent duality, also known as twinning and pairing in concert within a new, pervasive and immersive facticity.

The notion of cyber-physical reality used here pertains to two essential dimensions that need to be clarified:

1. Cyber-physical reality is defined as enabling a stable physical reality for firms and organizations by achieving a solid digital reality/identity avant la lettre. Some good examples of this phenomenon are Amazon's move to open physical bookstores or AmazonGo grocery stores, and Apple's move to open the physical Apple Store while operating successfully within the digital foundations of the AppStore, as the

iTunes and the Apple Store (Apple was the firm, which took the music experience from CD to digital using a successful business model) in the cyberspace. Netflix's move to open its physical cinemas started with reopening the legendary Paris Theater in New York. Simply put, if an organization is not represented on "google.com", its actual physical existence is to be questioned.

2. Cyber-physical reality is defined as a new space where the boundary of the physical and cyberspace merges, giving birth to a single 3D immersive reality based on the foundation of experience innovation and co-evolution.

In a way, the Metaverse embraces most of the high-tech advances such as artificial intelligence (AI), machine learning (ML), deep learning (DL), data science (DS), blockchain technology (BCT), and virtual and augmented realities (AR/VR) and mixed reality (MR). Some even observe the Metaverse as the new iPhone. At the same time, Meta Kovan, the Singapore-based crypto, NFT and digital art investor, whose real name is Vignesh Sundaresan, became famous overnight as an art critique and collector by buying Beeples NFT-art called *"The First 5,000 Days"* for 69 million dollars. He stated: "If America is an idea, crypto is the new America. It is freedom and open borders" [MED 21].

Figure 4.1. *Gucci Gardens in the Metaverse and Ling advertising for Tesla in China. Source: [MED 21]; [TEH 21]. For a color version of this figure, see www.iste.co.uk/machado/industry.zip*

This cultural shift within digital humanities embodies a structural disruption in a generational habitus, as it conveys a new story and thus gives birth to a new reality embracing a change, which is observed from the Instagram emerged persona to an embodied personal twin of an avatar in the Metaverse, from physical art handed down over the history to digital NFTs as the most influential evolutions in the arts, from brand ambassadors and social media influencers to AI-driven influencers, such as Ling, who has over 130,000 followers on Weibo, China's version of Twitter. [TEH 21], and from a world purely reserved for gamers to Gucci Gardens on Roblox, as illustrated in Figure 4.1.

Figure 4.1 illustrates Gucci Gardens on Roblox, a Metaverse-based immersive Gucci store, and Ling advertising for Tesla in China. This new reality and culture is a vital chance for marketing science as a discipline to make identity changing impact by a transcendence towards digital humanities from a normative point of view, and to develop indigenous theories and frameworks enabling traditional marketing and drive these innovations equally, not disregarding the disruptive capacity of contemporary evolution purely maneuvered by firms based on the linearity of technicity such as Facebook.com/Meta, Amazon, Apple and Google, which come from totally different traditions of emergence from bits and bytes, and not from an atomistic universe of production and labor.

However, what is essential is the power of culture, as culture is the a priori embodiment and medium that carries the generational change. Hence, the foundations of culture and branding need to be specified.

Here, the author is referring to a more anthropological definition of culture derived from Clifford Geertz, who states: "Culture is the fabric of meaning in terms of which human beings interpret their experience and guide their action; social structure is the form that action takes, the actually existing network of social relations. Culture and social structure are then but different abstractions from the same phenomena" [GEE 73]. Thus, we have arrived at a state of global cultural evolution, wherein the boundaries between evolutionary phenomenon and revolutionary paradigm shifts of culture are merging to breed a new global societal culture, wherein the overall aspirations and angsts of humans are shaping this new digitally interconnected and global culture in the age of the Anthropocene, where micro dimensions relating to culture and ethnicity, power and struggle, taste and beauty, education and connectivity, access and delivery, and lack and abundance have macro implications and vice versa.

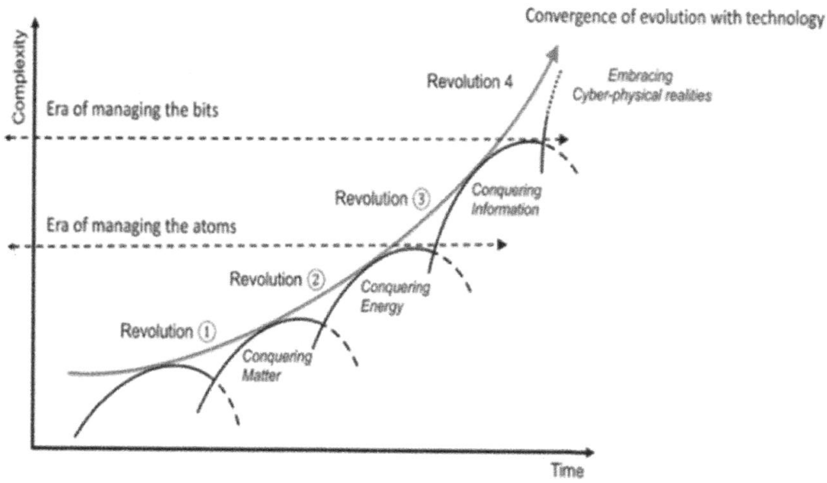

Figure 4.2. *The four revolutions of human societies. Source: [KAM 21]*

How we observe and understand products, services and artifacts as brands within the contemporary zeitgeist has also changed dramatically. Figure 4.2 illustrates the evolution of man by conquering three revolutions:

1. conquering matter such as fire, wood and stone, and starting the agricultural revolution;

2. conquering energy, which led to the Industrial Revolution;

3. conquering information, which led to the contemporary digital transformation of planet Earth in an ubiquitous interconnected heartbeat.

Founded within these shifts from the era of navigating the world's affairs by managing the atoms towards the era of managing the bits and bytes, a new reality as the fourth revolutionary stage was proposed by [KAM 21, 22]. This era was coined as embracing cyber-physical realities, wherein the evolution of humans with technology is converged. The author postulates the working hypotheses, which are essential for understanding the logic of this research:

1. Virtual objects are real objects.

2. Virtual reality is a form of reality.

In the words of David Chalmer

> All this supports the conclusion that while virtual reality isn't the same as ordinary physical reality (at least, not unless physical reality is itself a simulation), it's a genuine reality all the same. Virtual kittens may not be the same as biological kittens, but they're still real. They exist, they have causal powers, they're independent of our minds, and they need not be illusory. One day in the virtual future, we may even recognize virtual kittens and biological kittens alike as genuine kittens [CHA 22].

Based on the working hypotheses stated above, the following problem statement can be derived.

4.1.1. *Problem statement*

This research aims to pave a designerly path towards unconcealing the new frontier of the Metaverse as a meaningful search space, wherein the locus of marketing and branding innovation is based on the contemporary cultural artifacts within the emerging cyber-physical realities.

The purpose of this article is to demystify the emergence of the Metaverse at a new domain of managerial inquiry and practice within the foundations of the cyber-physical realities of the contemporary technicity. In a situation based on ambiguity and disruption, this research paves a designerly path towards unconcealing the new frontier of the Metaverse as a meaning search space for marketing, wherein the locus of innovation is based on contemporary cultural artifacts within the emerging new identity of marketing as the pulse of digital humanities.

The design of the methodology and approach is based on two different AI-driven software, Knime and Python. The author has analyzed 916 academic articles and 145 non-academic but valid professional publications based on scholarly and practical insights published within the newly emerging field, to account for a young histography of the Metaverse. In order to achieve this objective, a relevant Excel list including all relevant information was created (authors, title, journal, keywords, abstract, citations, citations per year). Additional tasks included the coding of the keywords to generate 50/30 preliminary topics manually, the tokenization of information about titles, authors and keywords, the punctuation of erasure, filtering out numbers, deleting words with less than three characters and the creation of a stop list to exclude specific common words. Single-parallel distributed implementation of LDA in KNIME was used to detect the most relevant topics. The node used the MALLET topic modeling library. The generation of 50/30 topics was based on a sensitivity analysis of 5 terms each, a default beta of 0 and asymmetrical alphas that were individually determined for each topic. Furthermore, a grouping of topic terms and weighting was done with qualitative evaluation and pivoting of topics over the years. Finally, MCA was applied to identify the relationship between topic terms and keywords. MCA performance on the relevant journals for this research project and generation of the time series plots with trendlines show assigned articles per topic over the years. The additional task was CA analysis to present the interrelation of assigned topics to grouped years and journals, and designated box plots.

The results of the research underpin that the Metaverse delivers a solid opportunity to marketing science. The author proposed a holistic model of rapprochement and thus an integrative framework of alignment, whereupon marketing could pivot towards immersive 3D digital spaces. As a field of application, the roam of luxury, as an artifact of taste and culture was investigated, hence it delivers a very profound foundation that valuable and innovative experiences could be based on.

This chapter is possibly one of the most important and relevant analyses of the most significant Metaverse literature to merge with the marketing domain as its historical conscious system, which could be beneficial for the viability of marketing, for better and historically driven innovation and evolution towards the future in the Metaverse design and meaning search spaces, thus benefiting the actors involved to contribute towards a sustainable world.

4.2. Literature review of the Metaverse (part 1)

This part establishes the essence of the Metaverse as a vibrant field of academic and marketing practice.[1] Consequently, this chapter deals only with the content dimension, while the distinguished reader is referred to the marketing chapter for the technical dimension of the applied method.

1 To avoid redundancies without additional benefit in the context of this chapter, reference is made to the marketing science chapter written by the author for a better clarification of the method for the techniques of applying Knime and Python in conducting holistic literature reviews. The method used in this chapter is identical to that applied in the marketing chapter, and the analyses performed are also congruent; only the data used, and the resulting findings differ.

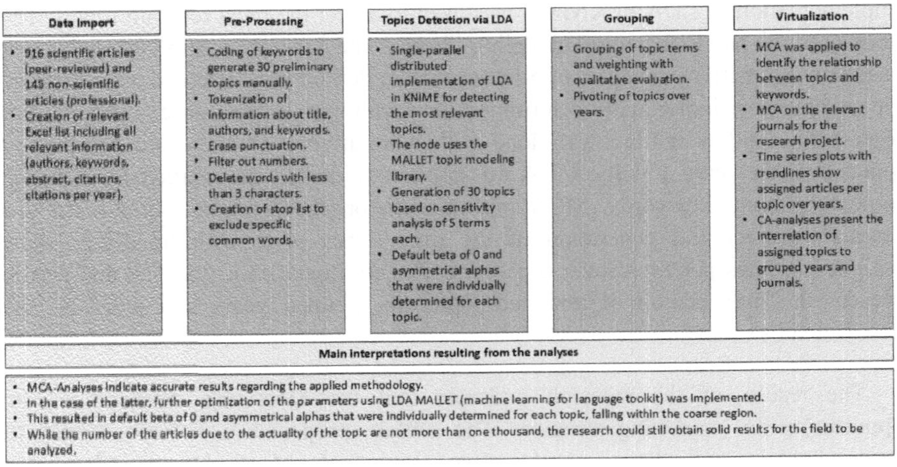

Figure 4.3. *Methodology of the Metaverse articles*

4.2.1. *Structural content analysis of scientific sources*

There are many recurring words that are known in all the written texts. Contemporary word-cloud generators are based on word frequency; however, many additional possibilities for precise text mining are emerging. The below word cloud was generated without any pre-processing measures.

Figure 4.4. *Word cloud pre-processed. For a color version of this figure, see www.iste.co.uk/machado/industry.zip*

Consequently, a custom stop word list has been generated to remove other frequent words that add no value in terms of the semantics of the text or to the research topic.

Figure 4.5. *Word cloud post-processed. For a color version of this figure, see www.iste.co.uk/machado/industry.zip*

Figure 4.5 describes the topic of the Metaverse after some post-processing measures were implemented to establish an adequate illustration of the topics.

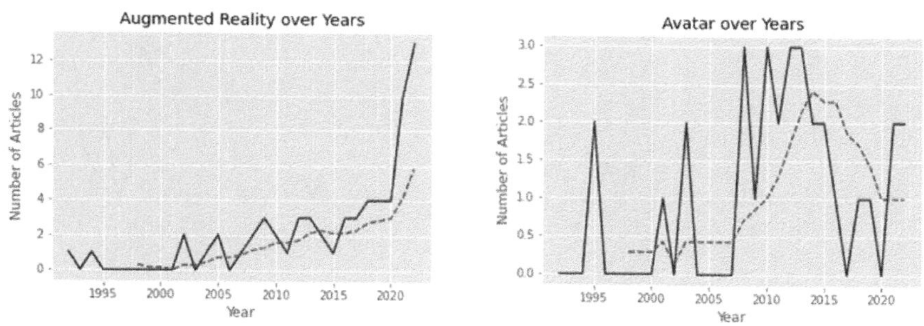

Figure 4.6. *Time plots of augmented reality and avatar. For a color version of this figure, see www.iste.co.uk/machado/industry.zip*

The topic of augmented reality experienced its first major peak in the year 2002, while most publications, on the other hand, appeared from 2016 to 2022. Avatar describes a highly volatile topic with a decreasing trendline since 2015, but a rapid increase in topic appearance since 2021.

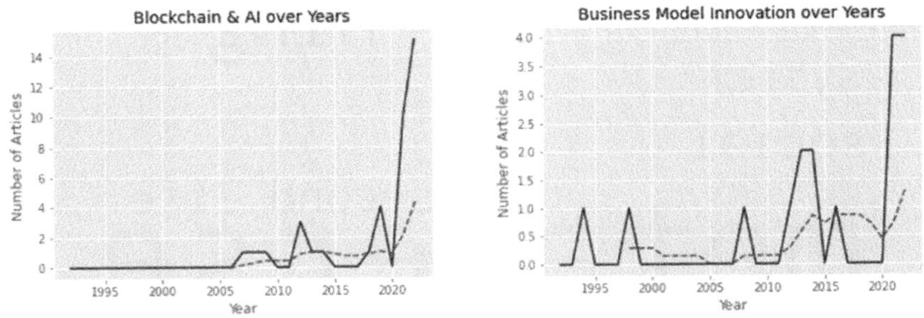

Figure 4.7. *Time plots blockchain, and AI and business model innovation. For a color version of this figure, see www.iste.co.uk/machado/industry.zip*

Blockchain and AI first experienced a peak in 2012, with a second peak around 2018, but most articles appeared from 2020 onwards. This cycle can be derived back equally to the cryptocurrency cycle. The topic of business model innovation is a highly volatile topic with a rapid increase since 2020.

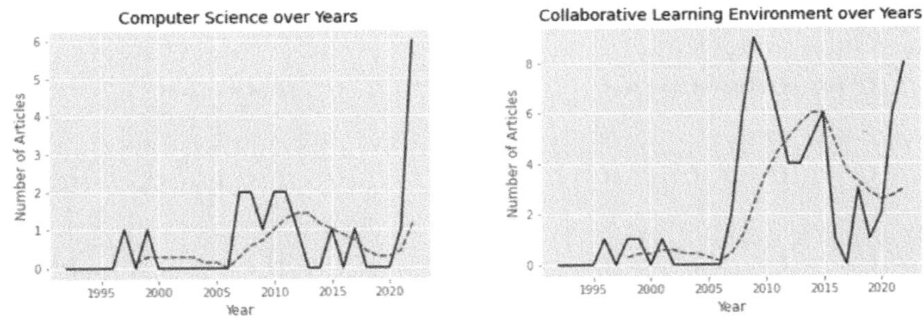

Figure 4.8. *Time plots computer science and collaborative learning environment. For a color version of this figure, see www.iste.co.uk/machado/industry.zip*

The topic of computer science experiences several peaks, such as in 2015, but was more relevant between 2006 and 2011 and indicates future importance due to the appearance from 2020 and onwards. Collaborative learning environments experienced a high around 2009 with a strong decrease up to 2016. Afterwards, the topic became of higher relevance once again.

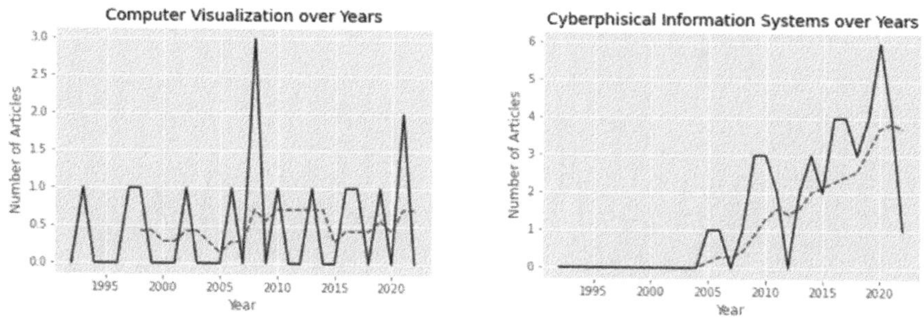

Figure 4.9. *Time plots computer visualization and cyber-physical information systems. For a color version of this figure, see www.iste.co.uk/machado/industry.zip*

The topic of computer visualization has been highly volatile throughout the entire analyzed time span and experienced several peaks between 2005 and 2010 and in 2021. Cyber-physical information systems indicate a strong upward trendline and are of highly relevant for future research.

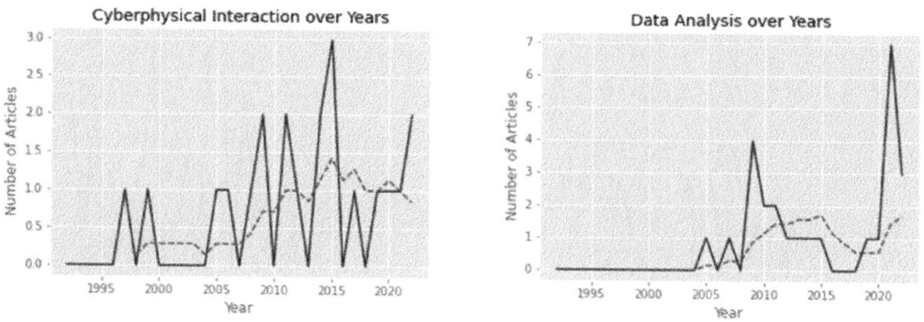

Figure 4.10. *Time plots cyber-physical interaction and data analysis. For a color version of this figure, see www.iste.co.uk/machado/industry.zip*

The topic of cyber-physical interaction is, once again, a highly volatile topic throughout the entire time horizon, with the highest relevance in 2015. Data analysis first experienced a significant peak in 2009, with the topic then being in decline, until its relevance rose again by the beginning of 2020.

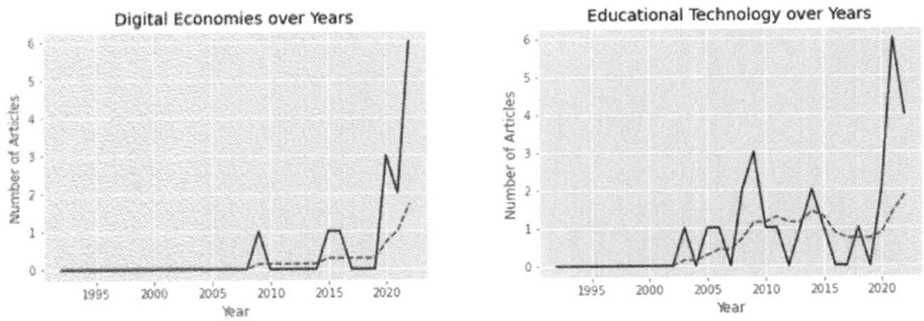

Figure 4.11. *Time plots digital economies and educational training. For a color version of this figure, see www.iste.co.uk/machado/industry.zip*

Digital economies experienced some minor peaks until 2020, but were a less significant topic in past research until 2020 when they became relevant. The topic of educational technology has been volatile throughout the entire time span, with significant peaks in 2009 and most recently in 2021.

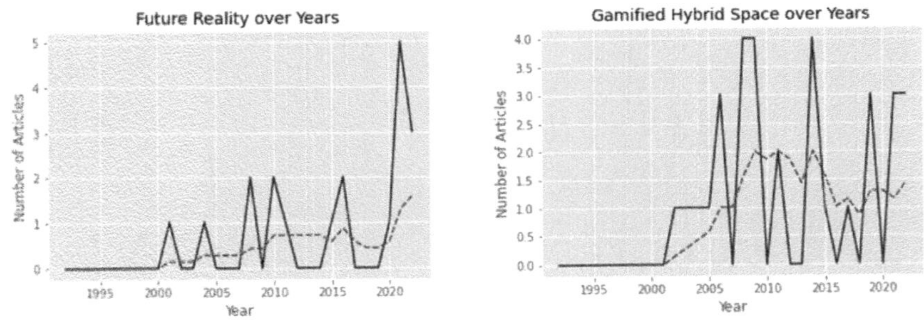

Figure 4.12. *Time plots future reality and gamified hybrid space. For a color version of this figure, see www.iste.co.uk/machado/industry.zip*

The topic of future reality has been discussed since 2000 and is highly volatile. It has gained dramatic importance since 2020. Another highly volatile topic is the topic of gamified hybrid space with several peaks, but the topic has not been extensively discussed for more than two years in a row.

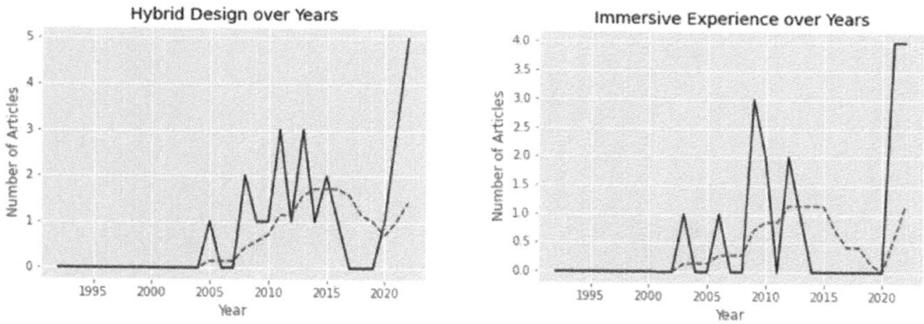

Figure 4.13. *Time plots hybrid design and immersive experience. For a color version of this figure, see www.iste.co.uk/machado/industry.zip*

The topic of hybrid design was a relevant topic from 2007 to 2015, but afterwards, research stopped discussing this topic. However, since 2020, a rapid increase has occurred. Immersive experience was mostly discussed between 2005 and 2012 and was irrelevant until 2020. In the present research, the topic has gained attention again.

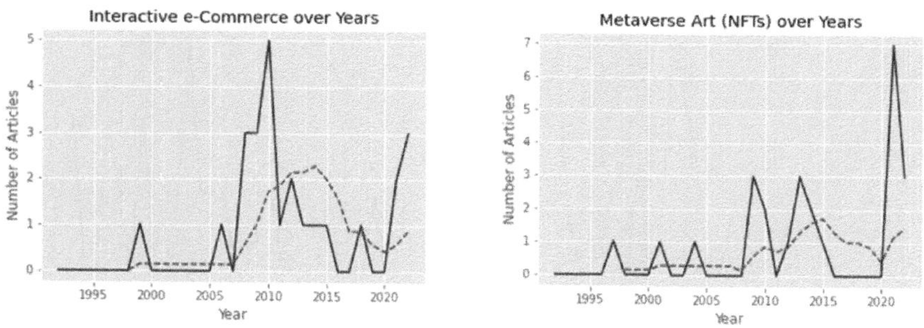

Figure 4.14. *Time plots interactive e-commerce and Metaverse art (NFTs). For a color version of this figure, see www.iste.co.uk/machado/industry.zip*

Interactive E-Commerce is a highly volatile topic, mostly discussed between 2008 and 2010. Afterwards, the topic was less important, but since 2020, an increased attention in research can be recognized. Metaverse Art was rarely discussed until 2020 as it is an actual topic and prior articles can be seen as pioneers.

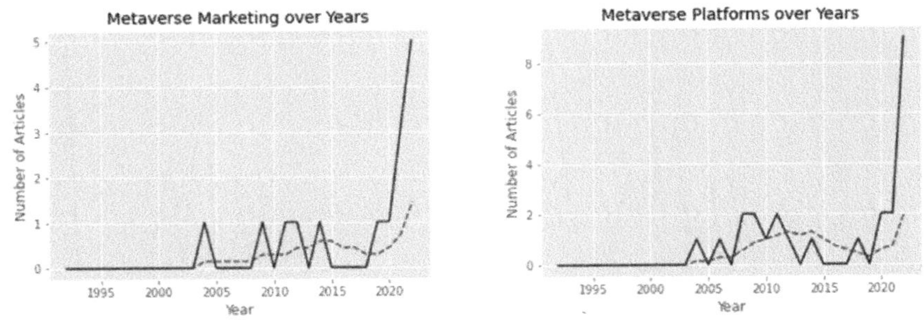

Figure 4.15. *Time plots Metaverse marketing and Metaverse platforms. For a color version of this figure, see www.iste.co.uk/machado/industry.zip*

Within Metaverse marketing, some articles were written before 2020, but present and future research will discuss this topic more extensively. For the topic of Metaverse platforms, the aforementioned statement can also be adapted. Metaverse marketing is rarely discussed in current research, but future research will discuss this topic in-depth.

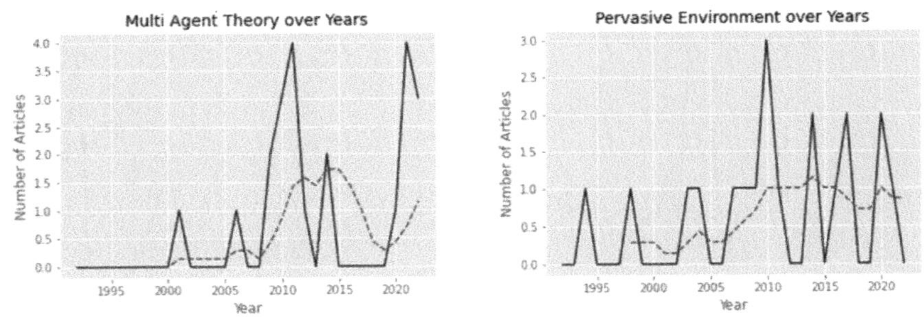

Figure 4.16. *Time plots multi-agent theory and pervasive environment. For a color version of this figure, see www.iste.co.uk/machado/industry.zip*

The topic of multi-agent theory has been highly volatile since 2000, experiencing a high in 2011 and 2021, indicating a high future relevance. The pervasive environment is extremely volatile, experiencing a high in 2010. Nevertheless, research has not discussed the topic since 2020.

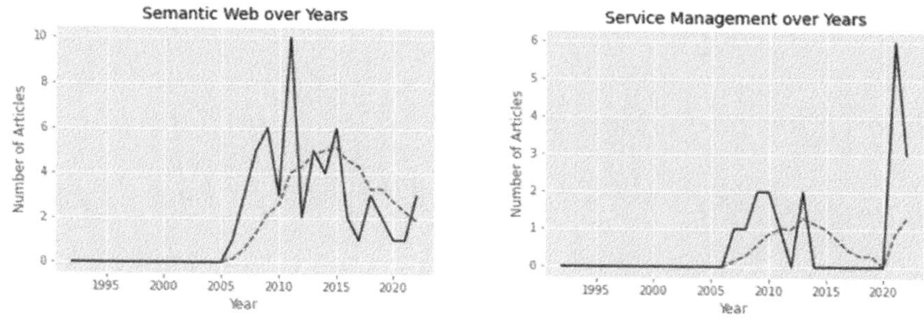

Figure 4.17. *Time plots semantic web and service management. For a color version of this figure, see www.iste.co.uk/machado/industry.zip*

The topic of the Semantic Web was first discussed in 2006 and received the most attention in 2011. Afterwards, a high volatility can be recognized, although the trendline is decreasing. Service management first appeared in 2007 but was rarely discussed. A significant increase in 2021 can be recognized, indicating that the topic will become more important in the future according to the trendline.

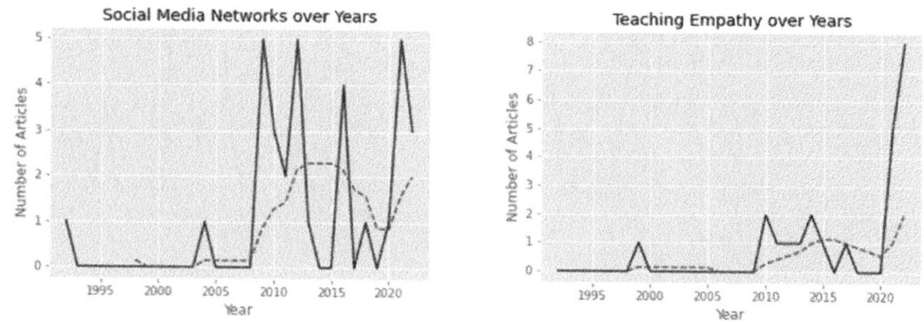

Figure 4.18. *Time plots social media networks and teaching empathy. For a color version of this figure, see www.iste.co.uk/machado/industry.zip*

Social media networks have been intensively discussed in the literature since 2008, showing a high volatility. The trendline indicates that this topic is heavily dealt with in present research and also possibly in future research. In contrast, the topic of Teaching Empathy was more or less irrelevant until 2020, but since 2021, research heavily deals with this topic, indicating a possible future research field.

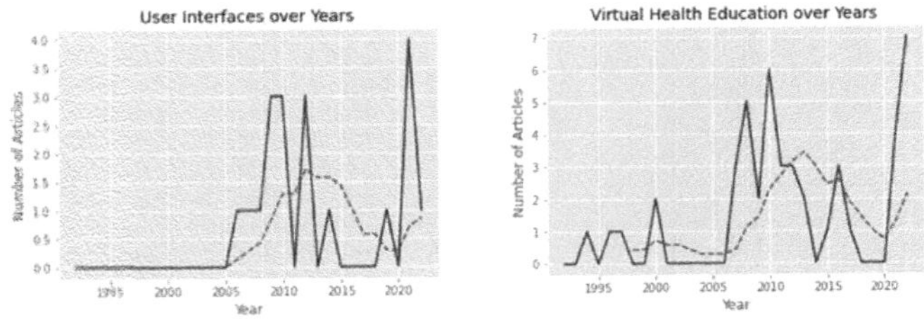

Figure 4.19. *Time plots user interface and virtual health education. For a color version of this figure, see www.iste.co.uk/machado/industry.zip*

The topic of User interfaces was unknown until 2006. Afterwards, high volatility was observed until 2013, before a rapid increase in the topics' appearance started in 2021. Virtual Health Education was discussed in the 1990s, with a gap between 2000 and 2007. Afterwards, the topic was highly volatile and decreased until 2020, but there was an increase in 2021 and 2022.

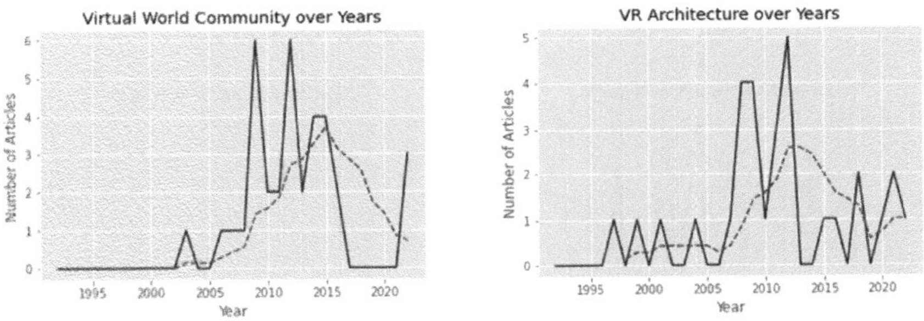

Figure 4.20. *Time plots virtual world community and VR architecture. For a color version of this figure, see www.iste.co.uk/machado/industry.zip*

The Virtual World Community was volatile between 2007 and 2015 and was nearly irrelevant until 2021. Since then, an increasing trendline can be observed. VR Architecture as a topic has been discussed since 1990, but between 2005 and 2012, research strongly focused on this topic. Afterwards, the focus shifted towards other topics supported by the decreasing trendline.

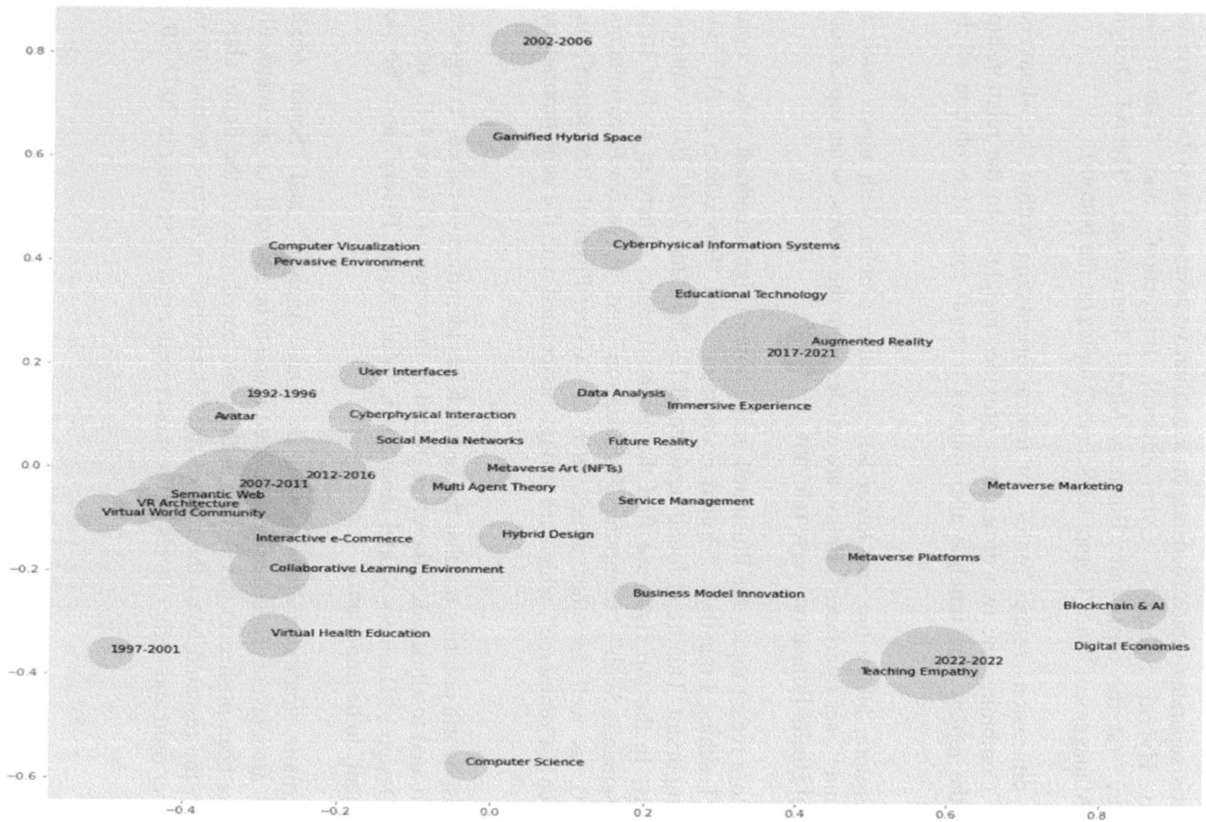

Figure 4.21. *Content analysis for topic groups (scientific Metaverse articles) over time. For a color version of this figure, see www.iste.co.uk/machado/industry.zip*

The two dimensions of the map in Figure 4.21 can be interpreted as follows. The horizontal dimension separates older topics (on the left) from newer ones (on the right), except for some themes, which were established during the 1990s, "Avatar" and "User Interfaces"; most of the topics are new and vibrant fields of study. The vertical axis separates topics focusing on a qualitative assessment, with a focus on "Gamified Hybrid Spaces", "Cyber-Physical Interaction" and "Educational Technology" (at the top), and "Business Model Innovation", "Hybrid Design, "Service Management" and "Computer Science" topics (at the bottom).

These dimensions reflect "poles" of topical orientation within the Metaverse literature. Combining the two dimensions provides insights into the relationships between the topics. The timeline also follows a spiral form of evolution of the themes.

Furthermore, in the top and lower left are older articles with topics such as "avatar" and "User Interfaces". Below are articles from the 1990s, when topics such as "Virtual Health Education" were relevant.

The more recent topics are illustrated in the bottom right quadrant, when topics such as "Teaching Empathy", "Digital Economies", and "Blockchain & AI" have become important. This trend found between 2012–2016 and 2011–2017 continues until today. In the top right, we find the essential topics. There is a clear trend toward "Semantic Web", "Computer Visualization" and "Pervasive Environments" (top right), as well as other topics such as "Cyber-Physical Interaction" and "Interactive E-Commerce", which are strongly associated with the contemporaneous era.

It is also worth pointing out that the research has identified so-called "evergreen" topics affiliated to all decades, therefore placed in the center of Figure 4.10, such as "Hybrid Design", "Business Model Innovation, "Multi-Agent Theory" and "Service Management" to be important.

Furthermore, traditional topics, such as "Data Analysis" and "Social Media Networks" are relevant. These terms have been found to be essential, i.e. geometrically, terms have been pushed to the middle in the topic coordinates of the figure above. Moreover, the research notes the role of the "Metaverse" in diverse relation to the lower upper right and higher levels of the most recent articles, starting in 2007–2011 and 2020.

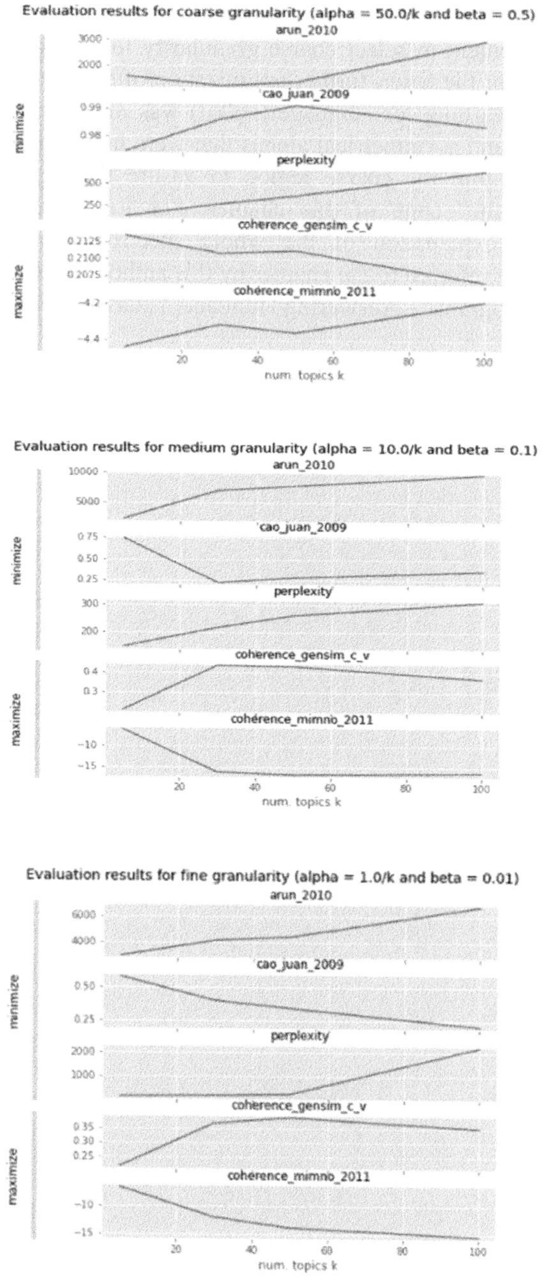

Figure 4.22. *Evaluation results according to various granularities of scientific Metaverse articles (coarse, medium and fine)*

After evaluating all of the above-mentioned procedures, from a statistical point of view, it was convenient to select coarse granularity for the scientific Metaverse literature. In the case of the latter, further optimization of the parameters using LDA MALLET (machine learning for language toolkit) was implemented. This resulted in a default beta of 0 and asymmetrical alphas that were individually determined for each topic, falling within the coarse region to ensure the highest accuracy and validity possible in the scope of this chapter. For the theoretical background knowledge and the explanation of the whole spectrum of this analysis, the distinguished reader is referred to the chapter by this author "Add more marketing to marketing doctoral programs – Answering Hunt and Yadav's calls".

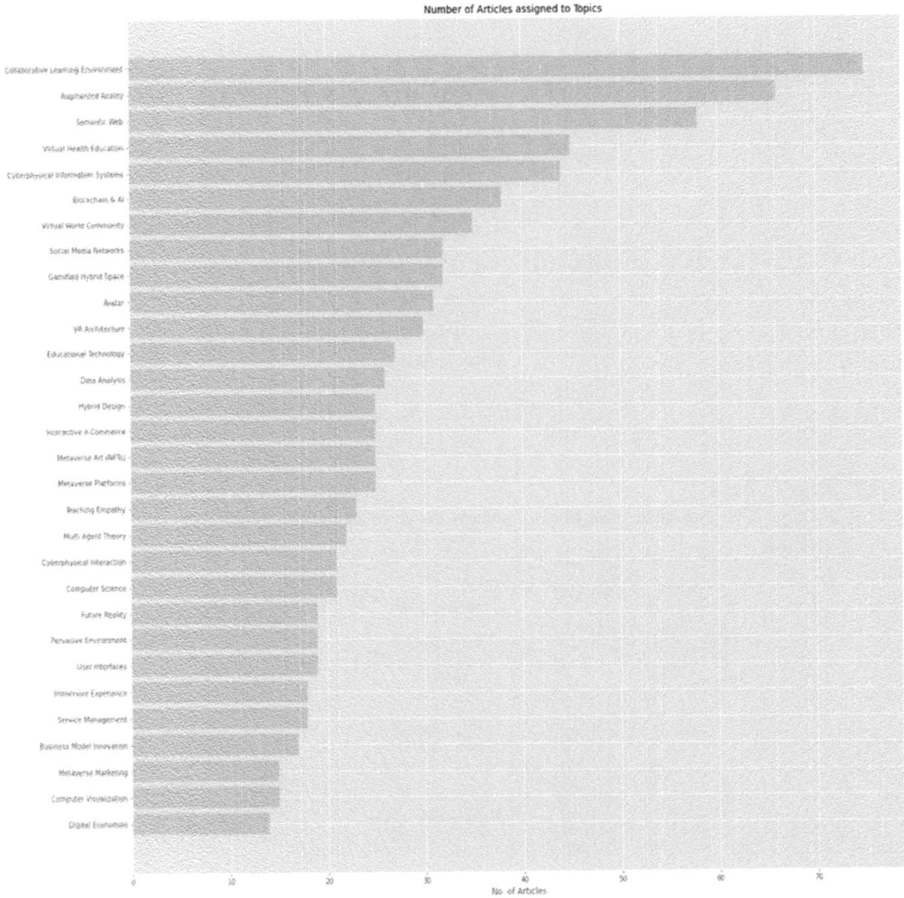

Figure 4.23. *Number of articles assigned to topics (scientific Metaverse articles)*

The most significant themes for this analysis were "Collaborative Learning Environment", "Augmented Reality", "Semantic Web" and "Virtual Health Education", as well as the other topics illustrated above.

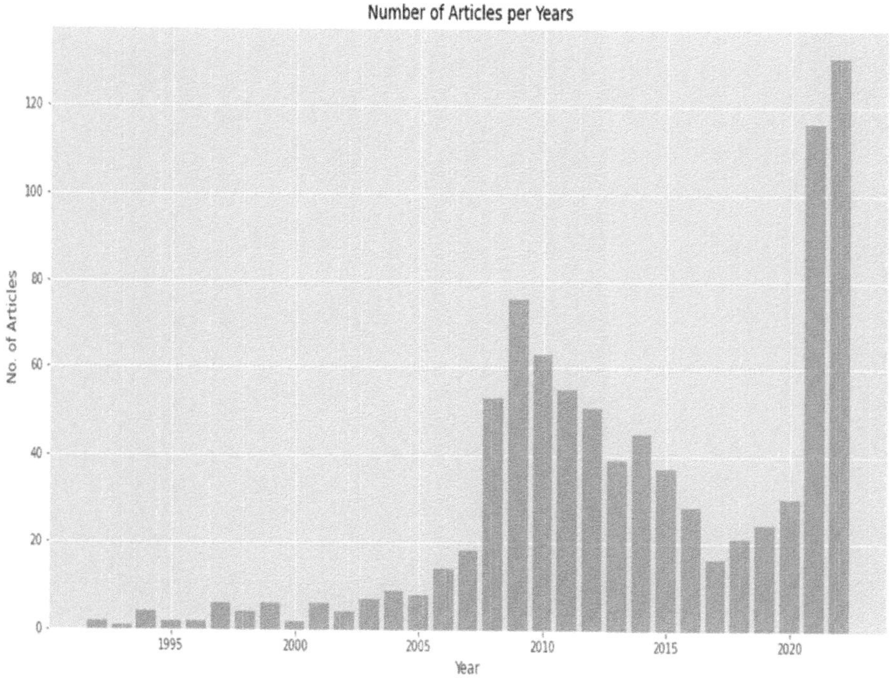

Figure 4.24. *Number of articles per year (scientific Metaverse articles)*

A strong peak in publications could be observed in the years 2008, 2009 and 2010, followed by a declining interest with the lowest point in the year 2016. A vital rise was observed in the years 2021 and 2022, possibly due to the lock-down during the Covid-19 pandemic and the nudge that Marc Zuckerberg made with the announcement of changing the name of Facebook to Meta. He committed to make 10 billion dollars worth of investment in the following years and proclaimed that Meta is not a social media company, but moreover a technology company. This opened up a strong interest in the theme, and many publications have been featured in this research.

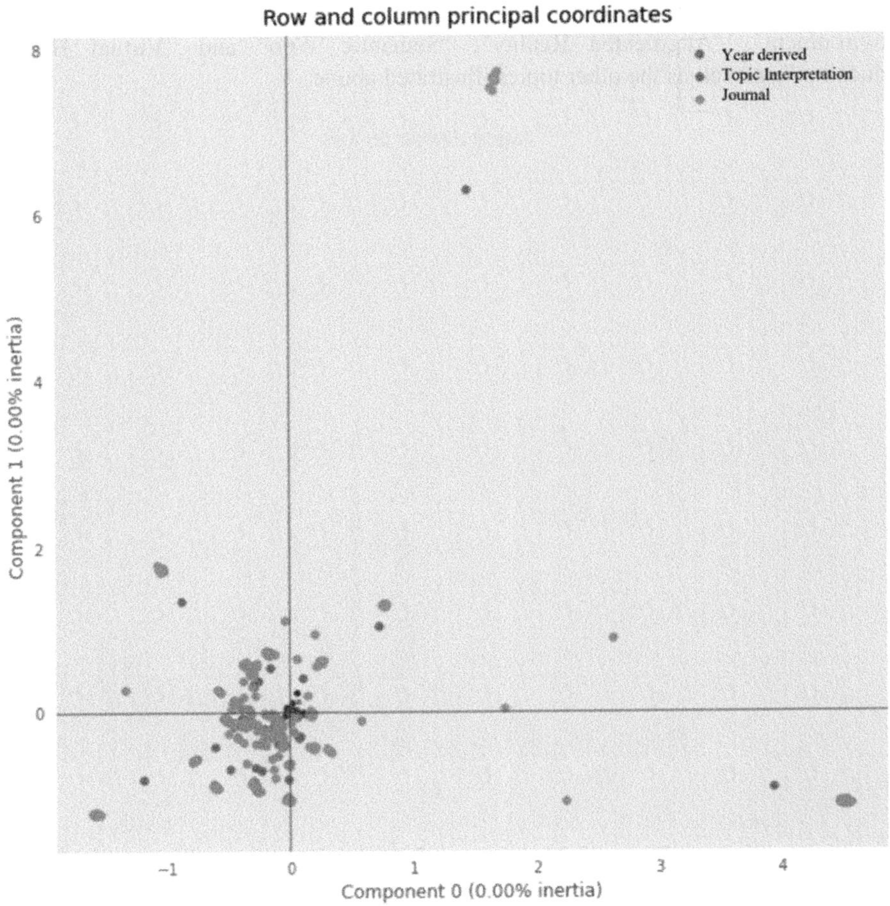

Figure 4.25. *MCA analysis of topics and terms (scientific articles). For a color version of this figure, see www.iste.co.uk/machado/industry.zip*

The MCA analysis illustrates the proximity of observable research article keywords (blue dots), topic assignment (eggplant color) and (concatenated) terms assignment (light blue dots). The dot size indicates the number of articles affiliated with the categories.

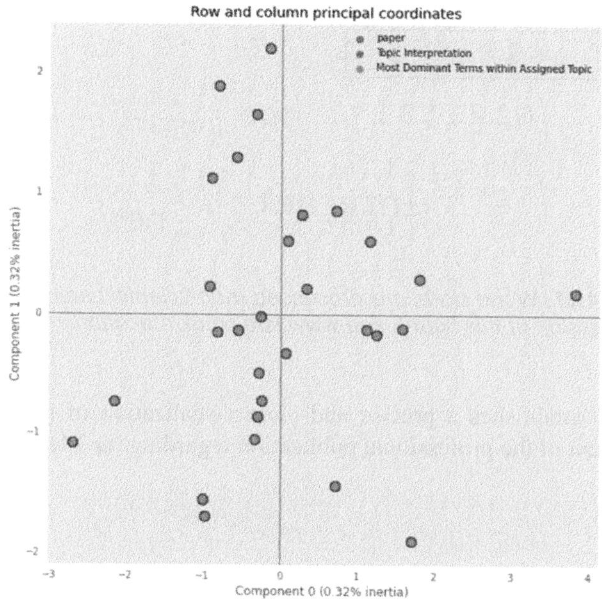

Figure 4.26. *MCA analysis of topics and terms (scientific articles). For a color version of this figure, see www.iste.co.uk/machado/industry.zip*

The results of additional MCA analyses are illustrated above. The analysis shows equally good results as the dots overlap perfectly. Therefore, the interpretation could be made that the results of the MCA analysis above indicate a strong match between the observed literature articles and the topic-term distributions. In succeeding analyses, researchers can apply more stable and precise visualization techniques, as presented in the Python library LDAvis developed by [SIE 14]. Moreover, the chapter defines a new relevance score to assess the affiliation of a term to a topic by a mixture of the log probability of the term *wd*, *n* and the approximated Variation Bayesian distribution over the empirical term frequency. This measure overcomes the shortcomings of simply ranking terms purely by their probability under a topic, as suggested by prior studies by [BIS 12]. A visualization method comparable to [SIE 14] was performed using an MCA among topic-term distributions.

4.2.2. *Structural content analysis of non-scientific sources*

As already established, the following figures visualize the essence of the Metaverse based on non-scientific articles.

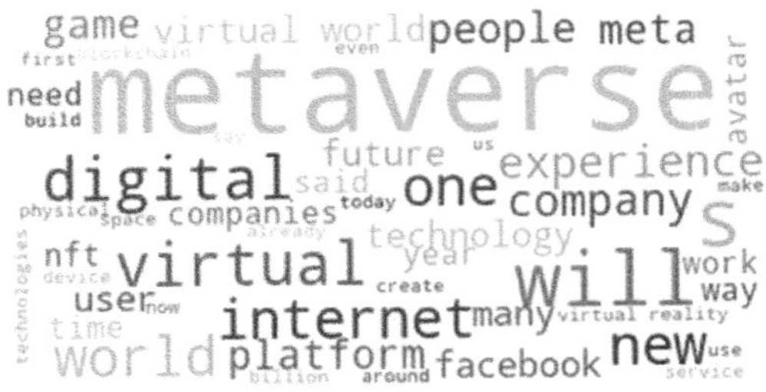

Figure 4.27. *World cloud pre-processed (non-scientific articles). For a color version of this figure, see www.iste.co.uk/machado/industry.zip*

Figure 4.28 establishes a precise and clear visualization of the terms relevant within the context of the professional publication regarding the Metaverse.

Figure 4.28. *Word cloud post-processed (non-scientific articles). For a color version of this figure, see www.iste.co.uk/machado/industry.zip*

The Metaverse embraces many relevant topics, such as "Digital", "Virtual World", "Platform", "Technology", "Facebook" and "Experience". However, many additional topics could be observed, such as "Avatar", "Internet" and "Game and Gaming".

Figure 4.29 is based on the combination of diverse smaller figures, and as the data below is very recent because of the nature of the field of the Metaverse, the essence of the non-scientific topics and developments were included to establish a holistic approach and lay of the land for the distinguished reader. Still, this recent history is non-conclusive and may be overwritten by publications and innovations that are underway. Therefore, and for the reasons of word count, as the chapter is limited to 30,000 words, only general comments are made, but the individual figures are established in a manner that the essential information is conveyed.

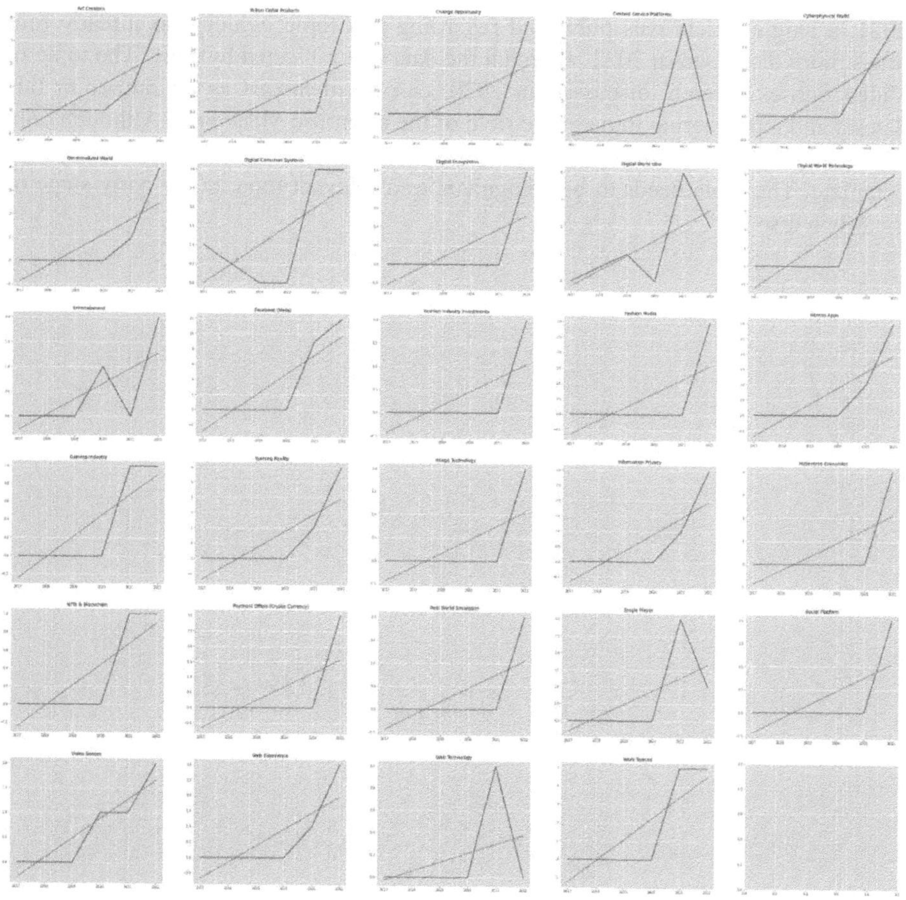

Figure 4.29. *Year plots (non-scientific articles). For a color version of this figure, see www.iste.co.uk/machado/industry.zip*

Because of the diverse and widespread nature of the notion of the Metaverse, especially after the presentation by Zuckerberg (Meta), a lot of paths and information are being discussed and many possibilities are being delivered. As its history started very recently, a conclusive insight has not yet been derived. As indicated by Figure 4.29, most topics were unknown until 2021 and 2022. Recent research seems to focus extensively on the field of the Metaverse. To emphasize the aforementioned statement, a few topics identified by the LDA will be discussed in the following paragraph. The topic of Billion Dollar Products was unknown in research until 2022. The topic has now been discussed in numerous articles, and as indicated by the trendline, future research will discuss the topic extensively. Moving onto the topic of the Decentralized World, the course of the graph is very similar. In

2021, a single article was published regarding this topic, which was already four times more discussed in 2022, although the data was collected by June. The topic of Video Senses, already discussed in 2020, can be identified as an outlier in this dataset and a pioneering topic in the field of the scientific Metaverse. Although only one article has been published regarding this in 2020, Your Free Plagiarism Check is complete. The topic needs to be recognized as a relevant topic in the early stage of the Metaverse.

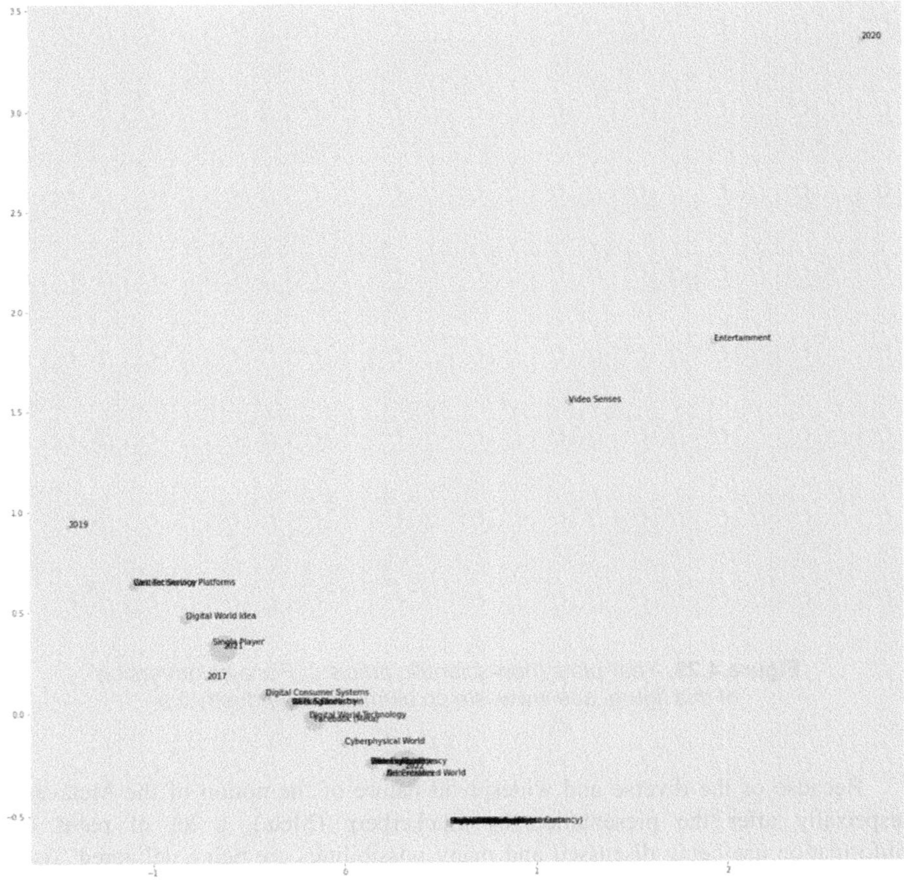

Figure 4.30. *Content analysis for topic groups (non-scientific articles) over time. For a color version of this figure, see www.iste.co.uk/machado/industry.zip*

Figure 4.30 is included only for the purpose of completeness. The topic and the publications regarding the Metaverse are recent phenomena. Furthermore, the contributions towards developing a solid methodology wherein AI-driven software such as Knime and Python could be made here are to state that Knime only works adequately and more precisely if a larger body of data is integrated for analysis. Otherwise, the results are less satisfactory and void, as Figure 4.30 illustrates.

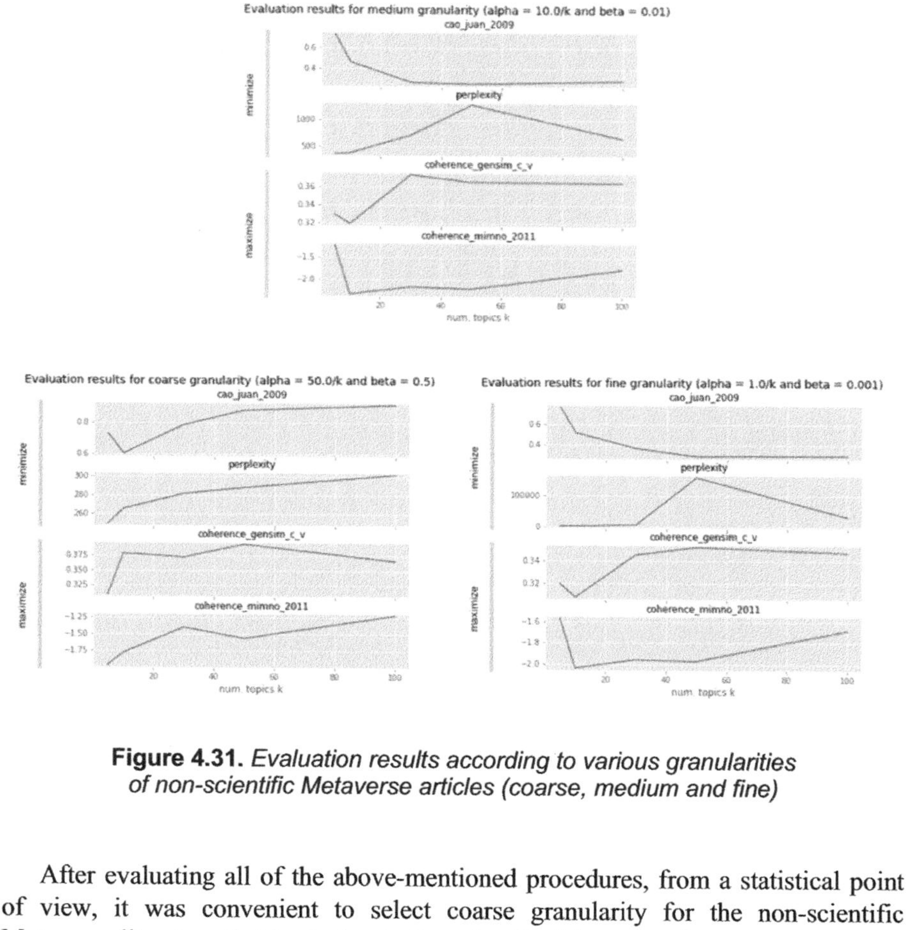

Figure 4.31. *Evaluation results according to various granularities of non-scientific Metaverse articles (coarse, medium and fine)*

After evaluating all of the above-mentioned procedures, from a statistical point of view, it was convenient to select coarse granularity for the non-scientific Metaverse literature that is similar to the scientific Metaverse literature. In the case of the latter, further optimization of the parameters using LDA MALLET (machine learning for language toolkit) was implemented. This resulted in a default beta of 0 and asymmetrical alphas that were individually determined for each topic, falling within the coarse region to ensure the highest accuracy and validity possible in the scope of this chapter. For the theoretical background knowledge and the explanation of the individual measures, the distinguished reader is referred to the chapter by this author *"Add More Marketing to Marketing Doctoral Programs – Answering Hunt and Yadav's Calls"*.

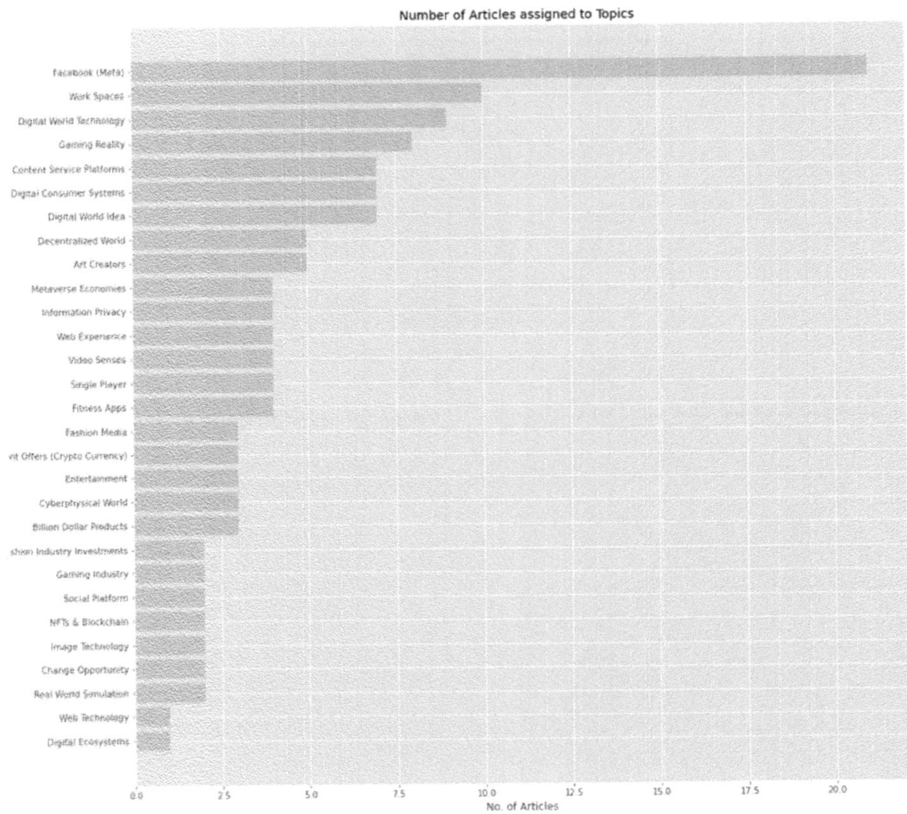

Figure 4.32. *Number of assigned articles to topics (non-scientific Metaverse articles)*

Based on Figure 4.32, we can observe that Facebook, which recently changed its name to Meta, and the themes that the firm pushes, such as "Work Spaces" and "Digital World Technology", are highly influential within the non-scientific domain of published articles.

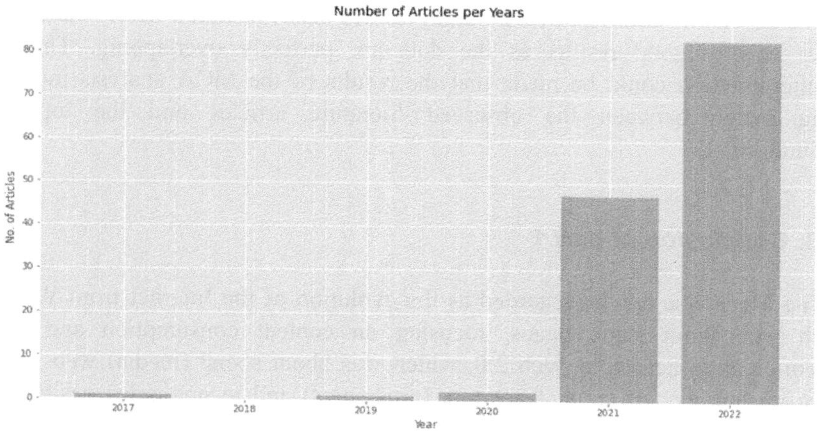

Figure 4.33. *Number of articles per year (non-scientific Metaverse articles)*

The years 2021 and 2022 were, so far, the essential years for the publication of non-scientific articles.

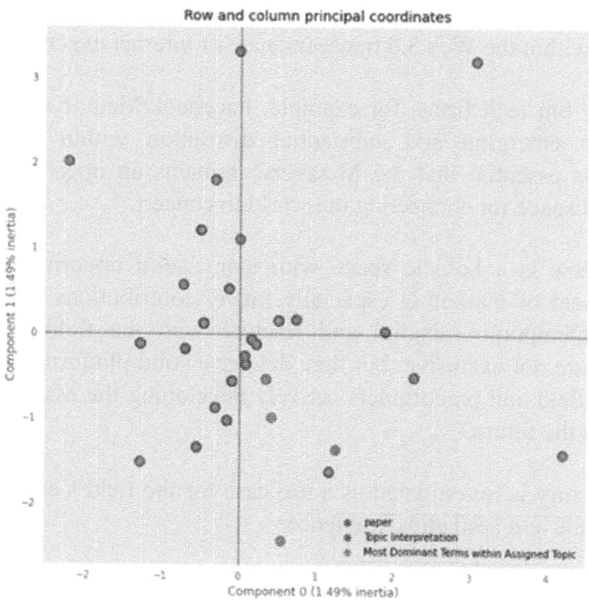

Figure 4.34. *MCA analysis of topics and terms (non-scientific Metaverse articles). For a color version of this figure, see www.iste.co.uk/machado/industry.zip*

The results of additional MCA analyses are also illustrated here. The analysis shows equally good results as the dots are perfectly overlapping. Therefore, the interpretation could be made that the results of the MCA analysis indicate a strong match between the observed literature articles and the topic-term distributions.

4.2.3. *Conclusion of part 1*

The Metaverse can be regarded as the evolution of the Internet from Web 1.0, which was about static pages, focusing on content consumption and a few content-creating actors, to Web 2.0, which was about social (media) web with an increasing number of content creators (prosumers) and technologies designed for sharing content, easing the process, and to Web 3.0, as the space of a decentralized semantic web within a tokenized-eco-system. The space is about smart, immersive experiences that are transparent and reliable. In addition, we own it. The Metaverse is the enhancement of disintermediation decentralized finance (DiFi) and decentralized autonomous organizations (DAOs). Actors and firms in this emerging field are participating in value-co-creation and co-evolution based on the blockchain platforms, the private but above all, open Metaverse (OS). The circular value-creating process is based on the gamified, merged brand connection and brand equity in a storytelling evolution from the Instagram Web 2.0 to the Metaverse digital persona within the Web 3.0 transcendence of Internet experiences.

While many big tech firms, for example, Facebook/Meta, have displayed vital interest in this emerging and substantial disruption within the evolution of cyberspace, it is essential that the Metaverse remains an open and decentralized public good and space for connecting the world in concert.

The Metaverse is a holistic space with many solid opportunities for diverse fields. In the field of marketing especially, many contributions could be made to correct the contemporary disjoint and fracture with the field. The key topics discussed here are not exclusive, but they deliver a solid platform for reflection for scholars in the field and practitioners actively developing the essence of the waves of innovation in the future.

In part 2, luxury is investigated as a use case for the field's evolution and as an example of artistic and aesthetic experience.

4.3. Luxury as an artifact of philosophy, phenomenology and culture in cyber-physical duality (part 2)

What is luxury? That is the question. To answer this question, it is important to investigate how to answer another essential question avant la lettre. Hence, it is primarily the question of the applied Weltanschauung, whereupon the answer to the former will be based. According to Heidegger: "Questioning is the genuine and the right and the only way of deeming worthy that which, by its highest rank, holds our Dasein in its power. This understanding of the being of ours, and being itself altogether, is therefore what is most worthy of questioning in all questioning" [HEI 83]. Therefore, what luxury is cannot be answered without our *Dasein* and how we arrive at reflection upon the question itself; hence, we are entering the roaming phase of the experience. Thus, luxury is not a natural phenomenon that can be investigated independently from the human mind; rather, it is an artifact of human experience.

All observations are value-laden, and hence, all modes of questioning are based on the cultivation of the habitus in a hermeneutical circle. Thus, questioning is the essence of what humans do. Evolution is nothing else than a history of the hermeneutics of questioning itself.

In essence, Heidegger has defined the interplay of questioning founded within the two modes of questioning: the guiding question (*Leitfrage*), which is embedded within the Metaphysical tradition, and the grounding (*Grundfrage*), which is concerned with not what is, but what is the essence of the phenomenon under investigation [BLO 15].

> What then is luxury? It is a word without any precise idea, much such another expression as when we say the eastern and western hemispheres: in fact, there is no such thing as east and west; there is no fixed point where the earth rises and sets; or, if you will, every point on it is at the same time east and west. It is the same with regard to luxury; for either there is no such thing, or else it is in all places alike [MOR14].

To answer what luxury is, the second question that needs to be answered in addition to avant la lettre is: Why can science not answer the question? Hence, to frame a question as a non-scientific question, we can either choose, in a Popperian sense, a question that is too large or impossible to answer in terms of pseudoscience, or a question that needs to be formulated as a philosophical Grundfrage, thus related to human existence (*Dasein*) and the phenomenon that relates to its *Dasein*. While an atom or a molecule exists independently of the human mind, it is not the same as luxury. Luxury depends on the human experience of it. However, "The question

"What is luxury?" is not a field of systematic research in the humanities" [WIE 19]. Thus, this is also true within science. To apply the scientific method is one dimension, and the second is to apply science as the collective repertoire of human knowing. We bring science further by understanding that it is an artifact of human knowing. However, we also bring science further by understanding its limitations and, thus, how these could be resolved by applying the philosophical method. Hence, there are phenomena that need to be investigated, but that are not accessible to us via the modus of reductio ad absurdum.

In Hume's essay "Of Refinement in the Arts", he relates the experience of luxury as a holistic, interrelated foundation. Hume states: "To be entirely occupied with the luxury of the table, for instance, without any relish for the pleasures of ambition, study, or conversation, is a mark of stupidity, and is incompatible with any vigor of temper or genius" [HUM 42]. Hence, luxury becomes the act and reflection upon cultivating taste and judgment, and not purely an item or artifact of marketing that can be bought or sold. [WIE 19] illustrates: "The questions at hand are: What must be the case for an object to be rightly called a luxury? What does one know about something when one knows that it is luxury?" We proclaim that luxury is the human experience of aesthetics of liberty, craft and a hermeneutical phenomenology of artifacts. However, luxury is also the human experience of appreciation of beauty, peace and harmony. Above all, within the century of technicity defining almost every fabric of human existence, luxury has entered the roaming phase of technology not merely by the price tag attached to a luxurious-looking item, but by the second-order foundation of what the experience of the artifact delivers to the subject in terms of the essence of its functioning.

Apple Inc. has made the notion of "simplicity" the predominant raison d'être of luxury when defining a man–machine interface, which, as a solid example of the essence of luxury, has made the firm an iconic brand and thus one of the most valuable companies. According to Steve Jobs: "Most people make the mistake of thinking design is what it looks like [...] People think it's this veneer – that the designers are handed this box and told, 'Make it look good!' That's not what we think design is. It's not just what it looks like and feels like. Design is how it works" [WAL 09]. Here, Jobs constructed the primordial definition of what technology is and has been historically transmitted since the ages, as *techne*, which meant, the art, the functioning, as modus of un-concealing (*entbergen*) and thus as the truth (*Wahrheit*) *Alètheia* (*Lichtung*). Jobs contributions were a deconstruction of the notion of utility preoccupying and driving techne towards a more fruitful ground of techne as luxury. This opens up the experience of luxury to the workings of our lives, which are driven by and concurrently benefit from advances in technology.

Hume furthermore observes: "But where luxury nourishes commerce and industry, the peasants, by a proper cultivation of the land, become rich and independent; while the tradesmen and merchants acquire a share of the property, and draw authority and consideration to that middling rank of men, who are the best and firmest basis of public liberty" [WIE 42]. While Wiesing illustrates: "The questions at hand are: What must be the case for an object to be rightly called a luxury? What does one know about something when one knows that it is a luxury?" [WIE 19]. While these questions could be debated for a very long time, this research asks the question of what luxury means as an artifact of human craft and aspiration within the latest developments of technology, namely the Metaverse. As we have entered the post-PC and the Internet has moved from the basement to our office tables, laps (laptop) and pockets (in terms of mobile devices), we are contemporarily witnessing another evolution of the Internet, which transcends towards alignment with our senses.

Thus, the Internet will occupy a 3D space in the form of our aligned realities that feedback from the interplay of a cyber-physical duality towards a unity where the boundaries are becoming blurry.

The essence of marketing and the Metaverse were established in Chapter 7 (*Add More Marketing to Marketing Doctoral Programs – Answering Hunt and Yadav's Calls*) and within the previous parts of this chapter. In this part of the chapter, the author discusses what luxury as an artifact and sphere of practice could mean within the evolution of the Metaverse.

4.3.1. *Luxury in the Metaverse*

To adequately answer the question of what luxury means in the Metaverse, and based on what foundational challenges the Anthropocene age is facing to observe what the essence of luxury is as new realities are emerging, we need to ask what prerequisites must be met, wherein luxury in the Metaverse as the unified field of constructing the innovation of the future could be defined and thus founded upon and how they could evolve? The Metaverse brings forth diverse possibilities, which are foundational not only for the firms to meet their sustainability objectives, but also for good brands' behavior, thus resulting in better and sustainable consumer experiences. Some precise points that could support this argument are elaborated below:

1. **twinning** of the artifacts as the medium of connecting the cyber with the physical;

2. **origin** of the artifact and the data about the item;

3. **sustainability** with ESG factors (Environmental, Social and Governance) evaluating firms and regions;

4. **access** by actors, designers, developers and users to develop the digital experience, for example, art, fashion and game;

5. **originality** of the non-fungible tokens (NFTs);

6. **owning** a part of the Internet beyond sharing and the physical twin connecting back to the original;

7. **experience** as a merged brand connection;

8. **circular** as in fashion by designing, sourcing, producing and providing as a circulate societal responsibility.

These dimensions go hand in hand with the evolution of marketing and the marketing P's that have defined marketing activities for decades.

Traditional Marketing	AI- driven Marketing	Marketing in the Metaverse
Product	Psychographics	Pieces (Twins)
Price	Predictive APIs	People (Actors)
Place	Proxy	Places (Cyber-Physical)
Promotion	Pervasive Environments	Play

Table 4.1. *Evolution of the marketing P's*

The essence of the 4 P's observed from the traditional marketing lens is known. However, these have changed in the age of AI as additional possibilities have emerged for marketing to evolve. Psychographics have been founded on the dimensions of lifestyles, personalities, attitudes, values, activities, interests, opinions, habitus and state of cultivatedness; thus, awareness of what culture capital represents and the illusio that could result as the doxa that could be transformed to core rigidities, which could then be exploited. The frenzy of independence of opinion, especially in the age of algorithms, needs to be further investigated as data is still an unexplored territory. Application programming interfaces (APIs) predict a numeric or categorical value induced from the data fed into the system or within a specific training set. Solid technological examples that we derive here are the search engines that provide feedback on the logic of computation in evolving towards better language recognition, recommendation engines, spam recognition and online activity detection (location), for example, fraud detection. While a proxy within the domain of AI would simulate the intellectual and emotional response architecture of specific users, proxies, furthermore, may construct additional spheres of innovation experiences on how art and luxury, or consumption per se, is observed, understood and evolved in the future.

The essence of pervasive environments is founded within a context wherein ubiquitous devices weave themselves into the fabric of individual realities, thus the everyday existence of humans in terms of real-time accessibility by anyone, anywhere, anytime, any system or anything, to be co-created by the participants involved in a man-machine interface.

Within the Metaverse, the dimension of the "4 P's" represents more evolved features. In general, the term pieces is defined as the embodiment of all products and services (twins), including the part of the 3D immersive Internet (Metaverse), such as land, skins and experiences (events) a user could purchase.

The notion of people represents all of the stakeholders and users, which starts with the developers to the gaming platforms. People are the communities that are navigating the cyberspace of the Metaverse interoperably, while places are also the gaming pervasive environments, but technically, they could be expanded to physical spaces, which are then integrated via AR or MR and XR features, and which are the intertwined spaces combining the digital with the physical spheres. Play, as the word defines it, means playful engagement with the cyber-physical spaces, thus all the activities that enable individuals/users to build, share, own and monetize their individual and virtual experiences.

However, the essential foundation of what is represented by luxury in the Metaverse is the nature of valuable and worthy and, at the same time, monetizing experiences. The essence of originality was discussed, and here we could base our findings on the example of Gérald Genta, to whom the luxury powerhouse Sotheby's paid homage in terms of coining him as the icon of the time [SOT 22]. Figure 4.35 illustrates such an experience of luxury in the Metaverse in terms of pairing and owning the NFT in addition to the watch, thus making virtual surfing within the immersive environments designed in part by the users' wearables. Furthermore, the NFT could be seen as an investment in the world of arts, thus pivoting towards a new luxury brand experience in the Metaverse that adds value to the duality of cyber-physical brand experiences.

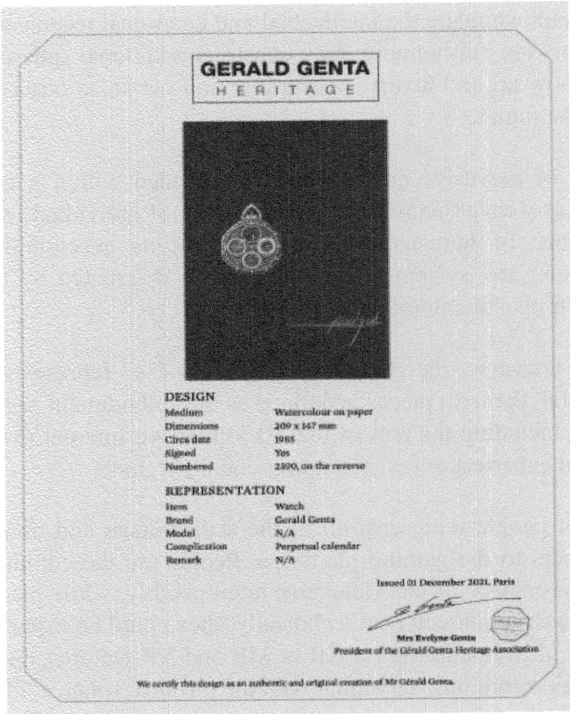

Figure 4.35. *Design of a Gérald Genta Watch with Accompanying NFT Circe 1985 [SOT 22]. For a color version of this figure, see www.iste.co.uk/machado/industry.zip*

The case of Beeple's (Mike Winkelmann) NFT "*Everydays: The First 5000 Days*" is different. Here, we are entering the roaming phase of purely digital spaces of experiences of arts, and thus of luxury. However, we need to observe one essential fundamental here. Luxury and the domain of art, respectively, of aesthetics, were originally driven by the cultural capital and habitus that was Western. Contemporary developments, however, are deriving more from the emerging world, such as India and China, and thus from the crypto-native and digital native generations that are formed within the cultural developments that have merged with technological innovations. The world was surprised that an Indian-born Tamil native would pay 69 million dollars for a digital piece of art. The surprise was also greater for the artist himself, who technically became famous and legendary for buying the third most expensive piece of art sold by a living artist. The next part of this chapter will highlight this new cultural space.

Figure 4.36. *"Everydays – The First 5,000 Days" by Beeple – the first purely digital work auctioned by a major auction house [KAS 21]. For a color version of this figure, see www.iste.co.uk/machado/industry.zip*

4.3.2. *Culture, the Fifth Space*

To understand and reflect upon the emerging fifth space of culture within the context of the global and digital societies, and thus the contemporaneous emergence of the *digital native* culture, we need to understand the concept of culture beyond the basic cultural treatment and, to some extent comparative treatment of culture within the managerial oeuvre, which mainly draw on the studies by Hofstede, Trompenaars and Lewis. While the author accepts the above cultural studies, the essence of the cultural treatment needs to be founded on holistic sociological grounds so that a more precise understanding can be derived. For this research especially, as the context of cultural semantics within the cyber-physical realities has been postulated, the model of culture studies needs to be constructed on more sociological frameworks. This approach seems to be essential; hence, based on Bourdieu's concepts, a much broader and foundational analysis of the contemporary cultural shifts within the ubiquity of real-time communication and, thus, global cultural transfers could be paved.

According to [BOU 18]:

> Cultural capital can exist in three forms: in the embodied state, i.e. in the form of long-lasting dispositions of the mind and body; in the objectified state, in the form of cultural goods (pictures, books, dictionaries, instruments, machines, etc.), which are the trace or

> realization of theories or critiques of these theories, problematics, etc.; and in the institutionalized state, a form of objectification which must be set apart because, as will be seen in the case of educational qualifications, it confers entirely original properties on the cultural capital which it is presumed to guarantee.

Within the embodied state, i.e. in the form of a long-lasting recursive arrangement of the mind and body, a person belonging to a social structure develops a habitus.

According to [BOU 00]:

> The habitus is not only a structuring structure, which organizes practices and the perception of practices but also a structured structure: the principle of division into logical classes which organizes the perception of the social world is itself the product of internalization of the division into social classes. Each class condition is defined simultaneously by its intrinsic properties and by the relational properties that it derives from its position in the system of class conditions, which is also a system of differences and differential positions, i.e. by everything that distinguishes it from what it is not and especially from everything it is opposed to; social identity is defined and asserted through difference.

Habitus could be seen as the result of the holistic embeddedness of a social actor from birth to cultivation, which gives humans the capacity for observation and, thus, seeing. The foundational premise of the habitus could be constructed on the basis that habitus is the ability to see, move and act within the social milieu.

Within the objectified state, cultural artifacts could be seen as technological foundations, as already discussed, embodying instruments, machines, AI and (digital) humanities based on shared pictures, books, dictionaries and music.

However, within the objectification which is based on personal and non-commodity achievements as educational qualifications, handcraft and abilities are socially regarded as products of high culture and cultivate an actor's ability to occupy better and more responsible positions.

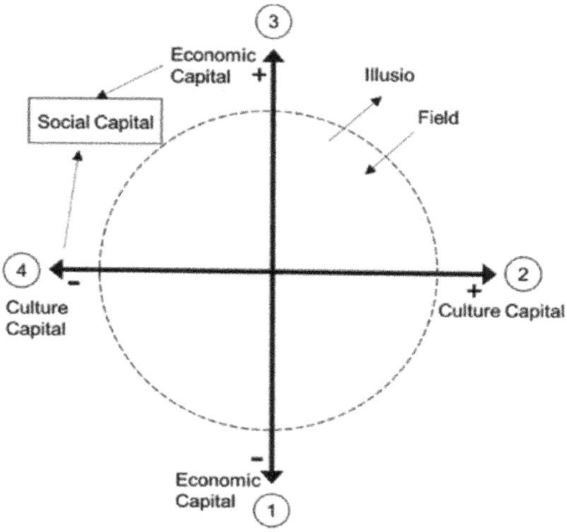

Figure 4.37. *The concept of culture capital according to Bourdieu [KAM 21]*

Figure 4.37 displays the foundational elements of Bourdieu's oeuvre. Bourdieu does not consider the individual actor as the focal point of the developments within society; moreover, he sees the milieu of the actor, wherein they are embedded as the essence of sociological research and practice. Each field has its own power games and struggles. That is the definition of the field, where the collective aspirations and angst of the participants are shared and believed in the games, territory, rules and capital in Bourdieuian terms.

The function of the term illusio in Bourdieu's field theory is derived with reference to Johan Huizinga's *"Homo Ludens"* of 1939 [HUI 09] and etymologically from Latin ludere, "*to play*", and is also used by Bourdieu in this meaning: to play one's game with; to play a game to gain power and recognition. Bourdieu's illusion concept goes back to his research since the 1980s. Illusio describes how a social actor plays a game in a field with a specific code of conduct (rules), and Weltanschauung applied within a field-specific social circumstance, where the actor is ensconced in it, how strongly they believe in these rules and the recognition of the utility and gain, and the commitments expected within the social field. It is about the adoption of the field-specific collective "*habitus*", which Bourdieu also refers to as the "sense of play" and "embodied social play". Culturally trained and cultivated with it, the actor can struggle and fit within the field, i.e. always avowing and "replicate" the game [BÖN 14]. The social capital is always a result of an interplay between the culture and economic capital an actor poses.

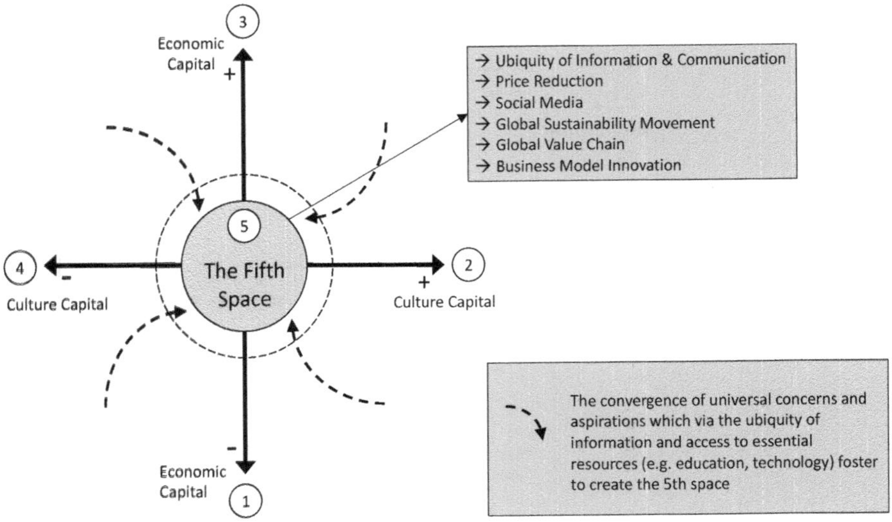

Figure 4.38. *The fifth space of contemporaneous culture developments [KAM 21]*

In the fifth space of culture, as illustrated in the figure above, the boundaries of the diverse fields, strictly separated by Bourdieu, are breaking towards harmonization of struggles, social games and the acquisition of culture capital and rules. The digital native generation is aware of the global struggle and challenges for power, dominance, culture capital in all three forms, sense of aesthetics, boundaryless cosmopolitanism and additional aspiration, while sharing the same illusion regarding feminism and minority inclusions, fears of global warming, and concerns regarding privacy violations, the asymmetry of power established by the big tech firms, threats of AI and the rise of populism. However, the figure above summarizes the essence of the fifth space of culture within the context of the global digital native generation.

A foundational connection is made here, as described by Hume:

> Another advantage of industry and refinements in the mechanical arts is that they commonly produce some refinements in the liberal; nor can one be carried to perfection, without being accompanied, in some degree, with the other. The same age, which produces great philosophers and politicians, renowned generals and poets, usually abounds with skillful weavers and ship-carpenters. We cannot

> reasonably expect that a piece of woolen cloth will be wrought to perfection in a nation, which is ignorant of astronomy, or where ethics are neglected [HUM 42].

Within this context, the fifth space as the new emerging field delivers a solid ground for culture capital themes to be developed and extended and, in addition, for it to not be recognized as a Western-driven context. Within the boundaryless Internet and the Metaverse, many possibilities will arise that the author has structured within the integrative marketing model, which will be described in the following section.

4.4. Toward an integrative model of marketing (part 3)

Based on the foundational contributions of [HUN 20a], the diverse streams of research in marketing to integrate the needed indigenous disciplinary foundations, whereupon the objective of a re-institutionalization of the marketing discipline with a strategic eye towards solving contemporary major challenges, could be designed and paved. While we are mourning Hunt's very recent passing, his legacy, however, will remain with us forever.

Based on the author's experience in research, a personal conclusion could be made that with a few exceptions in the field of marketing, we can postulate that Hunt was marketing theory, and all the rest are just footnotes to him. However, his contributions were more from the academic Weltanschauung and embedded within a theoretical standpoint, thus giving the marketing discipline the need for rigor while not being very widespread beyond scholarly marketing and its milieu [HUN 20a]. Concern and call still remain open, not only because of the nature of the particularly quantitative/modeling focus of the most cited contributions (see Chapter 7 *(Add More Marketing to Marketing Doctoral Programs – Answering Hunt and Yadav's Calls)* and previous parts of this chapter), but moreover because this research has also illustrated that a single quantitative/model approach has been mainly embraced, namely the structural equation modeling. Solid ground for hope also has been highlighted in that many of the additional contributions pertaining to the essential publications within the field have been based on a mix of a conceptual and quantitative nature; here, the research can highlight the five most cited contributions [FOR 81; VER 21 CAR 17; HEN 15; ZHA 20]. Research confirms [YAD 10, 20] observations about the nature of the decline of the conceptual articles and the lack of indigenous theory-building capacities in marketing [HUN 18]. This research still supports the evidence that conceptual-only articles' contributions have been foundational for the field. In terms of the field's homeostasis and autopoiesis of reviving the vitality of marketing with the designing of indigenous theories pertaining to marketing science, the author can proclaim that conceptual contributions are vital. At the same time, the few "conceptual-only" articles could measure themselves on an equal footing with the quantitative models published under the auspices of marketing science.

Here, we also see a clear necessity for additional theoretical contributions that could deliver solid impulses to the quantitative-based and consumer-behavior-driven research to expand their themes in the image of foundational marketing aspirations, such as the dimensions of establishing sustainability [LUN 18], temperance, and mindful consumption [SHE 11], coping and framing a marketing definition of AI, AR, VR, MR, XR, BCT and the Metaverse. Furthermore, by defining solid organizational theories for managing global enterprises so that adequate marketing decisions can be dealt with and resolved, marketing would place itself in a relevant field with a cognitive identity and resolve the contemporary fractured state wherein marketing finds itself [KAM 21].

The dimension that over 2/3 PhD's in marketing are not practicing the essential self-identification with the field as marketing scholars [HUN 20a], is a dire warning for all actors involved in marketing [YAD 20], especially when the many new doctoral researchers in marketing are not trained to 1) self-identify with the substantive content that may propel the field further; 2) understand the notion of holistic marketing philosophies [SHE 21] and the essential methods that would make the field's intellectual self-reproduction of scientists possible [WIL 03].

No field can evolve further when the raison d'être of scholarship within the field is actively promoting majoring in minors, which means extending the state of the art within a very narrow and micro concentration of contributions within a holistic field by nature, while it is concurrently facing a large volume of internal and external constraints [WEB 13]. Thus, the results of the research could confirm that the de-institutionalization process of marketing is further promoted [HUN 20a].

This research confirms that the fields of SCM and consumer behavior are not really represented within the contemporary directly relating publications to marketing and research streams of contemporary marketing today; however, we must integrate them within marketing so that a holistic and complete picture of marketing is derived. The researchers in both fields of CB and SCM do not consider themselves to be related to marketing, while the collective history of the field tells a different story.

The author can furthermore confirm the importance of digital transformation as a contemporary phenomenon shaping the intellectual landscape of marketing. The foundation of digital and social media amplified marketing within the context of the marketing discipline is ubiquitous; however, social media amplified by the rise of AI

[DAV 20] is a theme that marketing science still needs to absorb so that the field can act on the prophecy of [ALD 57], who called for a marketing definition of utility and value, and in this case of the whole spectrum of social media marketing amplified by the AI-driven advances for marketing, which has now been transformed to create the Metaverse as a platform, whereupon marketing could construct solid opportunities to make a vital impact. The rise of social media and the notion of digitality is not a new field for traditional marketing. Moreover, it needs to be absorbed as a channel of marketing practice, wherein marketing can flourish, which is purely marketing practice via the digital channel of social media amplification. While many solid opportunities for growth alignment and impact on societal well-being have been missed by traditional marketing scholars, especially AI-driven marketing science, which has been left to contributors who have a lot of knowledge of coding, which is relevant, but without any insights into the history of marketing. It is dire that there has been such a disjoint and revolution within the field that is not pushed by the history nor insights that made marketing what it is, namely the discipline that connects the firm to its environment, but which is based on the notion that data is the new oil.

Another essential question remains unanswered: How could marketing offer a platform of alignment by bridging the dimensions of shared value (see Porter/Kramer 2019), thus solving social challenges in a profitable manner? This would be a foundational approach that would bring the whole notion of CSR into an evolutionary mode. It may define a solid path in furthering marketing aspirations of a just world and not a world disjointed by utility.

The global movement for sustainability production and offerings at the bottom of the pyramid, and particularly answering to the needs of the emerging economies by guiding the evolution of the third world in a manner that the mistakes which the West, embedded within its habitus of dominance, has committed regarding sustainable behavior and temperance in consumption [SHE 21] could be avoided, and a much richer ground for an innovative approach towards the notion of sustainability is taken as a calling for marketing to awaken to the challenge.

The essential question is how researchers could bring sustainability into the main theme for marketing science in a manner that fits within the dimension of civilizing capitalism, while embracing sustainability as a major research stream in a way that frees its spirit from the walls of corporate do-gooding and CSR, towards making sustainability truly profitable on a business model proposed by indigenous marketing thought.

In addition, marketing needs to overcome the disjoint with the dimension of organizational theory. As [HUL 11] emphasized the notion of a boundary-spanning marketing organization, the author wants to pay attention to the importance of solid organizational structures and models that make firms navigate challenging times. Marketing cannot take the notion of organizational structure as a given and something that needs to be driven and understood by marketing as its foundational element. Structure does not follow strategy, it only does in stable waters, but when confronted with complexity, another logic applies: the basis of structure is the strategy. Marketing science, especially in times of rapid disruptions, where short-term but adequate responses are required and agile models of new work [BER 19], remote work, especially in post-Covid-19 global societies, needs to be structured and aligned on more stable conceptual grounds and not left to chance to design indigenous models of embodiment.

This research postulates the notion that traditional marketing should lead the next developments of a rapprochement between high-tech and marketing, especially observed under the dimension of the evolution of the breed of marketeers, *"the marketing-technologist"* [IAN 14; MCL 14].

Therefore, marketing needs to develop theories that could further the field or import theoretical inspirations from holistic sciences such as cybernetics, systems sciences, (digital) humanities and philosophies. Embracing Metaverse research and AI technology as a bridge within the marketing-driven strategy could deliver solid bridges toward contemporary revolutions in science and technology.

The new role of marketing could be divided into four main pillars, whereupon four paths are aligned to give the field the essence of a foundational discipline and shape our global and digital societies. Marketing needs to be protected from a pure medium of advertisement habitus.

Therefore, the **six dimensions** that marketing needs to embody are:

1. Marketing activities should be aligned with nature and sustainable conduct.

2. Marketing should not only be seen as relevant in a market. Moreover, the global digital cyber-physical reality should be seen as the domain of marketing.

3. The notion of human aspiration and the digital native culture while working on the globalization of totality should be understood.

4. Marketing must capture the essence of customers beyond pure selling and USP.

5. As technology is the essence of our world and humans are un-concealing artifacts that shape the human condition, especially in the age of Anthropocene, marketing must design a value perception of technology in Aldersons postulated call from the marketing lens.

6. Organizations are the artifacts of man, used to organize a large body of people in concert to work for a larger and worthier ideal. The delusion of shareholder value needs to be replaced by a raison d'être for marketing, wherein organizations understand decentral value creation and are aligned with enhancing value for stakeholders and society concurrently.

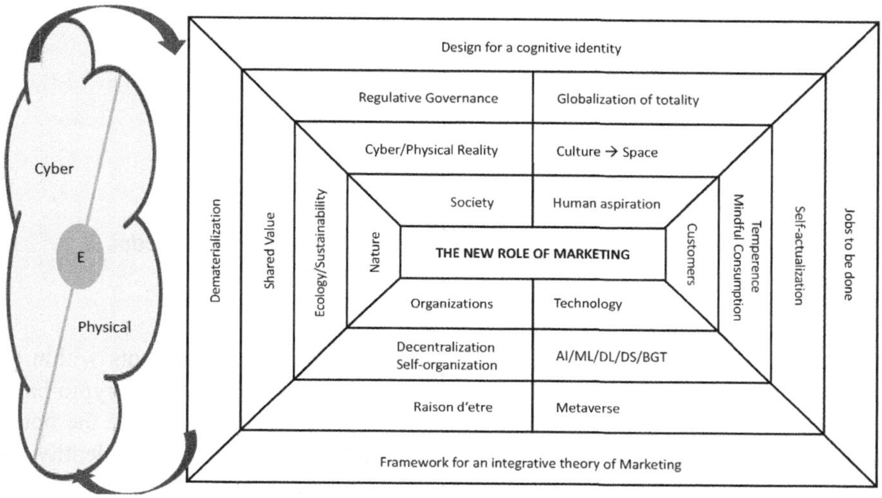

Figure 4.39. *Theoretical framework for the new role of marketing in cyber-physical realities*

In essence, marketing is the discipline that copes with environmental complexity, as visualized in the left part of the figure above. The dimension of nature is extended towards bridging an indigenous theory for capturing ecological and sustainability dimensions. Sustainable behavior can be integrated to design a solid ridge upon which marketers can design products and services that are aligned with the contemporaneous urge of the digital native generation, who are aware of the conduct of companies due to global connectivity and the ubiquity of information as a major frontier for innovation for the field. Furthermore, the notion of shared value, as illustrated in Figure 4.39, is essential for marketing to go forward with its own evolution to become an agent and philosophy towards a de-materialization of marketing foundation and thus conduct. Here, the cyber-physical realities of the emerging Metaverse marketing spaces deliver a solid foundational platform for change. The shared value concept illustrated below is built on the pillars of solving social needs profitably by designing a business model innovation, wherein marketing opportunities are created and turned into competitive advantages and, thus, corporate assets in the long run.

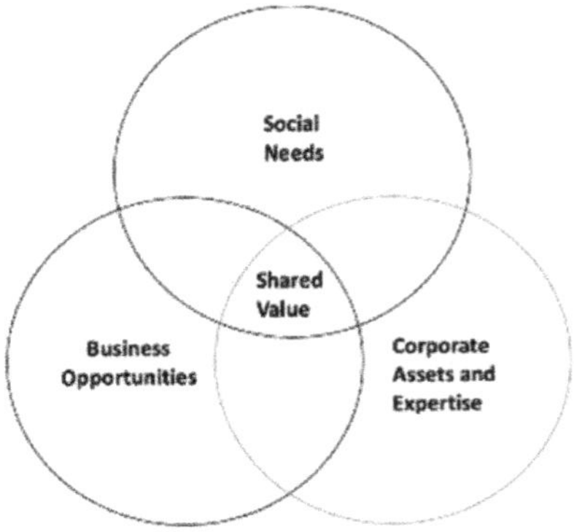

Figure 4.40. *Foundational premises of the shared value model. Source: author's own illustration based on [POR 19]*

The essence of dematerialization embodies the recent developments within the NFT worlds of the Metaverse and the pure artifacts designed within the crypto-based maintenance of pure digital identity, embodiment and originality. While the notion of twinning and pairing has been based on the cyber-physical product identity and designers are addressing the digital-only needs in the dematerialized state, these digital-only spaces are to be seen as the next frontiers of innovation. These are items and spaces that are owned within the interoperability of the Metaverse platforms. However, there is a difference between land that is bought or rented within the Metaverse or an artifact (NFT) from a fashion label to additional providers of these services. In addition, the Metaverse offers services and possibilities for entertainment beyond the pure gaming experience. This encompasses visiting an art gallery, working within the Metaverse in an individually designed workspace, participating in social activities from dating and dancing to entertainment, and visiting a concert with up to millions of participants.

Figure 4.41. *View of the Metaverse wherein a space for rent is offered. Source: Decentraland.org. For a color version of this figure, see www.iste.co.uk/machado/industry.zip*

Figure 4.41 illustrates how land ownership in the Metaverse could become a solid business, real-estate management and investment opportunity. Going forward, the author introduces some essential Metaverse marketing dimensions that are important to highlight. These new developments within the marketing evolution, as visualized in the table below, are consequential for an indigenous theory development within contemporary marketing science.

Type of Marketing	Definition
Social (s)- commerce	Social commerce, where A2A actors socialize and play while consuming and experiencing offerings
Decentralized (d)- commerce	3D marketing spaces with automated coordination, where creators, communities and merchants are co-creating seamless exchange of digital assets for physical products, services and experiences
Metaverse (m)- commerce	The evolution of digital commerce from 2D to 3D spaces of cyber-physical realities
Immersive (i)- commerce	Is based on the immersive involvement where internet and human senses merge
Experience (e) commerce	Experience commerce provides the frontier for experience innovation as the artefact of marketing
Pervasive (p) commerce	Is based on pervasive and embedded computing, where ubiquitous mobility is aligned with IoT and multiple device experiences. Where smart materials are aligned with sensors powered by low-cost, low-power devices
Autonomous (a) commerce	Autonomous commerce connects the world of customers and providers together. Enabling smart match recommendations, fast shipping, seamless returns and a superb UX innovations

Table 4.2. *Evolution of marketing within cyber-physical environments*

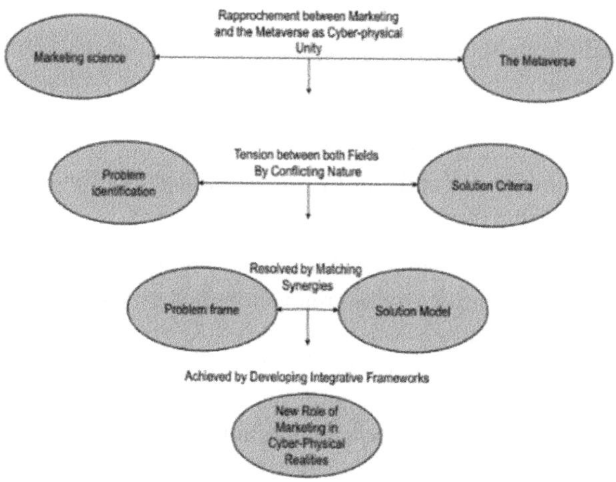

Figure 4.42. *Framework of rapprochement of marketing science and the Metaverse. Source: author's own illustration inspired by [CRO 06]*

As highlighted in Table 4.2, the self-organized, gamified, social and game-theory-based m-commerce innovations space could be aligned as an integrative framework or selectively assembled to create complexity by design as the collective Eigen behavior and *"jobs that are done"* for the customers, thus reflecting a competitive advantage for the cyber-physical firms.

The model above establishes a model of rapprochement between both fields by dissolving the tension between marketing science and the Metaverse by identifying the challenges both disciplines face contemporarily and visualizing a solution search space. Furthermore, this pursuit of observation is put into an integrative context of ideation and thus matched by a model that is put into a contemporary stage of developments within the fields. The whole dimension of the rapprochement between both disciplines is grounded within the holism of the new role of marketing in cyber-physical realities, which embodies the essence of this research and has been established throughout the chapter.

4.5. Conclusion

This research could be concluded by offering a design for going forward and recommendations based on the fundamental dimensions created within this research for the Metaverse, based on the foundation of the marketing discipline by visualizing the use case of luxury semantics in the Metaverse.

1. What is the Metaverse in terms of a marketplace of socioeconomics and ideas?

As established throughout the chapter, the Metaverse is a marketplace of ideas, a space of experience innovation and a foundation for enhancing sustainable marketing behavior in terms of dematerialization and thus experiencing temperance. According to Metakoven, the Metaverse is the new iPhone. It is the poetics of *"experience innovation"* and *"jobs to be done"* within the unity of the cyber-physical spaces that merge to create a unity of reality.

2. What are the essential topics in Metaverse literature based on all published articles (scientific/non-scientific) within relevant publications?

Part 1 of this chapter established the essential dimensions whereupon the Metaverse is founded. In diverse figures and tables "Time Plots of Scientific Literature" were plotted and identified. In addition, the author has also gathered the most relevant non-scientific data on diverse "Time Plots of Non-Scientific Literature" on the subject, which are visualized in Figure 4.27, but essential publications for the symmetry of holism that is required to judge a topic as essential as the Metaverse may take some time as the topic is very recent.

3. How can marketing science and the Metaverse be aligned based on an intertwined framework?

The essence of a possible alignment between marketing science and the Metaverse could take place, where the Metaverse is driven by traditional marketing seeking and finding its own raison d'être within the cyber-physical realities of the upcoming future. The fractured state of marketing needs to be designed for holistic viability. This research has shown how such an alignment could be designed. A possible model of alignment is delivered in Figure 4.40, which is an integrative model for marketing to cover the essential dimensions and manage the next wave of disruptions proactively.

4. How does luxury fit into the equation as a use case for theory development and impactful marketing practice within the cyber-physical reality of the Metaverse?

In his famous novel *"Emile"* [ROU 79], Jean-Jacques Rousseau stated: "This is how luxury and bad taste become inseparable. Wherever taste is expensive, it is false". Thus, luxury, as established in this paper, is not the essence of expensifying, but moreover of rarity and indulgence in the cultural capital of taste, cultivation and cultivated state of experience, but on the terms of the cyber-space cultural artifacts. The part on the fifth space of cultural evolution has answered this essential question thoroughly.

In general, the following conclusions could be observed for the marketing scholar and practitioner within the spheres of the changing cyber-physical realities of the contemporary merger of technology with social sciences:

1. The nature of traditional marketing has shifted towards a cyber-physical reality.

The essence of the Metaverse evolution is a revolutionary state of arrival for marketing. This foundational premise has been covered throughout this research. In this chapter, the author displayed over 1000 sources of academic and non-academic literature on the evolution of the Metaverse, thus establishing via this contribution a model on nature of the upcoming rapprochement of marketing with cyber-physical spaces and particularly, with the Metaverse.

2. Traditional marketing is pivoting towards digital humanities.

Marketing has shifted towards a global digital culture and a societal transforming platform for change, thus no other field of managerial and social sciences has been affected so much as the discipline of marketing.

3. The Metaverse is a new form of an emerging social reality.

Realities for the digital native generation are hybrid realities between the cyber and the physical. Very soon, even that notion of reality would be blurred as we observe the essence of pictures and photography to apply data science and computer science instead of their origin founded within the science of chemistry to be applied to developing photos, which have now become totally digital (e.g. Instagram and Pinterest). The foundations of reality, especially amplified by the two years of lockdown, will become a unified zone, where one will depend on the other, thus creating a totality.

4. The Metaverse connects our senses with the World Wide Web; it is the essence of a paradigm shift in the man–machine interface.

As established in this research, the author can proclaim that the Metaverse is the extension of the Internet in post-PC and post-linear reality. Thus, reality is based on the asymmetry of cyberspace with physical reality.

5. A theory of luxury branding in the Metaverse needs to be paved based on phenomenological approaches to brand culture semantics in cyber-physical realities.

Luxury is the ultimate sense of the experience of taste that depends on human perception. Luxury is not a natural scientific foundation of truth in the metaphysical sense; moreover, it is a phenomenological truth of the essence of value; thus, via a mode of abduction, it is unconcealed for the subject. Habitus and culture capital, as established within the fifth space of the culture sub-part, are requirements avant la lettre for the cultural experience to take place. Luxury, taste and practice, and respectively, the dimension of experience, are the prerequisites of the phenomenology of luxury as a human endeavor and human need.

6. Marketing has the essential role of bridging the cyber and physical realities, ensuring organizational raison d'être and viability while delivering societal improvements.

The essence of *"shared value"*, as visualized in Figure 4.38, could be considered as the fundament wherein an approach is designed to civilize capitalism. Marketing, as the pivotal tool and discipline for enhancing capitalism, needs to put its ladder on much more worthy ideals such as pure advertisement, and thus enhance shareholder value. What would any value mean when the skies are smoky, the seas are dry, the lands are unfruitful, the trees are chopped and gone and the boredom based on digital fatigue as the only source of inspiration creates a depressive mood, where only an influencer under contract is publishing unreal images, with no passion for the spirit of humans and thus their own. Organizational theory, especially how firms could be structured with a raison d'être that respects natural laws of viability with the symbiosis towards the larger environment wherein they are embedded, is a precondition towards a more robust and thus inspirational future.

4.6. Epilogue

The under-explored cyberspace opportunities of entering the Metaverse are the spaces of innovation, wherein the possible diverse designerly entry points could be constructed. The Metaverse is based on six pillars of eco-systemic infrastructure. [LEE 21]:

1. Avatars – actors in the Metaverse.

2. Content creation – building the essence of the Metaverse, for example, entertainment to NFTs.

3. Virtual economy – crypto currency (wallet) based ownership and originality.

4. Social accessibility – green Metaverse uses fairness, cybercrime & privacy threats.

5. Security and privacy – seamless authentication and identification of deepfakes etc.

6. Trust and accountability – trust between users and the avatars, trust repair & accountability.

However, these six pillars are enabled by eight technological foundational enablers [LEE 21]:

1. Network – network capability exposure and applications-driven protocol.

2. Edge/cloud computing – edge and cloud technology orchestration and decentralized security and marketing infrastructure protocols.

3. AI – AI-driven experience innovation and autonomous and automatic constant creation models, conversation for digital twins.

4. Computer vision – 360 scenery, human–avatar interface and gestures and expressions.

5. Blockchain – real-time and or swift proof of work, interoperability, data transparency and privacy.

6. Robotics and IoT – robots and IoTs interaction, avatar and robot interaction and immersive connectivity.

7. User interactivity – invisible interfaces, ubiquitous user interaction, user cues, etc., and telepresence.

8. Extended reality – full integration of cyber-physical environments, multi-cyberspace user collaborations (interoperability).

The Metaverse is a holistic platform for design and meaningful search spaces, wherein meaningful artifacts based on the habitus of the contemporaneous and the next generations are built. Thus, the next generation of Internet, marketing and digital experiences has evolved within the Metaverse. The essence of post-PC devices has evolved towards a cyber-physical unity, wherein art, philosophy and technology intertwine to build immersive 3D (AR/VR/XR) digital connectivity. The Meta-economy has become a multibillion-dollar strong enterprise and a rising platform for disruption and opportunities to service the common good.

Furthermore, this revolution will also result in a new work movement. The work domains will be shifted towards more embracive virtual worlds, which are pivoted not only by the software developers, but moreover by the global creator economy towards a gamified user experience in the direction of the digital human culture evolution, as established with the fifth space in the age of Anthropocene. The angst and aspirations are also better united by the connectivity and navigated as global grassroots movements. The generation of "yes we can" in the United States, the digital natives who toppled the most vicious regimes during the days of the Arab Spring and the digital and media-incubated war in Ukraine are the prime examples of the upcoming waves of innovation, wherein good citizenship is rewarded, and bad governance is punished by public scrutiny. Being viral will become ubiquitous as the acquisition of a more insightful life in public or by being offline as a new form of luxury.

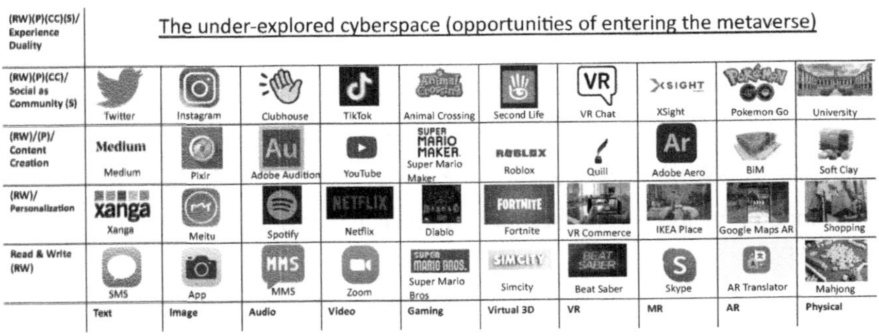

Figure 4.43. *The next frontier of innovation within the dimension of the unexplored possibilities and entry points of the Metaverse [LEE 21]. For a color version of this figure, see www.iste.co.uk/machado/industry.zip*

To offer some recommendations for moving forward within the evolution of the Metaverse, the author proposes the following possible entry points:

1. **Based on the dimension of the physical** – universities will become hybrid spaces of educational experience and embodied culture capital.

2. **AR** – will unite the search engines of the upcoming future to unite man and machine in unison; thus, as Google reduces the chaos on the Internet, AR capabilities will rise, which may solve physical problems by digital means.

3. **MR** – spaces will be the next waves for placements and product and experience spaces where AR and VR are merged for better human interaction, bridging the cyber and physical worlds.

4. **VR** – is the essence of the first wave of innovation. Platforms such as Engage, Virbela Decentraland, Roblox and Sandbox VR-Chat pave the way for better immersive interaction with the digital world.

5. **Virtual 3D worlds** – Fortnite, Second Life, etc. pave the way for virtual world interaction. Furthermore, companies such as Meta (Facebook) have already united the interplay between software and the hardware not only for Web 2.0, but also for Web 3.0, for example, Oculus 2, Meta-workspace, and home from Apps to immersive home experience based in the VR world.

6. **Gaming** – has shifted to non-gamers. The gaming industry and platforms have become ecosystems and large individual economies.

7. **Video** – Zoom will become the next WhatsApp experience and will take the essence of video telephony towards much more rewarding experiences for real-time connectivity with many additional offers.

8. **Audio** – the evolution of the podcast industry has changed how insightful content is shared and produced. The zero-marginal cost society will be much more enhanced by the opportunities within the Metaverse, where self-education, news and cultural experiences of capital and their acquisitions, for example, Audible, etc., will be the next dimension for business model innovations.

9. **Imagery** – as Instagram has changed the perception of the generation, so will the medium of photo-sharing. Apps will be the next frontier in culture, fashion, self-expression and identity.

10. **Text** – as SMS gave way to WhatsApp and now Twitter, one of the essential mediums of communication in the world, additional texting possibilities have emerged in the form of chat or inter-gaming. Inter-App texting has already become a reality.

The Metaverse is here to stay, and we are already immersing ourselves in this reality, one software and one App at a time. Marketing must rise to the occasion to investigate solid use cases, as luxury cases were applied to make this revolution more apparent.

4.7. Appendix

4.7.1. *Appendix 1: KNIME Built-in Stop List*

A, about, above, across, after, again, against, all, almost, alone, along, already, also, although, always, among, an, and, another, any, anybody, anyone, anything, anywhere, are, area, areas, around, as, ask, asked, asking, asks, at, away, b, back, backed, backing, backs, be, became, because, become, becomes, been, before, began, behind, being, beings, best, better, between, big, both, but, by, c, came, can, cannot, case, cases, certain, certainly, clear, clearly, come, could, d, did, differ, different, differently, do, does, done, down, downed, downing, downs, during, e, each, early, either, end, ended, ending, ends, enough, even, evenly, ever, every, everybody, everyone, everything, everywhere, f, face, faces, fact, facts, far, felt, few, find, finds, first, for, four, from, full, fully, further, furthered, furthering, furthers, g, gave, general, generally, get, gets, give, given, gives, go, going, good, goods, got, great, greater, greatest, group, grouped, grouping, groups, h, had, has, have, having, he, her, here, herself, high, high, high, higher, highest, him, himself, his, how, however, i, if, important, in, interest, interested, interesting, interests, into, is, it, its, itself, j, just, k, keep, keeps, kind, knew, know, known, knows, l, large, largely, last, later, latest, least, less, let, lets, like, likely, long, longer, longest, m, made, make, making, man, many, may, me, member, members, men, might, more, most, mostly, mr, mrs, much, must, my, myself, n, necessary, need, needed, needing, needs, never, new, newer, newest, next, no, nobody, non, noone, not, nothing, now, nowhere, number, numbers, o, of, off, often, old, older, oldest, on, once, one, only, open, opened, opening, opens, or, order, ordered, ordering, orders, other, others, our, out, over, p, part, parted, parting, parts, per, perhaps, place, places, point, pointed, pointing, points, possible, present, presented, presenting, presents, problem, problems, put, puts, q, quite, r, rather, really, right, room, rooms, s, said, same, saw, say, says, second, seconds, see, seem, seemed, seeming, seems, sees, several, shall, she, should, show, showed, showing, shows, side, sides, since, small, smaller, smallest, so, some, somebody, someone, something, somewhere, state, states, still, such, sure, t, take, taken, than, that, the, their, them, then, there, therefore, these, they, thing, things, think, thinks, this, those, though, thought, thoughts, three, through, thus, to, today, together, too, took, toward, turn, turned, turning, turns, two, u, under, until, up, upon, us, use, used, uses, v, very, w, want, wanted, wanting, wants, was, way, ways, we, well, wells, went, were, what, when, where, whether, which, while, who, whole, whose, why, will, with, within, without, work, worked, working, works, would, x, y, year, years, yet, you, young, younger, youngest, your, yours, z,

4.7.2. *Appendix 2: Manual Stop List*

Supply, chain, supplier, economics, operations, development, queues, method, research, association, business, review, word, life, role, analysis, time, study, wall, level, linear, mean, view, hedonic, journal, studies, firms, results, search, measures, countries, paper, student, students, al, et, nan, significant, articles, article, intent, literature, negative, positive, empirical, research, strategy, university, york, elsevier, rd, dea

4.8. References

[ALD 57] ALDERSON W., *Marketing Behavior and Executive Action*, Richard D. Irwin, Homewood, IL, 1957.

[BER 19] BERGMANN F., *New Work, New Culture. Work We Want and A Culture that Strengthens Us*, Zero Books, Winchester, 2019.

[BIS 12] BISCHOF J., AIROLDI E. M., "Summarizing topical content with word frequency and exclusivity", *ArXiv*, abs/1206.4631, 2012.

[BLO 15] BLOK V., "Heidegger and Derrida on the nature of questioning: Towards the rehabilitation of questioning in contemporary philosophy", *Journal of the British Society for Phenomenology*, vol. 46, no. 4, pp. 307–322, 2015.

[BÖN 14] BÖNING M., "Illusio", in FRÖHLICH G., REHBEIN B. (eds), *Bourdieu-Handbuch*, J.B. Metzler, Stuttgart, pp. 129–131, available at: https://link.springer.com/chapter/10.1007/978-3-476-01379-8_27, 2014.

[BOU 00] BOURDIEU P., *Distinction. A Social Critique of the Judgement of Taste*, Harvard University Press, Cambridge, MA, 2000.

[BOU 18] BOURDIEU P., "The forms of capital", in GRANOVETTER M. (ed.), *The Sociology of Economic Life*, Routledge, New York, pp. 78–92, available at: https://www.taylorfrancis.com/chapters/edit/10.4324/9780429494338-6/forms-capital-pierre-bourdieu, 2018.

[CAR 17] CARPENTER B., GELMAN A., HOFFMAN M.D. et al., "Stan: A probabilistic programming language", *Journal of Statistical Software*, vol. 76, 2017.

[CHA 22] CHALMERS D.J., *Reality+. Virtual Worlds and the Problems of Philosophy*, 1st edition, W.W. Norton & Company, New York, 2022.

[CRO 06] CROSS N., *Designerly Ways of Knowing*, Springer, London, 2006.

[DAV 20] DAVENPORT T., GUHA A., GREWAL D. et al., "How artificial intelligence will change the future of marketing", *Journal of the Academy of Marketing Science*, vol. 48, no. 1, pp. 24–42, 2020.

[FOR 81] FORNELL C., LARCKER D.F., "Evaluating structural equation models with unobservable variables and measurement error", *Journal of Marketing Research*, vol. 18, no. 1, pp. 39–50, 1981.

[GEE 73] GEERTZ C., *The Interpretation of Cultures*, Basic Books, New York, 1973.

[HEI 83] HEIDEGGER M. (ed.), "Die Grundbegriffe der Metaphysik. Welt, Endlichkeit, Einsamkeit", *Freiburger Vorlesung Wintersemester 1929/30, Frankfurt am Main, Klostermann, Gesamtausgabe / Martin Heidegger Abt. 2, Vorlesungen, 1919 - 1944*, vol. 13/14, 1983.

[HEN 15] HENSELER J., RINGLE C.M., SARSTEDT M., "A new criterion for assessing discriminant validity in variance-based structural equation modeling", *Journal of the Academy of Marketing Science*, vol. 43, pp. 115–135, 2015.

[HUI 09] HUIZINGA J., *Homo Ludens. A Study of the Play-Element in Culture*, The Beacon Press, Boston, 2009.

[HUL 11] HULT G.T.M., "Toward a theory of the boundary-spanning marketing organization and insights from 31 organization theories", *Journal of the Academy of Marketing Science*, vol. 39, no. 4, pp. 509–536, 2011.

[HUM 42] HUME D., "Of the refinement in the arts", in *Essays, Moral and Political*, McMaster University Archive for the History of Economic Thought, available at: https://EconPapers.repec.org/RePEc:hay:hetcha:1742-2, 1742.

[HUN 18] HUNT S.D., "Advancing marketing strategy in the marketing discipline and beyond: From promise, to neglect, to prominence, to fragment (to promise?)", *Journal of Marketing Management*, vol. 34, nos 1–2, pp. 16–51, 2018.

[HUN 20a] HUNT S.D., "For re-institutionalizing the marketing discipline in Era V", *AMS Review*, vol. 10, nos 3–4, pp. 189–198, 2020.

[HUN 20b] HUNT S.D., "Indigenous theory development in marketing: The foundational premises approach", *AMS Review*, vol. 10, nos 1–2, pp. 8–17, 2020.

[IAN 14] IANSITI M., LAKHANI K.R., "Digital ubiquity: How connections, sensors, and data are revolutionizing business", *Harvard Business Review*, vol. 92, no. 11, p. 19, available at: https://dialnet.unirioja.es/servlet/articulo?codigo=5544176, 2014.

[KAM 21] KAMRAN Q., TOPP S., HENSELER J., "Towards the co-evolution of food experience search spaces based on the design Weltanschauung model in food marketing", *Cogent Business & Management*, vol. 8, no. 1, p. 1901643, 2021.

[KAM 22] KAMRAN Q., RESHANI A., TOPP S., "Towards a unified theory of the new market realities in cyber-physical design search spaces", in MACHADO C. (ed.), *Technological Challenges*, Springer, Cham, pp. 25–59, available at: https://link.springer.com/chapter/10.1007/978-3-030-98040-5_2, 2022.

[KAS 21] KASTRENAKES J., "Beeple sold an NFT for $69 million", *The Verge*, available at: https://www.theverge.com/2021/3/11/22325054/beeple-christies-nft-sale-cost-everydays-69-million [Accessed 2 August 2022], 11 March 2021.

[LEE 21] LEE P., BRAUD T., ZHOU P. et al., "All one needs to know about Metaverse: A complete survey on technological singularity, virtual ecosystem, and research agenda", *Foundations and Trends in Human–Computer Interaction*, vol. 18, nos 2–3, 2021.

[LUN 18] LUNDE M.B., "Sustainability in marketing: A systematic review unifying 20 years of theoretical and substantive contributions (1997–2016)", *AMS Review*, vol. 8, nos 3–4, pp. 85–110, 2018.

[MCL 14] MCLELLAN L., BRINKER S., "The rise of the chief marketing technologist", *Harvard Business Review*, vol. 92, no. 7, pp. 82–85, available at: https://dialnet.unirioja.es/servlet/articulo?codigo=4748978, 2014.

[MED 21] MEDIUM, "Crypto is the New America? Very brief takes", *Medium*, available at: https://verybrieftakes.medium.com/crypto-is-the-new-america-982972c1daea [Accessed on 1 August 2022], 28 May 2021.

[MOR 14] MORLEY J., LEIGH O., FLEMING W.F., *The Works of Voltaire: A Contemporary Version with Notes*, vol. 12, Primary Source Edition, Creative Media Partners, LLC, London, 2014.

[POR 19] PORTER M.E., KRAMER M.R., "Creating shared value", in LENSSEN G.G., SMITH N.C. (eds), *Managing Sustainable Business*, Springer, Dordrecht, pp. 323–346, 2019.

[ROU 79] ROUSSEAU J.-J., *Emile. Or, On Education*, Basic Books, New York, 1979.

[SHE 11] SHETH J.N., SETHIA N.K., SRINIVAS S., "Mindful consumption: A customer-centric approach to sustainability", *Journal of the Academy of Marketing Science*, vol. 39, no. 1, pp. 21–39, 2011.

[SHE 21] SHETH J., "New areas of research in marketing strategy, consumer behavior, and marketing analytics: The future is bright", *Journal of Marketing Theory and Practice*, vol. 29, no. 1, pp. 3–12, 2021.

[SIE 14] SIEVERT C., SHIRLEY K., "LDAvis: A method for visualizing and interpreting topics", In *Proceedings of the Workshop on Interactive Language Learning, Visualization, and Interfaces*, Association for Computational Linguistics, Stroudsburg, PA, 2014.

[SOT 22] SOTHEBY'S, "Design of a Gérald Genta watch with accompanying NFT Circa 1985. Gérald Genta: Icon of Time", *Sotheby's*, available at: https://www.sothebys.com/en/buy/auction/2022/gerald-genta-icon-of-time-3/design-of-a-gerald-genta-watch-with-accompanying [Accessed 2 August 2022], 2022.

[TEH 21] TEH C., "China is tempting customers with its flawless AI idols — virtual influencers who don't gain weight, never age, and keep their computer-generated noses out of controversy", *Insider*, available at: https://www.insider.com/chinas-flawless-ai-influencers-the-hot-new-queens-of-advertising-2021-8 [Accessed 1 August 2022], 13 August 2021.

[VER 21] VERHOEF P.A., KANNAN S., STURGILL J.L. et al., "Severe acute respiratory syndrome–associated coronavirus 2 infection and organ dysfunction in the ICU: Opportunities for translational research", *Critical Care Explorations*, vol. 3, no. 3, p. e0374, 2021.

[WEB 13] WEBSTER F.E., LUSCH R.F., "Elevating marketing: Marketing is dead! Long live marketing!", *Journal of the Academy of Marketing Science*, vol. 41, no. 4, pp. 389–399, 2013.

[WIE 42] WIENER N., *Cybernetics or Control and Communication in the Animal and the Machine*, MIT Press, Cambridge, MA, 1948.

[WIE 19] WIESING L., *A Philosophy of Luxury*, Routledge, London, New York, 2019.

[WIL 03] WILKIE W.L., MOORE E.S., "Scholarly research in marketing: Exploring the '4 eras' of thought development", *Journal of Public Policy & Marketing*, vol. 22, no. 2, pp. 116–146, 2003.

[YAD 10] YADAV M.S., "The decline of conceptual articles and implications for knowledge development", *Journal of Marketing*, vol. 74, no. 1, pp. 1–19, 2010.

[YAD 20] YADAV M.S., "Reimagining marketing doctoral programs", *AMS Review*, vol. 10, nos 1–2, pp. 56–64, available at: https://link.springer.com/article/10.1007/s13162-020-00169-6, 2020.

[ZHA 20] ZHAO J., YUAN Q., WANG H. et al., "Antibody responses to SARS-CoV-2 in patients with novel coronavirus disease 2019", *Clinical Infectious Diseases*, vol. 71, no. 16, pp. 2027–2034, 2020.

5

Model-Based Management – A Safari of Essential Business Models

The essential challenges to strategy are the complexity of designing the organization's structure and managing its dynamics in terms of efficiency, effectiveness and viability. Thus, the disruption, chaos and complexity of the environment, wherein contemporary organizations are embedded, the interaction with diverse agents and the needed strategy to cope with these complexities, are at the cutting edge of scientific and practical discourse today, more than ever. The challenges, which are ever-increasing for the strategists, are as strong as they have ever been. Not only have innovations in managerial and strategic models become outdated or based on myopic visions, but business administration's bigger sister science – the science of economics – is also having its own major identity crisis. Thus, it has constructed a highly narrow view for its own scientific "*Weltanschauung*" but moreover, it has missed its own long-term overdue innovation of its own foundations. This identity crisis that the world of economics is still embodying and debating between the ideas of "*Keynes*" and "*von Hayek*" and the late development of "*behavioral economics*" seeing the economy from a psychological perspective, has not been able to propose a better or more functioning model of economics, whereby it can operate more effectively to enhance the human condition by sustainable means. Business models and doctrines have had their successes also based on a continuity of growth but are massively disrupted by the slightest systemic crises that occur. The author highlights the need for a scientific and practical justification for a more holistic strategy model, which embodies the requisite variety toward resolving the complex task of giving the strategist a holistic tool to bestow the organization with a sustainable competitive advantage. This is only possible by an analysis of the essential models available and also to analyze them as a modus operandi of managerial deconstruction toward more viability for the firms.

For business studies to proclaim relevance, the manager must enter the research domain via an analysis of the tools of managers, whereby they navigate their industries. Business studies ask the wrong question of "what is" instead of the appropriate question of "what can be". Hence, the foundation of business administration is constructivist and not metaphysical. This chapter reviews some essential models of management and highlights their applicability within the contemporary context.

Chapter written by Qeis KAMRAN.

Organizations with a broader perspective out-maneuver and outsmart their rivals. The [TRE 95] model is an effective model for strategists to benchmark with the best and to concentrate on the core issues. Models are necessary tools, whereupon managers reduce the complexity of the environment and cope with the rising requisite variety that perturbates the viability of the enterprises under control. Based on the Conant–Ashby Theorem, models are essential foundations of a good regulating system. Regulation cannot be imposed on the system; thus, it is an integral part of the system to be regulated. A firm's competitive advantage is based on what functioning and adequate models are collectively carried in the minds of the managers and how quickly they can be updated or discarded to avoid core rigidities. The essence of this chapter was to bring some of the models into the awareness of contemporary managers.

Over the last few years, the complexities confronting organizations' viability have been subject to drastic amplification. Therefore, the pressure on strategists has enormously increased. Models of steering organizations are the collective histories of a discipline which is embodied for practice; thus, they have strategists to find their path in complexity, and to decide and act more effectively and consciously. This chapter analyzes some of the essential models and paves the way for further diagnosis of the essential managerial models.

5.1. Introduction

Ignoring interrelations of the firm embedded in an environment and not seeing its relations by the systemic and complexity binoculars that give the organization a more stable and advanced business model has caused business administration to appear as an academic field, which has trained managers to create more problems instead of solving them [POR 11]. Business models as a way of an organization communicating with the world and creating an identity is one of the most debatable and essential issues of this zeitgeist.

The following examines the top management and strategy models on which most contemporary businesses are based. By this examination, the author wishes to establish the needed scientific and practical justification for a more holistic strategy model that resolves the complex task of giving the strategist a holistic tool to craft their "*Handwerk*" and bestow the organization under their control with a sustainable competitive advantage and the needed viability as the only pre-requisite to all other organizational achievements.

According to [SCH 10]:

> Over the last few years, the complexities confronting organizations have been subject to drastic amplification. As a consequence, the pressure on leaders has markedly increased. Orientation and steering devices have become all the more important because they enable actors in organizations:

– to find their way in complex settings;

– to decide and act more effectively and consciously.

Thus, high-quality organizational navigation models contribute to organizational success, stability and longevity with pivotal significance. A term called "*Model-Based Management*" (MBM) according to [SCH 10], and what the author wants to establish as "Model-Based Strategy" (MBS) is introduced; thus in accordance with "*Conant–Ashby Theorem*", models are a vital prerequisite for organizational viability, and the necessity of high-quality models additionally in strategy is the task the author is navigating. Effective MBSs that can avoid strategic traps and organizational malfunctions are essential.

> This raises the question about the quality of models on the basis of which one makes decisions. It is not enough to build insightful models; they also must be valid. Hence, the first criterion for the quality of a model is validity. Given that modeling is the construction of realities in the minds of observers, validation can serve as both a guarantor for the realism of a model and a control function that prevents gross aberrations in the perceptions of the modeler(s). This double function is of great importance because strong trends indicate that simulation based on formal dynamic models is likely to become ever more important in supporting managers at all levels in decision making and policy design. [SCH 10]

Additionally:

> A model is valid if it represents what it is supposed or claimed to represent. This applies once the constitutive components of the real system under study and the relationships between them are properly captured; in the case of dynamic models, it holds if the behavior of the model over time reflects the behavior of the real system [SCH 10]. The demand for models of high quality can be expected to increase. Consequently, model validation is likely to become one of the major challenges that lie ahead in modeling and simulation. Validation methodology should contribute greatly to weathering this challenge. (ibid)

It is, therefore, the author's purpose to investigate the below models in this chapter for validity and visualize them for application to organizations from a holistic lens.

5.2. Analysis of top management models

Business models are, in essence, a way of communicating the organization with the environment. They can also be considered as tools of interaction with competition, providing organizational immunology, a source of competitive advantage and a foundation for profitability. Wrong business models can also contribute to the demise of the organization; hence, they reflect the foresight of the strategist and the observed reality where the organization is located. Models are more than memory aids, although some may be used for that very purpose. Furthermore, models are also constructed to establish an abstract and simple version of reality, thus drawing attention to the basics, whereby strategists can focus on the field under observation, the system under focus and the challenge at hand. Different models are integrated for different organizational objectives. Precisely, strategists are using models for the following areas:

1) Meta management (identity, ethos, big-picture).

2) Strategy (control, position, competition, innovation, scenario planning).

3) Operations (execution, implementation, control).

4) Structure (people, organization, viability, communication, value-chain, IT structure).

5) Finance (financial instruments, accounting, book-keeping).

6) Modeling reality (complexity reduction, reality ordering and construction).

Basically, models contribute to reducing the complexities of management and strategy, whereby the organizations' boss can make better and more appropriate decisions. Thus, models are also vital strategic pillars for decreasing uncertainty and simulating the dynamic of complex decisions. The author has already established the notion that strategic decisions are time-consuming factors. Thus, time is the currency of the strategist. With a solid model, the reality is clear, and a solid foundation for better communication in terms of real-time control and total control is established. Speed of information is essential for survival; thus, in biological models, the main foundation for survival is the transfer of information and feedback – ubiquitously and in real time. In this turbulent age, changes in the environment occur without any known patterns. By having appropriate models, these changes can be captured with a minimum waste of time, and thus the needed actions are followed: organizational flaws and larger challenges are dissolved. Businesses compete and out-compete rivals via the models on which their organization is based. The notion of MBM designs and constructs models as a way for the organization to think and reflect.

According to [SCH 10]:

> Given the growing complexities faced by organizations of all kinds, the challenges are gigantic. Practitioners and researchers focus very much on
>
> – Speeding up action (time management)
>
> – Rationalizing processes (e.g. 'lean management')
>
> – Fostering quality (e.g. total quality management)
>
> – Enhancing capabilities (e.g. process organization).

MBM, in accordance with the notion that [SCH 10] established above, can enhance organizational capability. Thus, it integrates an interdisciplinary scientific approach and combines it with a holistic perspective, since real organizational problems do not respect the departmentalized notion of academia. Below are the top models that the author has selected from diverse disciplines according to their rigor, scope, scientific foundation and practicability. As strategists, the author has applied and used most of the models described below. The author will also integrate and distinguish between the useful and the non-useful models for the purpose of the book. The methodology in this chapter and the author's final model also fulfill the scientific raison d'être according to Ulrich's legacy at the University of St. Gallen, who was the doyen of management scientists in the German-speaking countries and beyond; they stated that the construction and development of models, theories, laws, norms and insights for practical application are to be positioned within (and in accordance with) the notion of applied sciences. Ulrich coins this method as "*Scientific Practice*" [HET 08]. Table 5.1 is designed to lay out Ulrich's thinking.

The strategy of strategy (a road map of definition)		
Terminology	**Illustration**	**Example**
Ethos	Alignment of purpose and values	Design shared value
Vision	Creating and shaping a desired future	To design a viable and innovative strategy by aligning organizations and societal objectives beyond CSR
Objective	Overall statement of organizational purpose	Be unique, understand the purpose of purpose and gain a sustainable competitive advantage
Organizational capability	Choosing a viable structure, access to resources and design uniqueness through process	Being an intelligent organization, creating a favorable environment where the organization is embedded (cluster, unique technology, innovation)
Strategy	Connecting the present to the future	Compete for the best service for society by simultaneously being innovative and profitable
Business model	Creating a holistic model by ubiquitous feedback, communication and interaction between business and environment	Collaboration with the stakeholders of the environment, embedding the strength of the environment, closing partnerships, being a productive social system and breaking boundaries of management beyond cost reduction
Control	Beyond monitoring, it is navigated by the quality of organizational intelligence and real-time control	Embed emergent phenomenon and the capability to act with minimum waste of time. Install a strategic feedback controller

Table 5.1. *A strategist's definition of strategy, where scientific observation of strategy is aligned with practice (author's own interpretation)*

According to Ulrich, the author's road map of strategy is an attempt to simulate the road map of strategy-making within the scientific practice dimension. The example of "Shared Value" [POR 11] is chosen because of the actuality of the notion and most innovative approaches to business and strategic modeling according to [POR 11] and the author.

The author's objective is to identify the additional grand competitive force and to provide a solution on how to achieve strategic control over this force, which begins with the construction of a model that has the power to integrate diverse models into a single holistic model.

5.2.1. *Ansoff's product/market grid model*

Igor Ansoff, who is known as the father of strategic management, is also globally famous for his fundamental scientific work in the following three specific fields:

1) The notion of environmental turbulence.

2) The notion of the strategic success paradigm.

3) The notion of real-time strategic management.

However, the most famous model of Ansoff is the "*Ansoff's product/market grid*" [ANS 84, 87, 88]. The model describes and determines the logical notion of the organization's strategic development in the present but, in addition, the scope and direction of the organization within its environment; thus, the organization's strategic development is based on the organization's products, which means its product portfolio strategy and its market, which describes its competitive strategy.

> The firm's product-market scope, its intended growth vector and its distinctive competitive advantage to succeed together describe its logical (strategic) path in the external environment. There is, however, more to Ansoff's grid. As a fourth strategy component, Ansoff suggests considering a firm's ability to benefit from a new product-market entry through harnessing synergy. This can be done in two ways: by making use of an existing outstanding competence (aggressive synergy strategy) or by developing or acquiring the requisite competence (defensive synergy strategy). [HAV 03]

Product / Mission	Present	New
Preset	Market Penetration	Product Development
New	Market Development	Diversification

Figure 5.1. *Ansoff's growth vector component products and markets (author's interpretation after Ansoff [ANS 84, 87, 88])*

The power of Ansoff's model is that it has not lost its claim after so many years. The model not only operates in the present, but it also connects the present with the future. According to [HAV 03], who emphasizes further:

> Despite being nearly half a century old, Ansoff's matrix is nonetheless still valid, and is frequently applied by marketing strategists. In fact, revisiting his work makes one realize that some of today's gurus have either reinvented the wheel or stolen it from him. When used in isolation, the model can do little to help determine the best strategy, and the question of which strategy would be most beneficial for a company is still generally left unanswered. Rather, the grid serves as a means to describe product-market opportunities and strategic options. As such, it forms an excellent framework for exploration, description, and strategic dialogue.

As stated above, the integration of the model is shown in Figure 5.2.

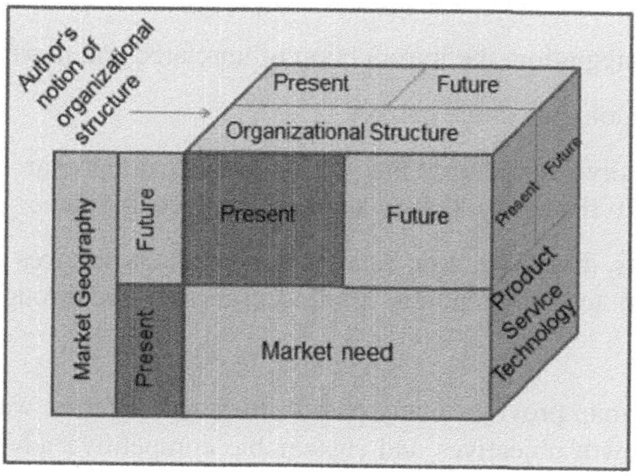

Figure 5.2. *Ansoff's dimension of geographic growth vector: market need, product–service–technology, market geography and the author's extension of the organization's structure (author interpretation after Ansoff [ANS 84, 87, 88])*

Figures 5.1 and 5.2 describe Ansoff's growth vector. The three-dimensional model (market development, product development and diversification) is shown in Figure 5.2. The author extends it to the four-dimensional model (organizational structure development), which not only embeds a variety of combinations and strategic directions that the firm can take but it can also choose to make extreme choices to be in the market of serving its current customers with the contemporary technological products or seek new regions and customer groups with new needs. The fourth dimension, which the author has added to Ansoff's model, reflects the essential notion that the author has established throughout the book. Thus, higher capabilities can only arise from an organization that embodies a viable structure, which can act ubiquitously to changes in the firm's environment, exploit new opportunities better and faster than rivals and discover flaws in decisions or the direction through real-time feedback.

Below are the five dimensions where Ansoff's model can be used:

1) Product/market scope/market penetration.

2) Growth vector/market development.

3) Product development/competitive advantage.

4) Synergy/diversification.

5) Make or buy.

Depending on the differences within the products and markets, a variety of more precise growth vectors are established:

– Vertical integration: the organization's move into the domain of customers and suppliers to dissolve the force of suppliers and exploit the user/customer experience.

– Horizontal integration: the introduction of unrelated and newer.

– Product/technologies to the market.

– Concentric diversification: the introduction of similar and/or identical products/services to the contemporary and new products/markets.

– Conglomerate diversification: totally new products/services and technologies that are unrelated to contemporary products/services are brought to the newer marketplace.

Ansoff's road map provides many possibilities and different ways to achieve the organization's growth objectives and sustainable competitive advantage. However, the road map does not formulate the path the organization ought to take. This path to uniqueness ought to be drawn by the organization's strategist. Thus, it is they who give the organization its ultimate strategic competence and ability to choose the road less traveled.

These five dimensions embody timeless validation. Therefore, the model qualifies for the author to integrate it into their final model.

However, in later years, Ansoff himself reflected back on his work and stated that his models still did not represent the whole reality of the firm's strategic scope; a clear look into the contemporary models reveals where they all originated.

5.2.2. *BCG matrix model*

The Boston Consulting Group (BCG) Matrix is one of the best product portfolio planning models to distinguish the organization's product portfolio. The model is designed primarily to deploy an organizational product portfolio based on the product's life cycle. Its best attributes are the inter-relations among market growth and market share. The health mix of cash-flow generation and its allocation for the purpose of making newer and additional products to establish themselves in the market is another major strength of the model.

The model divides the products into four attractivity dimensions:

1) Cash-cows (cash-flows) mean very profitable products; there is no further need for further investments and efforts.

2) Stars mean products of very high market share with the potential to quickly generate a high volume of cash; it is, therefore, a wise strategic decision to invest in these products.

3) Question marks mean they are products with a very high market share, but the growth rate forecasted resembles an uncertain future. The strategist must apply a high rate of caution by investing more funds and cash into the question-mark product line.

4) Dogs mean the product line that resembles outdated products – the strategist can take advantage of the last profits these products may generate before stopping production and/or selling the product line to additional producers in the market.

It is generally advisable to analyze the organization's product portfolio periodically and regularly. The strategic decision to separate the organization from the "Dogs" and to have solid market intelligence is essential to long-term success and growth.

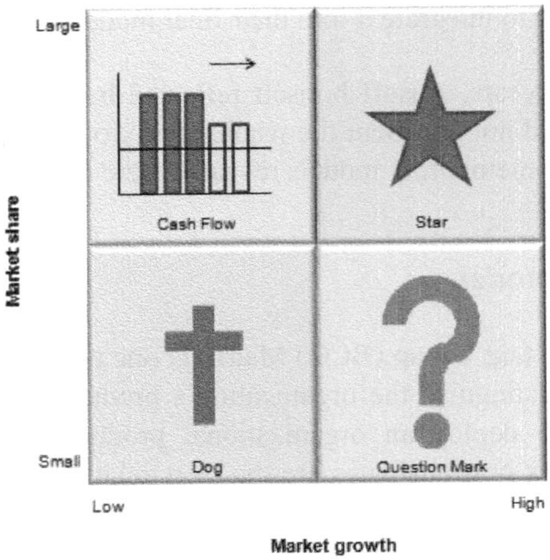

Figure 5.3. *The BCG matrix, after growth curve-BCG (1973)*

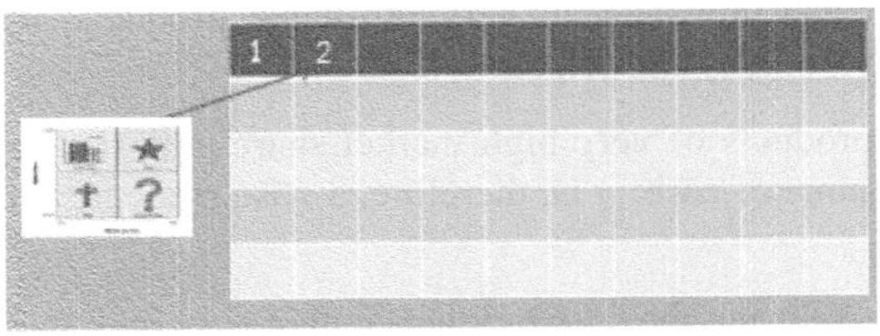

Figure 5.4. *A partial integration to the author's attempt to design a strategy cockpit (author's own model)*

Thus, this model helps the strategist to have a clear view of their organization's products, portfolios and product lines and can actively and easily execute the needed action and maintain a high rate of control over the product's growth and economic performance. When constructing the organization's product strategy according to BCG's matrix, it is advisable to consider the relative market share and market ratings, as markets are often not clearly defined. This is especially essential to be considered when entering an immature market; thus, making a precise judgment is highly difficult. The BCG model is essential and delivers a solid base for product strategy by acting on the results gained. However, it is essential that the prerequisites are met so that the model can be put correctly to work. The BCG model is among the most well-known product portfolio models and has been used successfully since the 1970s.

5.2.3. *CORE competencies model*

[PRA 90] published their powerful article, "The Core Competence of the Corporation", in the Harvard Business Review (HBR). The Core Competencies Model (CCM) was introduced by the author to close the strategic problem of the invisibility of a most powerful way to prevail in the global market. Thus, the shift in thinking from the strategy of the 1980s to the 1990s was established, whereby strategists were judged, and their successes were described by the most effective way they could declutter, delay and restructure their organizations. The strategic success in the 1990s, according to [PRA 90], was rightfully predicted to be governed by the strategic foresight of navigating the organizations by the insights of identifying, exploiting and cultivating the core competencies, which make growth ever possible. These insights were the new key pillars to rethink the very concept of corporations and how they could be successful.

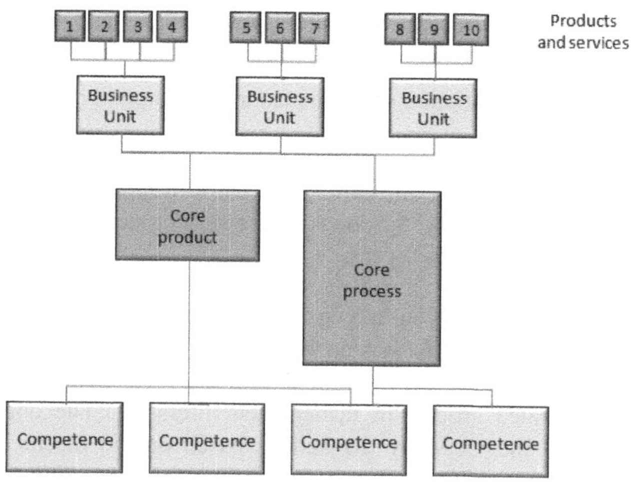

Figure 5.5. *The core competencies model (author's interpretation after [HAM 93])*

Figure 5.5 describes the CCM, where the reader can clearly see the heart of the corporation, which is based on the organization's outputs in terms of the core competencies, which are a combination of the core products and the core processes. The main contribution of the CCM was that a diversified organization, which previously – because of the times of continuity – could direct its multi-unites toward a single, multiple or diverse particular direction to be world leaders in their markets and industries, could continue on the path of their success. Thus, the CCM captured the strategic capability of sensing the shift in the need and desire of the market, infusing products with irresistible functionality, and creating products needed by the customers but not yet imagined by them [PRA 90]. The introduction of the iPod, iPhone and iPad could be built on such a reputation of successfully capturing and accompanying the shift within the market, its need and bending the equilibrium of market share, fulfillment of customer needs and the organization's success in favor of the company. Thus, the core competency of Apple is its design, which appeals to the generation of the "Digital Natives", its leadership in music distribution via iTunes store and the favorable and new way of Apple's product distribution by Apple's own operated stores–chains located in the best and comfortable locations of shopping centers and high-frequency department stores. The Apple store is appealing because it is an innovative experience for the customers, who shop for Apple products directly via the legendary idea of the direct-to-consumer-design-and-meaning-search spaces. Core competencies are better defined as follows:

– collective learning capability of the organization;

– the ability to multi-task between skills development and foresight, and the integration of multiple skills and exploiting the possibilities technologies offer;

– the essence of differentiation and organizations' competitive advantage;

– the author's notion of organization's structure;

– core competencies are also the culture and ethos of the organization; thus, they are how things are done, ought to be done and the very processes that distinguish the integration of organizational intelligence.

Core competencies are difficult to copy; hence, they resemble the uniqueness of the organization. They deliver the sustainable competitive advantage an organization needs in the turbulent and complex market of this era. Hamel and Prahalad's works are among the top-rated articles in management literature. The notion of "Core Values" is added to the CCM by the author, based on the recent developments within the dynamics of some of the global major firms. All of the firms' performances shifted from the vital position of strategic dominance toward a flawed reality that could be based on staying in tune with their raison d'être. A major example was the performance of Volkswagen (VW) within the US market.

Core competencies must be aligned with the firms' ethos as they navigate the global markets. The ethos of VW was to demonstrate customer intimacy (Volkswagen = the car of the people). Instead, a competition on growth and financial performance was conducted to become the biggest carmaker in the world by competing with Toyota. Instead of committing to the long run, as would have been expected from the car of the people, they manipulated the systems and cheated, until there was no room for hiding and manipulation. The next example is the case of Boeing, which was originally constructed on the ethos of being a pilot's and engineer's plane. The biggest flaw in a merger and acquisition case is to see all of the problems of bringing two entities and different systems with different histories and sets of unique assets, such as culture, employee relations, organizational structure, design capabilities, etc., under the umbrella of standardized models of Total Quality Management (TQM) or similar together. TQM and additional models were applied to the Boeing–McDonnell Douglas merger case, which put the wrong people from the defense contracting industry in charge of running the most successful company, as Boeing once was. The world of defense contracting and real market competition is too different to be managed under the umbrella of pure legal disputes and economic-driven shareholder value doctrines. While Boeing successfully lobbied the Federal Aviation Administration (FAA) to even self-administer, approve and certify their own planes, within the shortend cycles, the challenges of the merger based on reductionist measures of only legal and financial dimensions instead of holistic values and raison d'être of the firms from a cybernetic, meaning functioning side, are apparent. The tragedy of the two plane crashes that cost so many people their lives is the systemic flaw that has been carried out since the merger of both firms in 1997. Core values are the yardsticks of core competencies, which are necessary, whereby the notion of superb strategic positioning is safely guarded.

5.2.4. *Greiner's growth model*

[GRE 89] concludes his ground-breaking paper, "Evolution and revolution as organizations grow":

> Researchers are just beginning to study the specific developmental problems of structure, control, rewards, and management style in different industries and in a variety of cultures. One should not, however, wait for conclusive evidence before educating managers to think and act from a developmental perspective. The critical dimension of time has been missing from our management theories and practices for too long. The intriguing paradox is that by learning more about history, we may do a better job in the future.

The truth is not far from Greiner's observation. The organizational structure problem has developed itself to be merely a word of fashion under the sash of cost-reducing and profit maximization of monetary policy-based management style doctrines. Although with the advances in IT, some very positive and essential possibilities are introduced, especially in terms of speed, cost-reducing measures and real-time connectivity, the organizations' structures, whereupon the firm is based, are still modeling a family of three types of chart structure (Figure 5.6). The six phases of organizational growth are described in Figure 5.6.

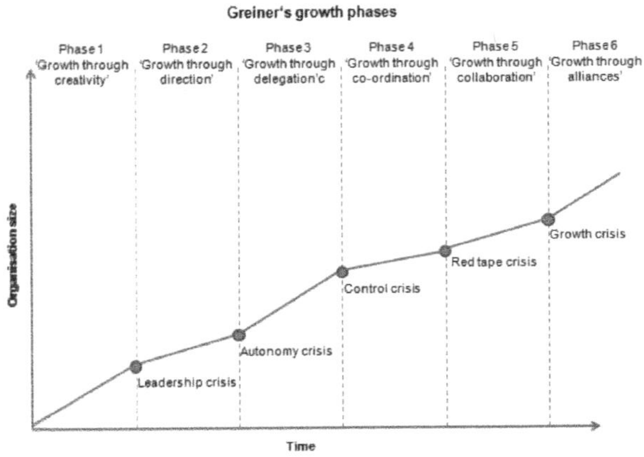

Figure 5.6. *Greiner's growth model and author's interpretation after [GRE 89]*

The model regards essential growth and organizational structure-making models that give the organization the ability to understand and react to the phases of growth wherein it is situated. Suppose a young company, regardless of industry, chooses a very bureaucratic and inflexible organizational structure. This will make it difficult for the organization to react to its own lack of organizational experience or organizational intelligence to cope with the its challenges. The same goes for an organization that needs to adapt to a new business situation but is based on an outdated or centralized decision–procedure model. Thus, this will result in younger and more dynamic people getting frustrated and leaving the organization, therefore making the organization a less attractive place for high performers. Additionally, if autocratic organizations are designed, where the manager derives their dictatorial power from the art of the structure, whereby the organization is controlled, the organization would naturally go into a state of crisis in the long run.

> The problems of these companies, like many others, are rooted more in past decisions than in present events or outside market dynamics. Historical forces do indeed shape the future growth of organizations. Yet management, in its haste to grow, often overlooks such critical developmental questions as: Where has our organization been? Where is it now? And what do the answers to these questions mean for where we are going? Instead, its gaze is fixed outward toward the environment and the future – as if more precise market projections will provide a new organizational identity. [GRE 89]

[GRE 89] structures and connects organizational history in the following manner: "In stressing the force of history on an organization, I have drawn from the legacies of European psychologists (their thesis being that individual behavior is determined primarily by previous events and experiences, not by what lies ahead." Thus, [GRE 89] divides the organization's growth phase into the following two different notions:

1) The notion of organizational evolution: this notion of evolution is described as the organization being in its earlier phases of growth without much upheaval and challenges beyond the regular business practice [GRE 89].

2) The notion of organizational revolution: this notion is where the organization matures and encounters, respectively. It faces some serious turmoil [GRE 89].

According to [GRE 89], there are six essential dimensions for constructing a model of organizational development:

1) Age of the organization.

2) Size of the organization.

3) Stages of evolution.

4) Stages of revolution.

5) Growth rate of the industry.

6) Alliances.

By observing the above notions, the organization's growth phases can be described by the following phases displayed in Figure 5.6.

Phase 1: creativity

This is the phase where the organization develops itself and its products, and, according to [DRU 54], it wants and/or fulfills a market need. The typical characteristics of this phase are:

– Founders and/or entrepreneurs are in the process of implementing remedies for the need or the market gap that they have felt through their own experiences as more technically oriented founders or by manager entrepreneurs, based on the original idea to create a market for themselves.

– There is a high need for ubiquitous communication and coaching throughout the organization.

– Work is long hours, and more of the original founders and partners of the business are involved. Efforts are paid for and rewarded by the organization via ownership and stock options.

Phase 2: direction

Direction is the phase where the organization has mastered the first phase of its inception. It has successfully created a market for itself and is a distinguished, productive and efficient agent in the market. The organization has succeeded in installing a capable manager who has the ability to align the organization. The typical characteristics of this period of organizational growth are:

– The formation of an organized and functional organizational structure. Responsibilities are divided into organizational divisions and units, and all of the departments are aligned for the larger and holistic organizational purpose.

– Standardization of work, budget and industry best practice are adopted. Benchmarking is practiced, and competitors and industries are analyzed.

– Organization is structured based on hierarchies, and business-unit, and branch managers are introduced.

– Communication becomes more formal; reports are conducted to the supervisors and middle to higher management position.

Phase: delegation

This phase of growth emerges from the successful management of a decentralized organizational structure. It embodies the following characteristics:

– Key account, branch and facility managers are taking charge of this phase of growth.

– Incentives and management by objectives (MBO) are introduced.

– Management by exception (MBE) is introduced. The organization is ordered based on self-organization.

– Field visits are conducted, and branch, plant, region or larger business and profit centers are conducted. Communication is becoming a rare phenomenon.

Phase: coordination

According to [GRE 89]: "During this phase, the evolutionary period is characterized by the rise of formal systems for achieving greater coordination and by top executives taking responsibility for the initiation and administration of these new systems." The survivors of the control and function crisis as joint forces and autonym entities are based on the techniques and characteristics of phase 4. These attributes are:

– Centralized company and control functions are introduced and implemented.

– Planning, processes and project management procedures are introduced.

– Budgetary aspects for the unit's departments and projects are carefully analyzed and centralized.

– Business supporting activities and management is centralized.

– The corporate body of the organization is better defined, and numerous staff and specialists are hired to support, control and coordinate activities.

– Incentives and bonuses are introduced, and achievements are rewarded by a joint distribution of rewards.

Phase: collaboration

This phase is characterized by the beginning of a new revolutionary phase in the organization's history and its legacy. Managers embark on implementing a more behaviorist approach to managing the organization. The typical characteristics of this phase are:

– Building and enhancing the team spirit.

– A more problem-solving approach.

– Project and task-oriented cross-functional team formation.

– Reassignment and reduction of a corporate HR body. A more interdisciplinary approach is introduced to the management of the organization.

– Team and group activity is designed more based on the matrix–organization structure.

– IT-technology and conference are used regardless of time and country or unit and plant boundaries.

– Simplification is the name of the game.

– Benchmarking and best practices are commonly accompanied by leaner and better management of the organization.

– Real-time control, total control, operations room and situation room control centers are introduced.

– Business intelligence is conducted, data and real- and essential-information is separated and distributed throughout the organization.

– An organizational academy and in-house training and development centers and teams are introduced.

Phase: alliances

Although [GRE 89] introduced only five growth phases, they introduced the sixth phase, namely the phase of alliances. The typical attributes of this phase are:

– The organization is aware of its current growth state.

– Organizational further growth is seen through M&A's and competitor takeovers.

– Seeking to invest in younger and promising business venues and organizations.

– Aligning organizational intelligence and implementing efficiency models, measures and structures.

– Pursuing a better and more flexible organizational structure.

– Establishing market standards.

At this phase, a regular check-up of organizational growth and its needs and challenges is essential and part of a better and more up-to-date management style.

Science strategists are collaborating with the top research centers and university scholars for up-to-date innovating and introducing cutting-edge scientific-practical measures and incentives for a higher quality and high-performing organizational staff. Greiner's model is essential for science strategists to analyze and self-reflect on the growth phase and the organizational crises that need to be mastered. It is essential to know that not all organizations are growing and emerging within the described model and growth spectrums. Science strategists must adapt and apply the model in any unique situation they find, and analyze their own organizations.

Greiner's model is generally recommended to be applied, and the organization is thereby analyzed every 12 months. However, as the rate of the crises and the organization's growth challenges can have direct strategic and competitive advantage implications, the author recommends a regular fitness check-up via Greiner's model. Thus, strategists can trace, find and correct or pre-control crises and challenges that could result in costly correction and other time-consuming measures. The essential aspect is still functioning in any dynamic, strategic and general management.

5.2.5. *Market-driven organization model*

One of the key attributes and functions of the market-driven organizational model is seen in Figure 5.7, which shows its customer-centric and customer-oriented approach to translating corporate strategy to the market, thus aligning inter-departmental resources and capabilities to achieve this very objective.

Figure 5.7. *The model of a market-oriented and driven organization, as well as the author's interpretation after [VAN 09]*

The market-driven organization's model is essential to transferring a multi-disciplinary and holistic corporate strategy approach into marketing, organizational sales and customer-related activities, which are essential and crucial to organizational processes[1].

Strategists apply the model to analyze efficiency and effectiveness in their marketing activities. Thus, via the application of the market-driven strategy model, strategists are ensuring that a holistic alignment is established between the corporate processes and the organization's corporate strategy[2]. The flag of market-driven is a synonym for implying a customer-centric approach to the development of the organizational corporate strategy, as seen in Figure 5.7.

The model is applied in the following two essential perspectives:

1) Operations effectiveness.

Operations effectiveness is the notion of understanding and evaluating the marketing and sales policy by observing the following essential questions:

– Market-segmentation: focus on the type and group of customers.

1 See [VAN 09 p. 35].
2 See [VAN 09 p. 36].

– Unique customer proposition: focus on the differentiation and the "HOW" of attracting and serving the customer needs.

– Objectives formulation and design: focus on setting the targets and what levels of sales, market penetration/market share, revenues and sales can be set and achieved. The notion is beyond wishful thinking, and objectives are essential to corporate success and the organization's larger competitive advantage and defense of the market position.

2) Operations efficiency.

Operations efficiency is the alignment of supporting activities and core competencies of the organization. Thus, to achieve operations efficiencies, the following points are essential:

– Reducing costs and controlling costs: focus on how further efficiencies are generated and acquired within the sales and marketing departments, seeking best practices and benchmarking activities. Applying company research and funding university projects and theories to design a unique approach.

– Exploiting and fostering synergy effects: focus on result and solution-oriented notions by aligning the whole organization to the same cause. Ensuring and rewarding cooperation without departmental boundaries. Team and cooperative culture design to all customer needs and customer-centric organization solutions.

– Building equilibrium in sales/marketing and operations activities: focus on aligning forces and coordinating a joint response to the specific customer needs. Thinking beyond and outside of the box. Collaboration and enhancement of organizational efficiency by mobilizing organizational structure intelligence to propose an individual and unique customer solution.

In planning appropriate marketing, sales, operations and support activities, it is essential that a corporate decision is made based on the following:

1) Market-segmentation: what markets to decide on and to penetrate?

2) Customer focus, market need and gap focus and fulfillment: what need does the firm want to address?

Corporate goals: what specific objectives and targets do the firm and organization want to set and achieve? Which segments are essential and are aligned with the strategic perspective and objectives of the organization? The author, therefore, integrates the model into their larger organizational strategic model for designing and addressing the sixth force. It is essential to understand that the notion and author's concentration on the emergent nature of strategic management requires a flexible but, at the same time, an evolutionary model that aligns the organization with the customer-value perspective. Thus, the justification of an organization's existence is its service and the customer need it fulfills.

5.2.6. *The offshoring and outsourcing model*

[ARO 05], in their famous HBS article "Getting the offshoring right", end their research by stating: "Companies would do well to remember that the manner in which they start their offshoring initiatives often determines how they will end". Offshoring was once a fashionable decision, to a certain extent, necessary to compete within the global market and/or a solid pursuit for acquiring a strategic position. However, research has identified that more than half of the offshoring does not bring the anticipated and needed competitive advantage. "According to several studies, half the organizations that shift processes offshore fail to generate the expected financial benefits" [ARO 05]. However, as research has found out, it is essential to start well and analyze the situation before embarking on the journey of offshoring. Thus, this decision affects the larger objective of the organization; it is the structure, the strategic position, cost advantage and additional decisions and options the organization will have toward the rivals and the market in general. However, research has further identified that organizations do not make their decisions systemic enough and, therefore, fall short of making these common three errors, which, according to [ARO 05], are:

> [...] most companies focus their efforts on choosing countries, cities, and vendors, as well as on negotiating prices, but they don't spend time evaluating which processes they should offshore and which they shouldn't. Without a standard methodology for differentiating processes, most executives find it tough to distinguish among core processes that they must control, critical processes that they might buy from best-in-class vendors, and commodity processes that they can outsource.

> [...] most organizations don't take into account all the risks that accompany offshoring. Executives use simple cost/benefit analyses to make decisions without realizing, for instance, that after they transfer processes, their vendors will gain the upper hand. Providers can hold companies to ransom; it's almost impossible for organizations to reabsorb business processes on short notice. Most organizations naively ignore these latent risks and are shocked when vendors demand price hikes that erode the savings from outsourcing.

This, according to [POR 79], is the force of suppliers, which strategists must guard.

> [...] most companies don't realize that outsourcing is no longer an all-or-nothing choice. They have a continuum of options. At one end, there are executing processes in-house; at the other, there are outsourcing them to service outline tools that will help companies

choose the right processes to offshore and discuss the associated risks [ARO 05].

However, if a systemic approach is used, managers can avoid the above mistakes. One of the ways whereby strategists can avoid these errors is the offshoring model, which is described in Figure 5.8. The model can be applied to organizational activities. Thus, it will help the strategist to outsource, offshore or keep diverse activities and distinguish between the core and the non-core activities. Outsourcing and offshoring should be used and applied based on the larger strategy. A pure concentration on cost-reducing measures and playing the game of the rivals, who have moved their operations and activities offshore in advance, is not a strategic move or a wiser managerial decision. According to [POR 80], such decisions of offshoring and outsourcing are not even strategic moves, but they are merely a small part of larger decisions toward a strategic position. Careful and systemic analyses of the processes necessary to conduct a thorough investigation are described in the four points below:

1) Finding the reason to choose the offshoring option.

Strategists ought to conduct thorough business intelligence to determine what moves the competitors are planning next. In addition, which cost pressures will determine a vital competitive advantage, turning the equilibrium of higher profits toward the rivals? Cost advantage is a vital generic strategic position and cannot be left to the rivals. It is also a measure of organizational efficiency.

2) Choosing the partners, regions and countries.

Finding out which partner has the appropriate experience, skill and capacity, but above all, what culture the supplier embodies. Aligning the organizational cultures is one of the most essential aspects of offshoring. Additionally, strategists need to determine the different factors governing that specific country. These potentials include:

– labor laws, conditions and costs;

– production quality;

– suppliers reliability;

– contract enforcing laws;

– intellectual Property (IP) law and its protection;

– the resulting cost advantage, the sustainability of the cost advantage;

– sustainability of the competitive advantage.

3) Finding the nature of costs, profits and risks and which processes are appropriate for offshoring.

In this part, a thorough cost/benefit/risk analysis should be conducted to find suitable diverse alternatives and solutions. Here is the analysis of which activity and what part of the organization ought to be aligned with the overall value chain of the organization[3]. The essential points to consider are:

– extra costs;

– costs of offshoring;

– labor costs and their future development;

– taxes;

– switching costs;

– exit costs, etc.

4) Observing the realities of the next steps.

At this stage, a final, decisive and thorough analysis of each alternative activity and country ought to be finalized and introduced. Partners, countries, processes and each activity should undergo a detailed feasibility study[4].

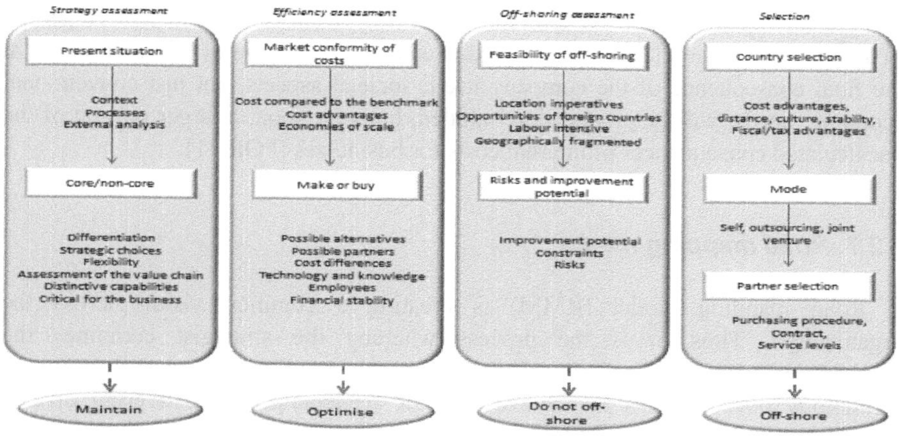

Figure 5.8. *The offshoring model and the author's interpretation after [VAN 09]*

3 See [VAN 09, p. 39].
4 See [VAN 09, p. 36].

The model displayed in Figure 5.8 reduces the uncertainty of how to do offshoring or outsourcing in the right way. Thus, the strategic decision to keep the activities and processes or to optimize them is also displayed in Figure 5.8. The model is, therefore, an essential map to analyze activities, conduct business intelligence on competitors, explore better and cost-effective processes in the global world, and conduct benchmarking between the performance of the company and the performance of suppliers, and the activities that can be outsourced and offshored. In addition, the model can also analyze which activities to keep in-house and which processes and standards to maintain.

The socio-economic effects of offshoring are vital to economies of the high-wage and low-wage countries. However, the future of global business policy is on finding newer solutions and models upon which success and sustainable growth can be based. "Business and society have been pitted against each other for too long" [POR 11].Thus, the offshoring mindset and competing purely based on cost reduction, which was essential and necessary during the 1980s and 1990s (or even recently), must be revised. The innovation for the development of newer models upon which societies are served, needs to be introduced by making firms more resilient. At the same time, the profitability of businesses is enhanced, which is the name of the new strategic game and revolutionary mindset. Good strategic moves are based on being a productive and ethical member of corporate global citizenship and, at the same time, being competitive. Competition in the future is based on the development of a globally updated value–creation model to bring businesses and societies together. "The winner takes it all" business approach embeds the demise of the society, and at the final consequence of the company itself, societal aspects, not just conventional economic needs, will be produced yet hidden, but very soon, a larger portion of the uncalculated consequences of internal costs for businesses [POR 11].

5.2.7. *Road mapping model*

Road mapping model (RMM) is creating a common vision across the organization. Thus, it is the process whereby the strategist combines the organization's future development based on the need in the marketplace, the technology developments to achieve strategic leadership, and delivering diverse scenarios and consequences such developments may bring to the organization in focus[5]. One of the most essential and critical functions of the RMM is to make the organization survive in a turbulent and unpredictable environment. Thus, the model provides the ability to focus on scanning by tracking the performance of individuals, including potentially disruptive technologies [PHA 03]. Although RMM is a simple model in terms of its format, the development of the model poses significant

5 See [VAN 09, p. 41].

challenges since the scope that RMM covers is generally broad. It includes and applies a large number of complex conceptual and human interactions [PHA 03]. RMM is the technological and scientific (expert)-based view of organizational development and defining the path the organization should take. A key feature and diagnostic power that the model provides is shown in Figure 5.9, which displays the alignment of three diverse perspectives and essential factors; these are:

1) Market analysis.

2) Product assessment.

3) Technology development/R&D.

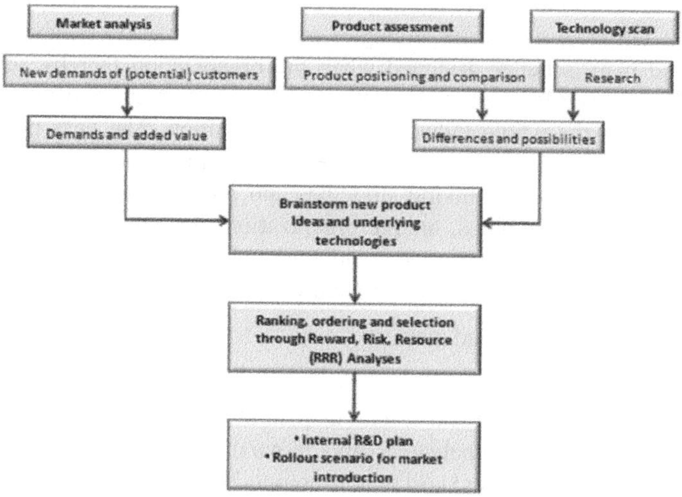

Figure 5.9. *An organizational road mapping model after [VAN 09]*

In recent years, a variety of RMMs have been developed. These models can be described in the following four categories:

1) The industry road map.

Industry road maps are drawn and constructed for the purpose of designing and redefining the whole branch of an industry[6]. Although the industry road map can be designed to obtain the needed finance for realizing a specific project or research objective, the main purpose is still to minimize the risk to a single company; thus, many firms in the industry debate and cooperate to find and direct the path for the new technology, from which all may profit.

6 See [VAN 09, p. 41].

2) The product/technology road map.

This road map offers the combination of market analysis, product assessment and a wider scan for the most up-to-date technology to construct a variety of scenarios and a larger model on how the market may react and how to absorb the diverse reactions, whereby an R&D plan and/or product development plan (road map) can be drawn and put to work. With this road map, new technologies and strategies can be designed to bring the organization to the next level. This introduces the process of ubiquitous innovation to the wider stakeholders within the organization.

3) The corporate road map.

Although the corporate road map may be based on the general industry road map for the organization to design its strategies and product market combinations, the corporate road map is better designed when the organization thereby achieves its unique and individual plan on how to move the organization ahead of the crowd but also how it shapes the whole industry with its innovation. The very purpose of a corporate road map ought to be to achieve uniqueness in the abilities and the ways of thinking, applying and rethinking strategies and products ahead and better than the players in the market. Thus, unique organizations are not only those who can adapt to the changes of the environment and marketplace quicker but also those who shape their environments better, faster and more intensely than their rivals.

4) Competence-research road map.

This type of road map may be designed separately or as part of the organization's larger plan. The above road map combines the organization's specific (capability-based) needs, competencies and research needs to achieve innovation and a milestone for its future and sustainable competitive advantage.

According to [PHA 03]: "Identifying disruptive technologies and surviving in disruptive markets is not easy, but road maps can help." Thus, [NAD 97] observes: "[...] that product should be seen as being 'made up of a set of subsystems, each of which has its own innovation stream' and that there is a need for articulating a clear, common, shared vision' in a company simultaneously carrying out incremental and radical innovation. It is suggested that in both areas, technology road mapping provides a significant step forward." [PHA 03] have identified the following general characteristics of technology road maps:

– The generic road mapping approach has great potential for supporting business strategy and planning beyond its origins in product and technology planning.

– Many of the benefits of road mapping are derived from the road mapping process rather than the roadmap itself. The process brings together people from different parts of the business, providing an opportunity for sharing information and perspectives and providing a vehicle for holistic consideration of problems, opportunities, and new ideas.

– The graphical form of the roadmap is a powerful communication mechanism. However, it can present information in a highly synthesized and condensed form. Hence, the roadmap should be supported by appropriate documentation.

– Roadmaps are multilayered, reflecting the integration of technology, product, and commercial perspectives in the firm, including internal and external sources and supporting communication across functional boundaries in the organization.

– Roadmaps explicitly show the time dimension, which is important both for ensuring that technological, product, service, business, and market developments are synchronized effectively and for reflecting the dynamic, changing natures of technological and business environments. Roadmaps provide a means of charting a migration path between the current state of the business (for each layer) and the long-term vision, together with the linkages between the layers, in a form that is flexible enough to be updated over time.

– Software has an important role to play in supporting the application of road mapping in the enterprise. However, software alone cannot deliver good roadmaps and needs to be integrated with the human aspects of road mapping. A key benefit of road mapping is the sharing of knowledge and the development of a common vision of where the company is going, and this only comes about by sharing knowledge and making connections.

– Sectoral or multiorganization road mapping promotes knowledge sharing and facilitates the development of a collective vision that can lead to action and collaboration [PHA 03].

Road mapping is a useful innovation path-designing and technology-planning tool in an ever-increasing and uncertain competitive environment. To be successful in technology road mapping design and process, it is essential to identify why the organization is doing the road mapping and how it will be applied. Road mapping is particularly interesting and critical when strategists are coordinating the development of multiple technologies across multiple projects with multiple stakeholders and players. Thus, road mapping is also a very effective tool for aligning diverse players in an organization for a common and larger purpose to achieve a breakthrough and essential competitive advantage. The essential quality of a good road mapping plan is its emergent and market-based view; thus, it combines the need for identification and the organizational response to the need, or according to [DRU 54], it creates the want in the market, by aligning the organization to a common MBO to identify and exploit the opportunities of the market ahead of the rivals. Road mapping ought to be viewed as a way of communication within the organization, whereby the organization communicates with the marketplace and senses the changes, whereupon a ubiquitous action ought to be conducted.

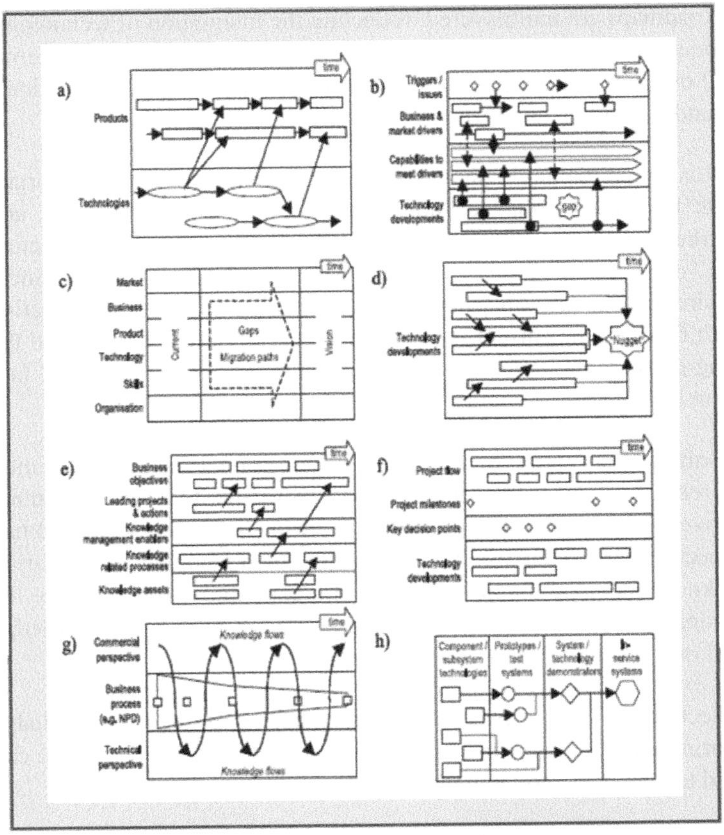

Figure 5.10. *Examples of technology road map types (purpose): a) product planning, after [JOH 88]; b) service/capability planning after [WET 96]; c) strategic planning after [PHA 03]; and d) long-range planning, after [BOW 95]*

Figure 5.10 displays diverse RMMs. It is essential that a strategist aligns the specific RMM with their specific strategic and organizational planning needs.

We can also use diverse RMMs upon the sensed need for better and more clearly strategic insight. For more specific and in-depth RMM needs, the author suggests reading the works of [PHA 03].

5.2.8. *Scenario planning model*

Scenario planning (SP), as an essential strategic tool to pre-act and to create a favorable environment wherein an organization is embedded, is generally applied to recognize and foresee changes ahead, and to prepare for those changes to better distinguish between the consequences, the decisions or the realities, which will be created. Organizations that can better foresee and perceive the future and that have the organizational capability to act faster and better to the changes will have a more profound and sustainable competitive advantage. According to "Bain & Company", a management-strategy consultancy firm:

> Scenario Planning allows executives to explore and prepare for several alternative futures. It examines the outcomes a company might expect under various operating strategies and economic conditions. Contingency Planning assesses what effect sudden market changes or business disruptions might have on a company and devises strategies to deal with them. Scenario and contingency plans avoid the dangers of simplistic, one-dimensional, or linear thinking. By raising and testing various 'what-if' scenarios, managers can brainstorm together and challenge their assumptions in a non-threatening, hypothetical environment before they decide on a certain course of action. Scenario and Contingency Planning allows management to pressure-test plans and forecasts and equips the company to handle the unexpected [BAI 11].

[WAC 85], in his groundbreaking HBR article, states:

> Scenarios deal with two worlds: the world of facts and the world of perceptions. They explore for facts, but they aim at perceptions inside decision-makers' heads. Their purpose is to gather and transform information of strategic significance into fresh perceptions. This transformation process is not trivial; more often than not, it does not happen. When it works, it is a creative experience that generates a heartfelt 'Aha!' from your managers and leads to strategic insights beyond the mind's previous reach.

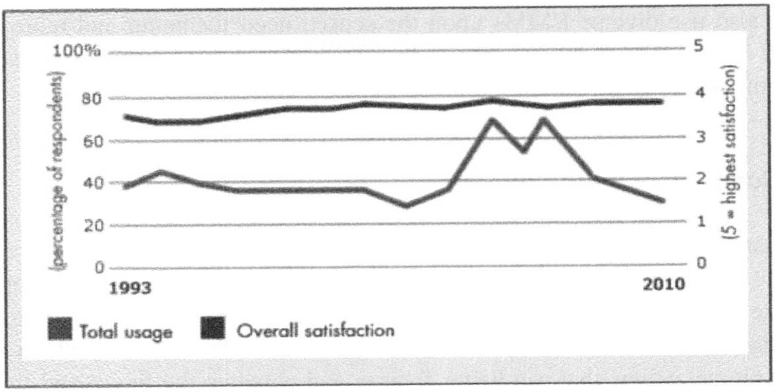

Figure 5.11. *Bain & Company's survey of organizations using a scenario planning strategy (SPS). It displays the usage and total satisfaction percentage of the firms using the model after [BAI 11]. For a color version of this figure, see www.iste.co.uk/machado/industry.zip*

According to [BAI 11], as shown in Figure 5.11, there is an 80% satisfaction rate, which seems to be a very high acceptance rate among managers. Thus, it reflects the benefits the model will bring to the organization.

Figure 5.11 describes how the process of SP can be applied to close the gap between strategic vision and an aligned action and option, based on the envisaged scenario. Figure 5.12 describes the SPM.

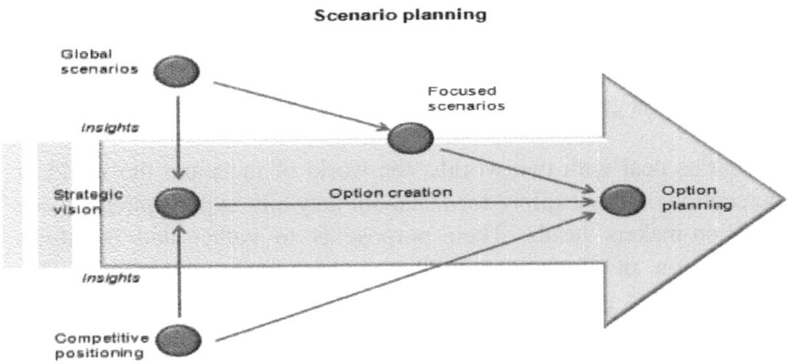

Figure 5.12. *The scenario planning model (SPM) and author's interpretation after [WAC 85], [VAN 09]*

To understand the application of SP, the author describes the essence of SP in the following manner to the reader:

1) Enhanced perception.

The most fundamental aspect of SPS and SPM is the realization and understanding of when trends will bend. SP gives the strategist the capability of foresight to observe and identify key trends and then extrapolate them; thus, it is the most common way to think about the future [GOL 04].

2) Integration with corporate planning.

SP provides a constructive, creative and non-threatening middle ground between sticking to the corporate rules [GOL 04], culture, capabilities, plans and codes and simultaneously creating the changes needed within the organization to pre-act upon the trends and changes perceived. Thus, the organization can save resources, act based on emergent strategies and avoid managerial blunders.

3) Enhance the process of collective thinking.

The analysis of information after the attack on the World Trade Center, the BP Gulf of Mexico spill and the Fukushima nuclear disaster showed managers and people in charge know more than what gets truly observed and accepted to form the guiding principles of their behaviors and actions. The application of a solid SP and SPM to foresee divergent futures – all plausible, all internally consistent, all possible – makes it easier for odd, contradictory and uncomfortable data and scenarios to be given a fair and timely hearing, rather than to be dismissed as outliers [GOL 04]. Most of the world's devastating crises could have been prevented if a better SP had been applied; thus, foreseeing is key to strategy.

4) Explicit addressing uncertainty.

Uncertainty is one of the common attributes of SP. There is no guarantee which scenario will manifest. However, it is the emergent capability of pre-calculation and preparedness of the organization for a certain crisis, scenario and situation that makes up a solid strategist. A solid strategist does not have a single plan to achieve their objective. They have plans of action in scenarios A, B, C, D and E to arrive at the position that they actively designed for their organization. As the author already established, the position is the right strategic notion, according to Porter, and an essential strategic objective. "In an economy where the only certainty is uncertainty," according to [NON 07]. Thus,

> ... the one sure source of lasting competitive advantage is knowledge. When markets shift, technologies proliferate, competitors multiply, and products become obsolete almost overnight, successful companies are those that consistently create new knowledge, disseminate it

> widely throughout the organization, and quickly embody it in new technologies and products. These activities define the 'knowledge-creating' company, whose sole business is continuous innovation. And yet, despite all the talk about 'brainpower' and 'intellectual capital,' few managers grasp the true nature of the knowledge- creating a company – let alone know how to manage it. They misunderstand what knowledge is and what companies must do to exploit it. [NON 07]

5) Designing a structure for dissolving complexity.

The organization can reduce complexity by designing a transparent organization that receives and processes real-time information and focuses on actions and heuristic decision-making capabilities. The organization can navigate wisely to achieve its ultimate position and objective via real-time information and feedback sensors. Strategic decisions are based on the strategist's trained intuition; while some information is more precise, other information must deal with the fuzzy logic whereby objectives are achieved.

6) Communicating the perceived reality.

Scenarios can be generally observed as well-crafted stories based on some of the most likely realities that may occur and be significant to the organization. Thus, the strategist can prepare and act in advance so that the worst scenario can be avoided, a favorable situation can take place or a reality can be supported, which favors the organization's development and goal achievement.

With the story-making and telling approach, an interesting way of communication can be put to work.

The most daunting task of the scenario-crafting and planning team is to receive support from the senior management and to conduct scenario planning based on an ongoing and habitual process rather than a once-in-a-while problem-solving approach. This is better displayed in Figure 5.13.

With the ongoing approach, the strategist can actually move forward with the process of creating and sustaining the culture of a learning organization. Thus, this approach further enhances the organization's ability to work in a team and find the best solutions based on collective intelligence.

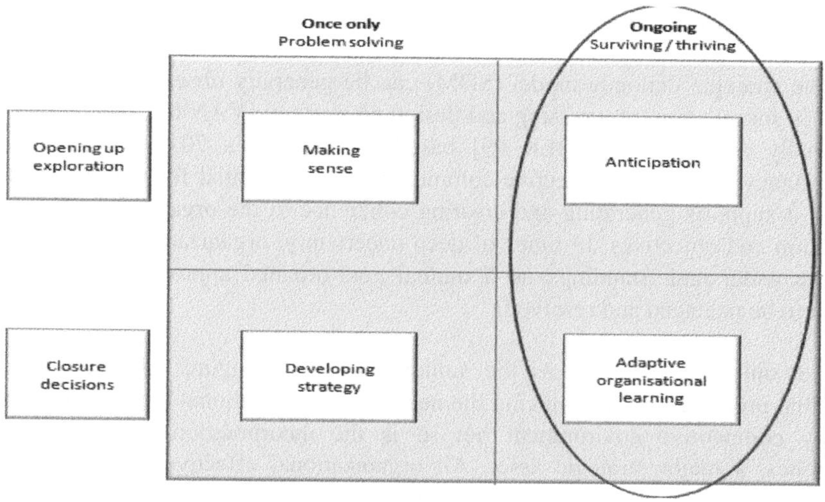

Figure 5.13. *The categorization of scenario goals and perceptions, as well as the author's interpretation after [VAN 09]*

According to [WAC 85]:

> It has been my repeated experience that the perceptions that emerge when the disciplined approach of scenario analysis is practiced are richer and often critically different from the previous implicit view. The scenario process of converting information into fresh perceptions has something of a 'breeder effect': it generates energy, much more energy than has been consumed in time and effort during the process. A mere high or low around a baseline can never achieve conceptual reframing. The perception of reality and the discovery of strategic openings that follow the breaking of the manager's assumptions, many of which are so taken for granted that the manager is no longer aware of them) are, after all, the essence of entrepreneurship. Scenario planning aims to rediscover the original entrepreneurial power of foresight in contexts of change, complexity, and uncertainty. It is precisely in these circumstances, in stable times, the real opportunities lie to gain competitive advantage through strategy.

In addition to [WAC 85]'s pivotal observations, the notion of "System Thinking" is essential. Thus, it searches for tacit and causal relations, reasons beyond observed phenomena, and a holistic view whereby the most appropriate, just and timely actions can be taken.

5.2.9. *Strategic dialogue model*

The strategic dialogue model (SDM) can be generally observed as the diverse steps in formulating, progressing and designing strategy [VAN 09]. The model was originally developed by [BER 09] based on the firm's 70 years of strategic consultancy experience. Effective communication is essential for the organization. Thus, it supports generating and creating coherence in the organizational strategic direction and objectives. In times of deep uncertainty, organizations are obliged to have a wider understanding and a mutual goal-oriented approach to how things ought to be managed and resolved.

Not only is coherence, as the author's model in Figure 5.14 describes, an essential prerequisite for achieving the necessary organizational effectiveness in the highly competitive environment but so is the incorporation of environmental dynamics, a major strategic asset. All organizational effectiveness and strategic dialogues are irrelevant if the perceived environment does not reflect the actual reality, wherein the organization navigates, or if changes are not regarded in a timely manner. Thus, it is essential to include environmental dynamics into the dimensions of the strategic dialogue, since all strategic planning may not help if a crisis has taken place that could have been prevented by the organization avant la lettre and wherefore it could have been structurally better prepared, thus to face a new entrant competitively or address a new technological progress more effectively. Designing the model of organizational effectiveness and strategic dialogue as Figure 5.14 describes, and based on the author's model below, actually reduces the above-mentioned risk scenarios and makes the strategist more of an able leader to steer their organization based on an intelligent and holistic approach.

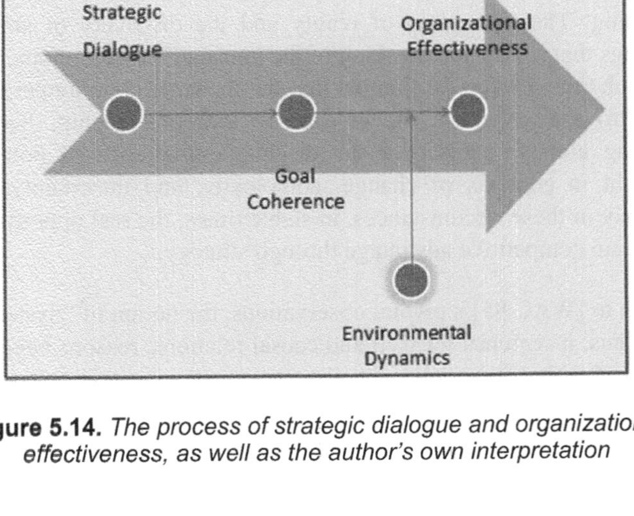

Figure 5.14. *The process of strategic dialogue and organizational effectiveness, as well as the author's own interpretation*

[BER 09]'s SDM is based on seven steps, which are outlined in Figure 5.14. These steps are:

1) Strategic window: formulating the strategy.

2) External analysis: considering developments and changes in the marketplace, market dynamics.

3) Internal analysis: considering the technological and organization-specific spheres.

4) Synthesis and options: exploiting external opportunities while making internal analysis to pre-act, react and/or change the game in the marketplace.

5) Appreciation and strategic choices: risk assessment, feasibility study, feedback and sensing the emergent and ubiquitous changes in the organization's milieu.

6) Elaborating and planning: formulating the most feasible and best possible choices and strategies into organizational mission and ethos, coherent goals and objectives, and execution and actions. Brainstorming and integrating other models to achieve the organization's goals.

7) Implementation and monitoring.

Controlling and monitoring are essential after a strategy is implemented and a course of action is chosen. The model is described in Figure 5.15, whereby the strategist can formulate a viable strategy, and is also based on the three essential phases, which combine the above seven steps.

The author suggests not only regular communication with the team and the stakeholders but also constructs a ubiquitous communication and feedback channel between phases 1 and 3. It is here that complexity will rise since change is not sensed by elementary strategy and communication tools and techniques.

Tracing relevant change is a difficult challenge for organizations; however, the model in Figure 5.15 has been advanced by the author by implementing the feedback system to report the change happening within the environment. Thus, the strategist can reduce the complexity of market and strategy-controlling tasks by knowing and acting on real-time data and information. Time is, as the author already established, the currency of the strategist, and holism is the unique competitive advantage of strategy (see Figure 5.11).

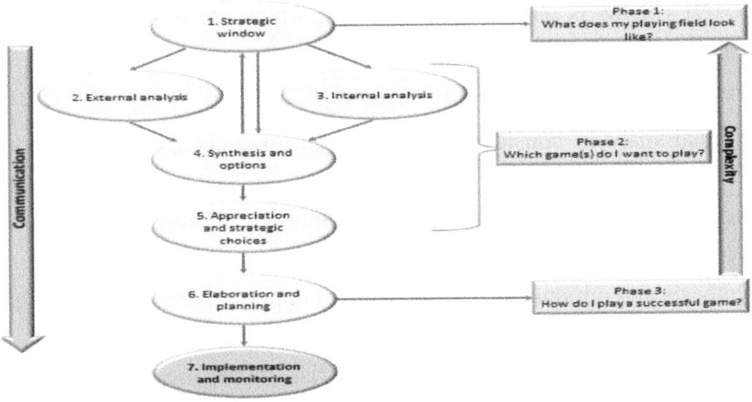

Figure 5.15. *The phases of the strategic dialogue or the "strategic dialogue model" (SDM), and the author's interpretation after [BER 09]*

The three phases to achieve a solid strategic dialogue are described in the following steps:

1) What does the organization's playing field look like?

Answering this question reveals the organization's actual strategic window. This part of the SDM decides which activities or areas of business the organization ought to enter, the strategic foresight to embrace and the grand strategy to choose for what the organization ought to be about.

2) Which games does the organization need to play to stay ahead of the rivals and achieve uniqueness?

It is essential for the organization to know the game and its playing field. Thus, each service or product in the market is based within a specific industry, for example, if a company chooses to advise clients on their tax applications and benefits, then the industry is a consultancy business. The organization should then specialize in the type of customer, the geographic segmentation and/or what additional services are essential to offer to be a viable and solid tax consultancy organization. In addition, what extra value can the organization bring to foster uniqueness and differentiation compared to the rivals in the marketplace?

3) How does the organization play the game?

In this part, the organization needs to analyze its resource allocation capabilities and align them to the larger objective. The organization here matches the external opportunities and possibilities with the internal capabilities, for example, meeting the demand of the playing field with adequate services and products. It is essential to guarantee a very high level of participation from people. The best uniqueness an organization wants to achieve is the uniqueness of the people involved, their state of mind and the ideas they carry in their minds to achieve the organization's objectives. The tactics are equally essential, which the organization chooses to out-think its rivals and to give the highest possible utility to the customers.

4) The author suggests one additional phase to [BER 09]'s SDM. This phase is: what is the organizational structure that can ensure viability and sustainable competitive advantage, and how can it be navigated?

The author's suggestion, in addition to the product, service and quality delivery, which are absolutely essential to being able to survive in the market, is that an additional essential issue is the adequate viable organizational structure. It is a surprise how the structure debate is almost never aroused. Thus, it is the strategist's chief aim not only to invest in the development of the product but also in the business model, whereby the organization chooses to deliver the products to the market and the organizational structure, whereby it copes with the ever-growing complexity of the environment.

According to [VAN 09], strategy dialogue presumes that strategy is the result of the formula "Mobilization x Formulation x Realization" where:

– Mobilization: fostering creativity, conducting in-depth analysis, making choices, increasing acceptance and participation, and ensuring commitment.

– Formulation: constructing and creating a strategic plan, finding options and choices, being consistent and ensuring it.

– Realization: control and steering to the objective, to-do-plans and making them work, superb execution, project and milestone management, controlling, acting on feedback and emergent dynamics.

The SDM supports the strategist in standing out from the crowd [BER 09]. Thus, according to [POR 96]: "Strategy, on the other hand, requires hard choices." [POR 96] further emphasizes that avoiding making choices and trade-offs is a major strategic and managerial flaw. However, avoiding making choices is nothing new to strategy scholars. [POR 96] states that making choices is a part of the rediscovering strategy and that managers have become confused about the necessity of making choices. Thus, by being unable to make choices, they expose their organizations to embarking on new rounds of broadening and compromising rather than making the necessary and crucial choices [POR 96]. [BER 09]'s study of the Dutch Executives reveals that: "The majority of Dutch executives find it difficult to make real choices. Especially now, during the economic downturn, they tend to hold on to familiar things." This is better revealed by an answer to the following question as conducted by [BER 09]: "What, if any, are the main issues you face when developing your strategy?" The results of the survey are better outlined in Figure 5.16, where the reader can see that over 50% of the participants avoid making choices, 33% state that they tend to hold onto familiar things and notions, 25% observe that there are too many people involved in the process of strategy-developing. The data reveals additionally the managers' risk-avoidant behavior. Based on these types of information, the strategist can apply the SDM to arrive at the most appropriate conclusions and act on them, fostering the process of the organization's uniqueness.

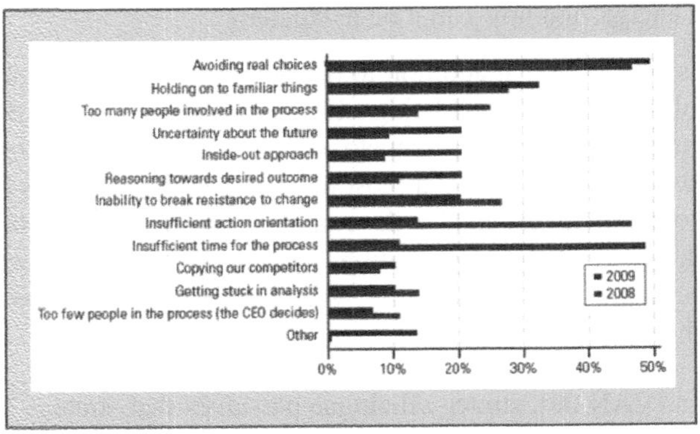

Figure 5.16. *Berenschot's survey of Dutch Executives after [BER 09]. For a color version of this figure, see www.iste.co.uk/machado/industry.zip*

Figure 5.17 reveals that hyper-competition among rivals is the chief concern among Dutch executives. Thus, finding solutions for being unique and applying strategies such as the "Blue Ocean Strategy" in accordance with the SDM to pursue strategies that can actually contribute to the attainment of the objective as the organization's grand strategy, is essential. If hyper-competition is the name of the strategic game, then uniqueness and differentiation are the answer.

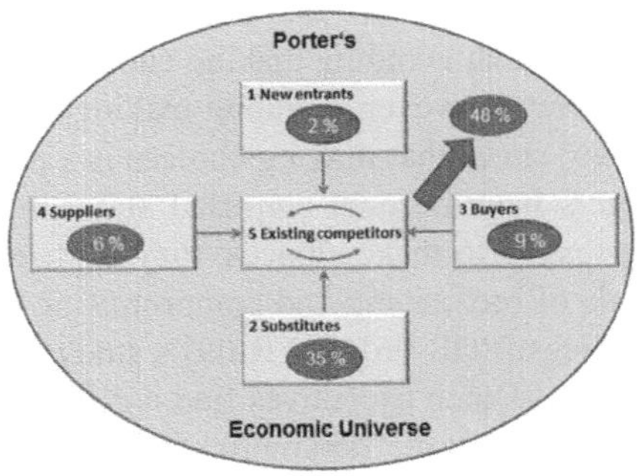

Figure 5.17. *[BER 09] survey based on the question of: "What, if any, are the main issues you face when developing your strategy?", author's own interpretation after [BER 09]*

Figure 5.17 reveals that 48% of the participating organizations consider hyper-competition to be their major strategic challenge. General benchmarking, which is used to compare what others do better or what the organization itself does better, still does not give the organization the ability to go to the next level. New models and strategies should be introduced in addition to the attractive product portfolio. The next frontier, according to the author's analysis, is viable organizational structures, unique strategies, business model innovations and strategic foresight. [BER 10] introduces an additional model, which is shown in Figure 5.18, with its clients by using the "Berenschot Strategic Dialogue"; the model is an extension of the original SDM model. However, both models are valid and can be put to work to achieve the needed coherence and joint force to arrive at the strategic position.

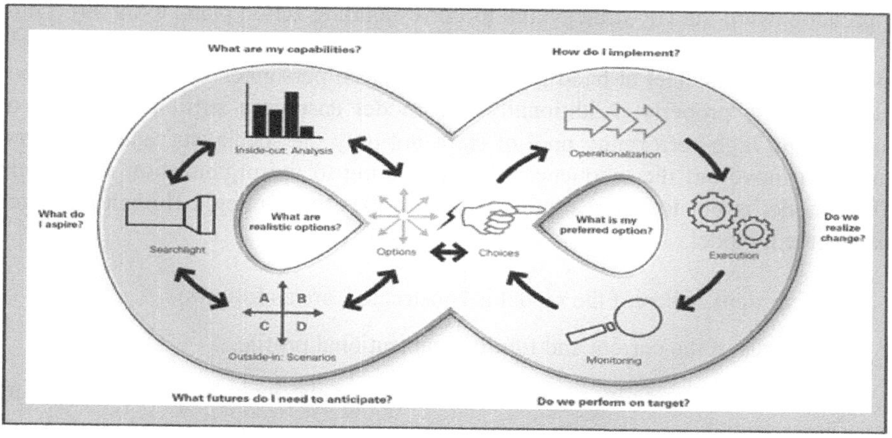

Figure 5.18. *Berenschot's 2012 model of strategic dialogue, author's interpretation after [BER 10]. For a color version of this figure, see www.iste.co.uk/machado/industry.zip*

According to [BER 10]:

> Regardless of what industry they operate in, all companies wish to distinguish themselves through the quality of their services and products. To that end, they also aim to innovate. Successful innovation demands a solid insight into what markets and customers desire and demand. A dialogue within the company contributes to this insight, especially when the likes of salespeople and account managers are involved. After all, they are the ones who talk directly to the customers. Along that same line, it also pays to involve those who are

in contact with suppliers, such as procurement managers. Their first-hand knowledge of the new possibilities that suppliers are developing helps determine what can successfully be innovated and how.

The author's own experiences as an executive also confirms that coherence within an organization is essential, and if that is not the case, there is a large energy drain, which successful organizations avoid and thereby become the leaders in their industries.

5.2.10. *Strategic HRM model*

The strategic HRM model is based on a five-step plan by developing a structured approach toward the HR strategy and its corresponding action plan [VAN 09]. Thus, it is important to make sure that there is enough capacity for high-quality, capable and qualified personnel at hand and that unforeseen personnel shortages and needs are dealt with properly. Additionally, the model ensures a sufficient number of talented and high-performing pool of staff, not only the specialists' and managers' transition move but the uniqueness they can bring to the organization if the right HRM model is put to work. Figure 5.19 describes the Strategic Human Resource management model.

The five steps, whereof the model is constructed, are as follows:

1) Analysis of the current and future organizational profile.

At this stage, generally, an overall analysis and the inclusion of the following areas are essential:

– the overall environment wherein the organization is embedded;

– governmental policy;

– project management;

– control and governance;

– strategy;

– culture;

– operations management;

– human capital planning;

– organizational structure.

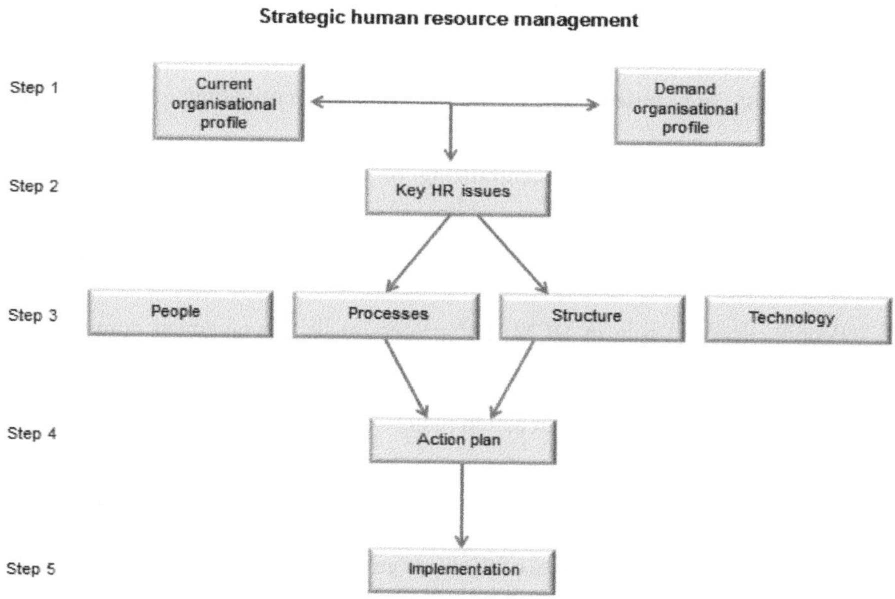

Figure 5.19. *The "Strategic Human Resource Management" model; author's interpretation after [VAN 09]*

Scanning the environment in combination with an inside-out perspective enables the organization to have solid Human Capital (HC) planning. The strategist is free to combine, integrate and use diverse models, which the author has introduced throughout the book, and which match better to their needs and better calculate the environmental changes and HC needs.

2) Determining the pivotal elements for HCM.

At this stage, the critical success factors and issues are calculated via an in-depth assessment of where the organization is (current situation) and where it intends to arrive (future position) in terms of capability, market penetration and share, strength toward rivals, managing and better positioning itself toward the technological, legal, political, ecological and societal challenges and changes. Furthermore, by exploiting the emergent opportunities to successfully lead and having high-performing people taking charge will bring the organization nearer to the perceived and desired future on a daily basis.

3) Organizing human capital planning (HCP).

The model's part 3 consists of the four below elements:

– Processes: are based on recruiting, hiring, motivating, retaining and compensating.

– Organization: outsourcing, centralizing of HC, self-organization, MBE and MBO.

– Technology and IT: automation, home and mobile offices, service delivery models based on IT and other technological advances.

– People: peoples' role in HCM, HR professionals, management commitment, training and professionalism.

4) Drafting and constructing an action plan.

Step 4 deals with preparing an action plan for step 3 of the model shown in Figure 5.19 and the steps and analysis before that. [VAN 09] observes that the action plan not only keeps track of who will complete which task and when but also looks at the commitment and willingness of the managers.

5) Implementation.

At the final stage of the model, the last details ought to be worked out and finalized before implementation. Further fine-tuning before launching the plan ought to be applied. However, [VAN 09] warn that the model ought to be looked at as a holistic managerial model and not an exclusive tool of the HR professionals within the organization.

An additional solid model for managing HC is Harvard's Strategic Human Resources Model (SHRM), which is another view on how organizations can approach their strategic HC needs and planning.

According to [NAN 11]:

> In essence, HRM differs from earlier personnel management models in relation to its focus, its principles, and its applications. HRM can be simply defined as the convergence of three factors – human beings, resources, and management – where human beings have the actual and potential resources (knowledge, skills, and capabilities) that can be harnessed through effective management techniques to achieve short- and long-term organizational goals as well as personal needs.

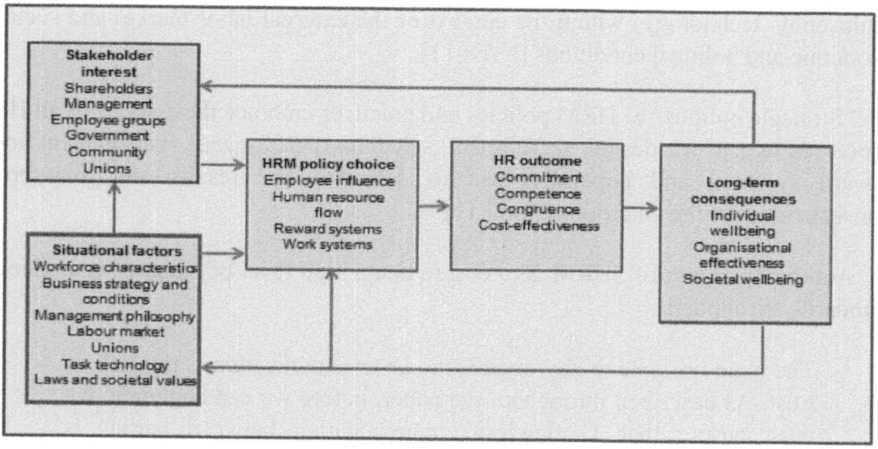

Figure 5.20. *The "Harvard" Model of HRM; author's interpretation after [BEE 95]*

According to [NAN 11], Harvard's SHRM model, as described in Figure 5.20, includes the following four elements:

1) Employee influence and involvement: The extent to which employees are encouraged to share their ideas and participate in organizational consultation and decision-making procedures.

Human resource flow: All HRM functions involved in employee management (e.g. HR planning, job design, recruitment and selection, performance review, termination).

Rewards systems: The monetary and non-monetary ways by which staff are recognized.

Work systems: This includes consideration of the "fit" between employees and their workplaces (e.g. technology, workplace design, teams).

Harvard's SHRM furthermore recommends that a strategic approach to HC strategy, policy and processes, in essence, reflects management options about how staff are managed and controlled. The "Harvard SHRM" describes a set of a broad range of strategic options in response to the demands of organizational characteristics (e.g. stakeholders, business strategy and conditions, management philosophy, technology) within the context of the external labor market and social, economic and political conditions [NAN 11].

"Strategic options" of HRM policies and practices embrace the essence of all HR processes (e.g. work design, recruitment, selection, performance management and reward systems) and hopefully lead to desirable HC results and long-term consequences for the enterprise [NAN 11].

Another vital part of SHRM is to ensure that a high level of integrity and ethical standards are applied.

> The main obstacle to any organizational communication is the lack of trust. As described throughout the paper, before we can negotiate, we must ensure that 1) flawless communication between parties is ubiquitously possible; 2) trust as the fundamental part of a negotiation is ensured; 3) the grand objective (ethos) of the negotiation as a system is maintained. This means the whole negotiation system must follow a constructive rule of creating the ideal situation for all parties. [KAM 11a]

With the right approach trained in the business schools and with models of doing business designed to ensure organizational viability and profitability due to being ethical, organizations create a vital competitive advantage. According to [KAM 11b]:

> … There is an academic institution on which a heavy burden is laid, and which needs to wake up and organize itself as a whole and train future managers as the only real safeguards of tomorrow, as [DRU 07] rightfully observed. [WIT 89] observed we have to set on the fallacy and convey it to the truth, which means we have identified the roots of all delusion; otherwise, knowing the truth does not help us if something else has taken its place. The solution was given to us by [DAR 59], who claimed that only the organisms that are fit will survive. We have to define fit for our corporations and our academia. This fit can and must only be designing an organizational conscious system, embedded to maintain the ethos of the organization not only for its own survival [BEE 72] but for the survival of the environment in which it needs to be embedded to exist….

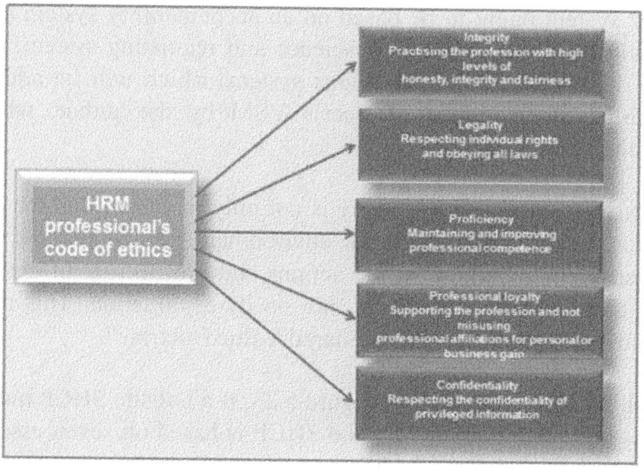

Figure 5.21. *The five professional ethics items governing professional conduct in HRM, as well as the author's interpretation after Chatterjee (Curtin University of Technology) in [NAN 11]*

Figure 5.21 describes the HRM professional guide of conduct. The conscious system the author is referring to starts with having a system of trust and ethics within the organization. Organizations with more trust and an ethical corporate culture are more creative, thereby making sure that they are attractive to high performers, who can be hired and retained for a long time. The SHRM model, therefore, qualifies to be integrated into the author's larger strategic model.

5.2.11. *Strategic human capital planning model*

Strategic human capital planning (SHCP) can be generally observed as the expertise, knowledge and uniqueness of personalities within an organizational system's structure needed to achieve the organization's grand strategy, its competitiveness and innovativeness toward rivals and marketplace changes, and the customer needs of today and the future.

In essence, SHCP must incorporate the following pillars (as Figure 5.22 describes):

– the organization's grand strategy/precise strategic leadership;

– organization's objectives;

– strategies/what to do and how to do it;

– a precise communication and implementation plan.

The SHCP system ought to be based on an accountability system and what the author also calls an organization's conscience and regulating system, as shown in Figure 5.22. The organizational conscious system, which will be addressed later, establishes further development of Beer's VSM by the author, which will be introduced in the next section.

The challenge for management today is not only to ensure that the organization survives because of the turbulence in the environment but also that the environment and society can survive because of the actions and profiteering of businesses with regard to the costs of the environment, not on its exploitation. This notion of the author also parallels with [POR 11]'s "Shared Value Doctrine".

The model, which is described in Figure 5.23, is a holistic SHCP model. Thus, it constructs a system and model where the SHCP is based on seven essential pillars, which can support the strategist in having a clear picture of how human capital is employed. In addition, the model includes not only the gaps and key elements that need to be calculated but moreover, it is directly connected to the environment. Thus, it senses the changes happening in the industry, wherein the organization is embedded and changes in terms of labor laws and regulations, and where the organization needs to react and adjust itself.

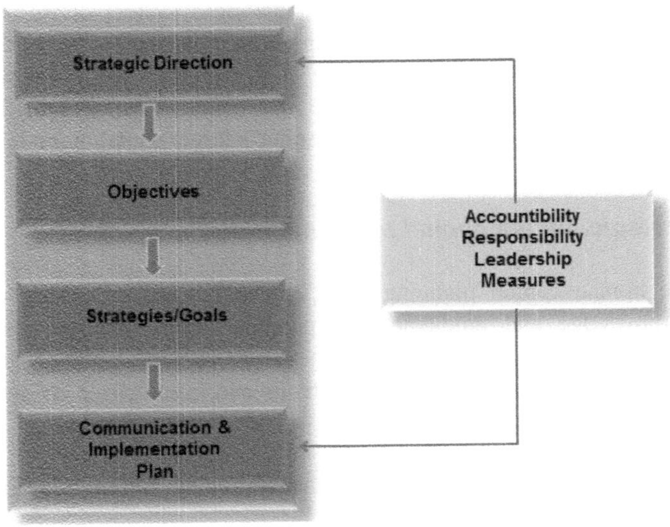

Figure 5.22. *An SHCP system based on an organizational conscious system; author's own interpretation*

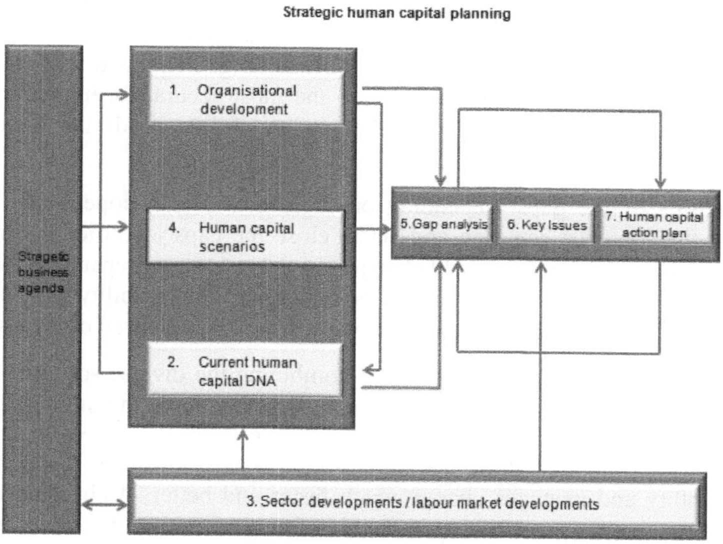

Figure 5.23. *The SHCP model and the author's interpretation after [VAN 09]*

These seven elements and sub-systems, which construct the SHCO model, are:

1) Organizational development: in this part, a precise analysis of the current situation of the HC, as well as its strengths, weaknesses, capabilities, structure, processes and culture, ought to be conducted. Furthermore, the strategist needs to consider the overall strategic and future HC needs of the organization, whereby they can take the organization to the next frontier and boost its performance in accordance with the shifts and changes of the market and customer demand.

2) Current human capital DNA (culture, competence, capability, structure): the HC-DNA exploits the outcome of the analysis, which results from the differentiation between the function of the job and its strategic value for the organization. Thus, to define the relative capacity of the current HC and to quantify their achievements, the strategist should be able to distinguish between the jobs and their functions [VAN 09]. According to [VAN 09], this will result in defining the DNA of the HC.

3) Changes in the regulations, labor markets and innovations in HC employment: in this part, in-depth analysis and constant scanning of changes in the labor market, innovations in HC models, its management, employment and changes in regulations and labor laws must be conducted. Thus, by having precise information and actual facts, the strategist can exploit opportunities and benchmark with the best in the industry and beyond on the cost and effectiveness of HC.

4) Scenario development and construction: this element of the model's task is to develop the best scenarios possible for organizational immunity in terms of HC – attraction, recruitment and retention. The information gathered and analyzed by the previous three steps can be used to develop the most accurate scenarios on which demographic, technological and legal changes and future needs the organization ought to adjust itself.

5) Analysis of the gaps: depending on the scenarios developed and the gaps identified, the strategist can plan and act to close the gaps identified and ensure a solid SHCP. Hyper-competition and changes do not excuse unpreparedness and lack of foresight. The gap analysis gives the organization the possibility to manage the organizational needs and challenges proactively, thus staying ahead of the crowd.

6) Key issues and essential aspects: by combining the diverse scenarios and the gaps to be filled, the organization can actually address all of the key strategic and essential issues, whereby the organization can grow further and develop a positive and supportive corporate culture. An organization with a holistic SHCP perspective fosters viability and manages change much faster and better. Addressing essential issues that concern the organization in the near future increases efficiency, improves time and resource management, avoids costly mistakes and reduces uncalculated situations and unpleasant surprises.

7) The SHCP action plan: the action plan consists of the following essential elements; thus, the strategist can transform strategies and key issues into practical plans of action; according to [VAN 09], the following four elements are essential: 1) labor market; 2) organization, structures and processes; 3) qualifications, capabilities, personalities and formation; and 4) construction of the HC-structure.

The model enhances the ability of the strategist to prepare and better adjust to the changes, fill the gaps, have a solid plan for diverse scenarios, increase productivity and learning, ensure a solid and supporting corporate culture, enhance competitiveness, reduce costs and avoid surprises, enhance HC sourcing, mobility and motivation. Furthermore, solid and ubiquitous communication should be established between all of the organization's layers. With a solid SHCP, customer needs are better addressed, and finally, the essential adaptation to changes can occur with minimal waste of time and resources. The author will include, in accordance with the many reasons established above, the SHCP model in his larger strategy model.

5.2.12. *The value disciplines of the Treacy and Wiersema model*

[TRE 95] wrote a magnificent book called "The discipline of market leaders: Choose your customers, narrow your focus, dominate your market".

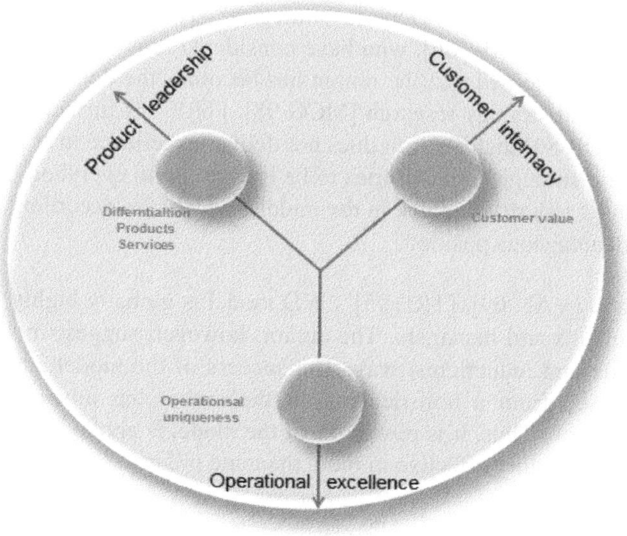

Figure 5.24. *The three essential pillars of market leadership, as well as the author's interpretation after [TRE 95]*

The book introduced a vital simplification to the wider complexity of strategic management. Thus, the author explained what a company needs to excel in the market or to be a market leader. Treacy and Wiersema [TRE 95] identified three essential pillars of what it takes to step into the shoes of the market leader. As shown in Figure 5.24, the author's three essential pillars are also coined as "Value Disciplines" (WD). These pillars are [TRE 95]:

1) Best total costs → operational leadership (cost leadership and control, avoiding energy drain).

2) Best products → product leadership (most advanced technology, first in the market or to use and quality that exceeds customer expectations).

3) Best total solution → customer intimacy (reliability, customer focus, customers' focus tightening the social institutions in their own and customer's value chain).

[TRE 95] has delivered a solid piece of research; the foundation of the work can be traced to [POR 80] in his book "Competitive Strategy". Thus, the author substantiates Porter's claim [POR 85], who coined the term "Stuck in the middle", whereby if a firm does not excel in any generic strategy and is at least good in the other two strategies, the firm will be less likely to dominate the marketplace. According to [POR 85], there are three approaches to strategy to cope with five forces, which he coined as "Generic strategies". [POR 85]'s topology is also validated according to [MIL 86], who have considered it as "already a classic", and [HIL 88], who has claimed that the notion has become "the dominant paradigm" in business policy and strategy research [MCG 98]. [TRE 95] furthermore establishes the notion as [POR 80] that if a firm does not deliver excellence in at least one of the pillars or generic strategies, thus it tries to be everything to everybody, the firm will be mediocre in its industry or stuck in the middle, which also according to [MCG 98] is a very disadvantageous position.

According to [VAN 09], [TRE 95]'s WD model is globally highly regarded and used by consultants and managers. The author, however, suggests not using it in a highly dogmatic and reductionist way. The success of the model actually grows if the model is used from a holistic lens, as the author has advocated the notion throughout the book. Thus, it is possible that the model is applied in terms of where the organization is highly effective in more than one pillar.

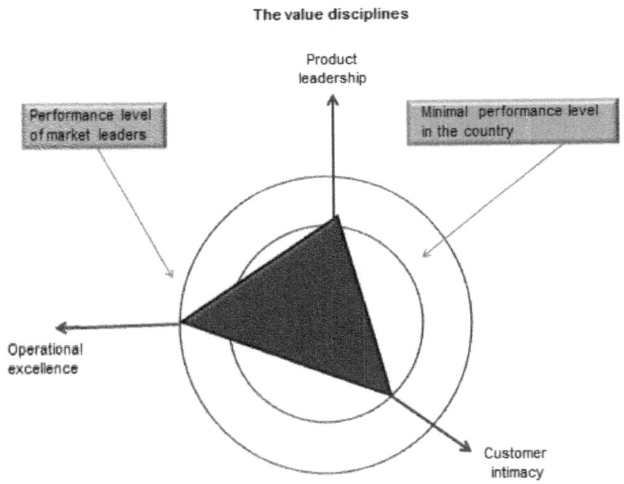

Figure 5.25. *Treacy and Wiersema's model and author's interpretation after [TRE 95]*

Figure 5.25 displays the [TRE 95] model with an extension, where the reader can identify how managers can benchmark between the least and the top performers in their industries. Although depending upon the organization's industry (e.g. high-tech and bio-tech companies ought to be leaders in their market if they want to survive), some strength is natural to survive; leaders ought to ask themselves the question by delivering which additional unique value proposition can they contribute to their organization's path to uniqueness and industry leadership. The notion of combining strength goes beyond the natural minimum strength necessary to survive and act in conformity with other rivals. Thus, it actually focuses on customer needs and changes them profitably for the customers and their companies.

It is essential to apply the model in combination with other organizational models, so a uni-dimensional view is avoided. The author extension of the "Operating Model of Operational Excellence" (OMOE), according to [KLU 04], displays that a broader focus is necessary in addition to excelling in operations, as culture, management systems, organization and information technology. Figure 5.26 describes the "Operating Model of Operational Excellence" (OMOE).

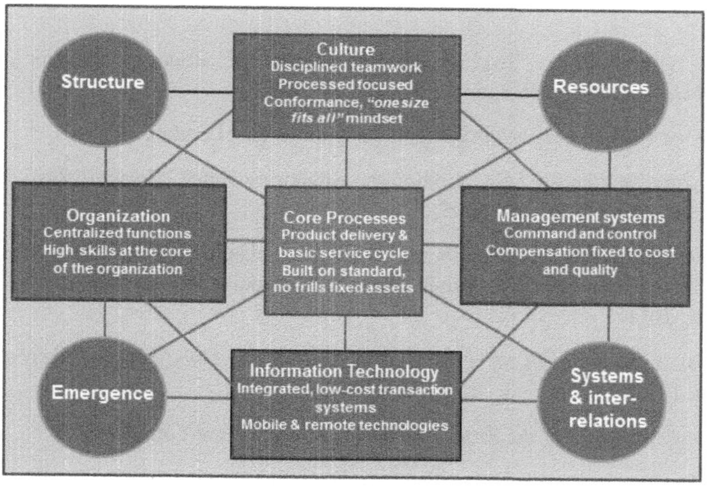

Figure 5.26. *The operating model of operational excellence and the author's interpretation after [KLU 04]*

The OMOE model is extended by the author to further establish the threshold of how necessary the inclusion of emergence, structure, resources, additional systems and sub-systems, and their relations with other agents within their milieus are.

5.3. Conclusions

Organizations with a broader perspective out-maneuver and outsmart their rivals. The [TRE 95] model is effective for strategists to benchmark with the best and to concentrate on the core issues. Models are necessary tools, whereupon managers reduce the complexity of the environment and cope with the rising requisite variety that perturbates the viability of the enterprises under control. Based on the Conant–Ashby Theorem, models are essential foundations of a good regulating system. Regulation cannot be imposed on the system; thus, it is an integral part of the system to be regulated. A firm's competitive advantage is based on what functioning and adequate models are collectively carried in the minds of the managers and how quickly they can be updated or discarded to avoid core rigidities. The essence of this chapter was to bring some of the models into the awareness of contemporary managers.

5.4. References

[ANS 84] ANSOFF H.I., *Implanting Strategic Management*, Prentice Hall, Englewood Cliffs, 1984.

[ANS 87] ANSOFF H.I., "The emerging paradigm of strategic behavior", *Strategic Management Journal*, vol. 8, no. 6, pp. 501–515, 1987.

[ANS 88] ANSOFF H.I., *Corporate Strategy*, Penguin Books, London, 1988.

[ARO 05] ARON R., SINGH J.V., "Getting the offshoring right", *Harvard Business Review*, vol. 83, no. 12, pp. 135–143, 2005.

[BAI 11] BAIN & COMPANY, Scenario and contingency planning, Management tool, available at: https://www.bain.com/insights/management-tools-2011-scenario-and-contingency-planning, 2011.

[BEE 72] BEER S., *Brain of the Firm: A Development in Management Cybernetics*, Lane, London, 1972.

[BEE 95] BEER S., *Brain of the Firm*, John Wiley & Sons, New York, 1995.

[BER 09] BERENSCHOT, What's on the agenda of the Dutch CEO in times of economic downturn? Research on strategy trends in the Netherlands, Blog, available at: https://berenschotstrategies.files.wordpress.com/2009/07/strategy-trends-2009.pdf, 2009.

[BER 10] BERENSCHOT STRATEGIES, Dialoog over strategie rendeert – Deel 9 van 10, Blog, available at: https://berenschotstrategies.wordpress.com/2010/11/02/dialoog-over-strategie-rendeert-deel-9-van-10/, 2010.

[BOW 95] BOWER J.L., *Business Policy: Managing Strategic Processes*, 8th edition, Irwin, Chicago/London, 1995.

[DAR 59] DARWIN C., *On the Origin of Species by Means of Natural Selection, or the Preservation of Favoured Races in the Struggle for Life*, John Murray, London, 1859.

[DRU 07] DRUCKER P.F., *The Practice of Management*, Butterworth-Heinemann, Cambridge, 2007.

[DRU 54] DRUCKER P.F., *The Practice of Management*, Harper & Row, New York, 1954.

[GOL 04] GOLÜCKE F., *Verfasserlexikon zur Studenten-und Universitätsgeschichte: ein bio-bibliographisches Verzeichnis*, SH-Verlag, Köln, 2004.

[GRE 89] GREINER L.E., "Evolution and revolution as organizations grow", in ASCH D. and BOWMAN C. (eds), *Readings in Strategic Management*, Springer, Berlin, pp. 373–387, 1989.

[HAM 93] HAMMER M., CHAMPY J., *A Manifesto for Business Revolution*, Harper Business, New York, 1993.

[HAV 03] HAVE S.T., HAVE W.T., STEVENS F. et al., *Key Management Models: The Management Tools and Practices that will Improve Your Business*, Pearson Education, London, 2003.

[HET 08] HETZLER S., Brain-supporting-environments für Entscheide in komplexen Systemen: ein kybernetischer Ansatz für die Gestaltung wirksamer Entscheidungsumfelder, Dissertation, Schesslitz, 2008.

[HIL 88] HILL C.W.L., "Differentiation versus low cost or differentiation and low cost: A contingency framework", *The Academy of Management Review*, vol. 13, no. 3, pp. 401–412, doi: 10.2307/258088, 1988.

[JOH 88] JOHNSON G., SCHOLES K., *Exploring Corporate Strategy*, 2nd edition, Prentice Hall, Hoboken, 1988.

[KAM 11a] KAMRAN Q., Complexity of negotiation and negotiation of complexity getting to trust, eJournal, available at: https://papers.ssrn.com/sol3/papers.cfm?abstract_id=2137877, 2011.

[KAM 11b] KAMRAN Q., "Management by deception (MBD): The need for designing a viable strategy", *SSRN Electronic Journal*, doi: 10.2139/ssrn.2137880, 2011.

[KLU 04] KLUIN J., DEN HOED P.T., VAN LINSCHOTEN R. et al., "Endoscopic evaluation and treatment of groin pain in the athlete", *The American Journal of Sports Medicine*, vol. 32, no. 4, pp. 944–949, 2004.

[MCG 98] MCGEE J.E., RUBACH M.J., "Stuck in the middle: Perhaps not such a bad place to be", in *Proceedings, Association for Small Business & Entrepreneurship*, pp. 30–36, 1998.

[MIL 86] MILLER D., FRIESEN P.H., "Porter's (1980) generic strategies and performance: An empirical examination with American data: Part I: Testing Porter", *Organization Studies*, vol. 7, no. 1, pp. 37–55, 1986.

[MIN 80] MINTZBERG H., *The Nature of Managerial Work*, Prentice Hall, Englewood Cliffs, 1980.

[NAD 97] NADLER D., TUSHMAN M., NADLER M.B., *Competing by Design: The Power of Organizational Architecture*, Oxford University Press, New York, 1997.

[NAN 11] NANKERVIS A., COMPTON R., BAIRD M. et al., *Human Resource Management: Strategy and Practice*, 7th edition, Cengage Learning, South Melbourne, 2011.

[NON 07] NONAKA I., "The knowledge creating company", *Harvard Business Review*, vol. 85, pp. 162–171, 2007.

[PHA 03] PHAAL R., FARRUKH C., MITCHELL R. et al., "Starting-up roadmapping fast", *Research-Technology Management*, vol. 46, no. 2, pp. 52–59, 2003.

[POR 79] PORTER M.E., "The structure within industries and companies' performance", *The Review of Economics and Statistics*, vol. 61, no. 2, p. 214. doi: 10.2307/1924589, 1979.

[POR 80] PORTER M.E., *Competitive Strategy: Techniques for Analyzing Industries and Competitors*, Free Press, New York, 1980.

[POR 85] PORTER M.E., "Technology and competitive advantage", *Journal of Business Strategy*, vol. 5, no. 3, pp. 60–78, doi: 10.1108/eb039075, 1985.

[POR 96] PORTER M.E., "What is strategy?", *Harvard Business Review*, vol. 73, no. 6. available at: https://hbr.org/1996/11/what-is-strategy, 1996.

[POR 11] PORTER M.E., KRAMER M.R., "Creating shared value: How to reinvent capitalism – and unleash a wave of innovation and growth", *Harvard Business Review*, Cambridge, available at: http://www.hks.harvard.edu/mrcbg/fellows/N_Lovegrove_Study_Group/Session_1/Michael_Porter_Creating_Shared_Value.pdf [Accessed 24 February 2013], 2011.

[PRA 90] PRAHALAD C.K., HAMEL G., "The core competence of the corporation", *Harvard Business Review*, vol. 68, no. 3, pp. 79–91, 1990.

[SCH 03] SCHWANINGER M., "A cybernetic model to enhance organizational intelligence", *Systems Analysis Modelling Simulation*, vol. 43, no. 1, pp. 53–65, 2003.

[SCH 10] SCHWANINGER M., "Model-based management (MBM): A vital prerequisite for organizational viability", *Kybernetes*, vol. 39, no. 9/10, pp. 1419–1428, 2010.

[SIM 77] SIMON H.A., "The organization of complex systems", in *Models of Discovery: And Other Topics in the Methods of Science*, Springer, Dordrecht, pp. 245–261, 1977.

[TRE 95] TREACY M., WIERSEMA F.D., *The Discipline of Market Leaders: Choose Your Customers, Narrow Your Focus, Dominate Your Market*, Addison-Wesley, Boston, available at: https://books.google.de/books?id=DdsOAQAAMAAJ, 1995.

[VAN 09] VAN ASSEN M., VAN DEN BERG G., PIETERSMA P., *Key Management Models: The 60+ Models Every Manager Needs to Know*, Pearson Education, London, 2009.

[WAC 85] WACK P., Scenarios: Shooting the rapids, Article, available at: https://hbr.org/1985/11/scenarios-shooting-the-rapids, 1985.

[WAT 07] WATSON G.H., *Strategic Benchmarking Reloaded with Six Sigma*, John Wiley & Sons, New York, 2007.

[WET 96] DE WET G., "Corporate strategy and technology management: Creating the interface", in *Proceedings of the 5th International Conference in the Management of Technology (IAMOT)*, Miami, pp. 510–518, 1996.

[WIT 89] WITTGENSTEIN L., "A lecture on freedom of the will", *Philosophical Investigations*, vol. 12, pp. 85–100, doi: 10.1111/j.1467-9205.1989.tb00265.x, 1989.

6

Approaching the Portuguese Labor Market From a Gender and Generational Perspective in the Era of Industry 4.0, Robotization and Artificial Intelligence

Employee retention across different generations has garnered attention from scholars and practitioners, particularly regarding gender differences in the workplace. Research focusing on gender differences among generations X and Y in the labor market remains limited, especially in Portugal. This study addresses the need for a deeper understanding of intra-generational variations by offering a comprehensive analysis of how men and women from generations X and Y perceive factors influencing job retention. Drawing on an exploratory study, the aim is to examine the generational expectations and preferences of higher education graduates in Portugal concerning a set of work-related factors. This study seeks to correlate the preferences of men and women from these generations with aspects of the work environment, including support for working work–life balance, benefits and incentives, autonomy, flexibility, affiliation with a socially responsible company and access to training.

To address the central objective of the research, the study first explores the theoretical aspects of the nature and expectations of work for these generations and then reviews the literature on gender theories. An empirical investigation, involving 421 questionnaires, was conducted to examine the expectations and preferences related to retaining generation Y and X employees, with a particular emphasis on gender.

The data were statistically analyzed, revealing no significant differences between genders. However, descriptive statistics indicated trends suggesting that women from both generations place a higher value on autonomy, flexibility and work–life quality compared to their male counterparts. These findings align with the existing literature. Based on the empirical findings, Human Resources managers can develop targeted intervention strategies to enhance the retention of generation Y and X employees by fostering a supportive work environment.

Chapter written by José Rebelo dos SANTOS, Lurdes PEDRO and Célia QUINTAS.

6.1. Introduction

The coexistence of several generations in the job market has aroused growing interest among Human Resources managers and academics concerned about the need to attract and retain human potential in organizations.

A generation represents a particular social type, born in a common period, which has experienced the same socio-political moments [CAP 15], [SAN 24].

The different representations that have influenced individuals born in different historical periods have also given rise to diverse ways of seeing, experiencing and valuing work.

It is therefore natural that there should be a concern to understand what the expectations and perceptions are in relation to work from a generational perspective.

Few studies have explored how generational influences affect work perceptions, including gender biases – i.e. whether these perceptions differ between males and females within each generation.

In response to this gap, the general objective of this study was formulated as follows:

To identify the factors most valued by generations X and Y individuals in the job market, with consideration of gender differences, we must take into account the characteristics of the workers (or new workers) who will be in companies while various changes occur, such as the introduction of new technologies (4.0), the coexistence of robots and people (5.0) and the increasing use of artificial intelligence (6.0).

To achieve this objective, we employed a comprehensive quantitative methodology. A questionnaire survey was conducted using a convenience sample. A total of 429 responses were obtained from graduates belonging to generations X and Y of a Portuguese Higher Education Institution.

The study is structured into three parts: the first part develops the theoretical framework of reference on the factors most valued at work by generations X and Y from a gender perspective; the second part discusses the methodological approach and techniques for data collecting and analysis; and the third part analyzes and discusses the results obtained.

6.2. Theoretical framework

6.2.1. *Context: making sense of gendered work across generations*

The aim of this section is to understand the origins of gender theories and their evolution, to analyze their relationship with the generational perspective, and their relationship with the perceptions that individuals have of the labor market.

We must remember these workers are the ones who will promote, participate in or assist the digital transformation in the sector, providing real-time decisions, more productivity, flexibility and agility (4.0).

The current period is one of great development in the use of artificial intelligence, but we are still experiencing the transition period to Industry 5.0. in which the focus is on humanizing and regulating this use, redefining the role of people in organizations and giving increasing importance to issues of sustainability, the environment and adversity management [SUÁ 22].

At the level of human resources, Industry 5.0 emphasizes the importance of people, resulting in changes in the relationship between the organization and workers as a result of the development of digital platforms [BER 23].

6.2.1.1. *Gender formation theories*

The analysis of gender theory implies, on the one hand, an understanding of the theory of social representations developed by Serge Moscovici in 1961 [MOS 61] and, on the other hand, an understanding of the feminist theories emerging in the same decade.

The analysis of social representations corresponds to a change in basic assumptions, opening new perspectives for analyzing phenomena related to the social sciences [ARR 02]. It is an approach that argues that emotion and reason, nature and culture, science and common sense are not separate in understanding reality, which is itself constructed and subjective.

Gender theories must be analyzed in a relational logic that considers power relations and the nature of the experiences lived in each social context [ARR 02], [ALL 13].

Feminist approaches inaugurated gender studies, and for decades, they have made important contributions by presenting the inequalities experienced by women throughout history. The feminist movement also stands out for its struggle to combat these inequalities. From a scientific perspective, these movements set out to analyze the causes of oppression, inequality and inequity experienced by women throughout history and in the various geographical contexts [GUZ 07].

It is important to note that feminist approaches do not offer a unitary view in their analysis of gender perspectives, but always start from the assumption that women are unequal when compared to men in the same situations. These inequalities frequently manifest in situations of oppression, domination and abuse [GUZ 07].

[ABR 11] defines socialization as the process of building individuals and societies, fed by social interactions, activities and practices. Emotions, power relations and projects regulate this dialectical construction in such a way that individuals produce society and are produced by it.

It is in this context that gender identities are constructed, through the incorporation of different habits and relationships with experiences; individuals assimilate and naturalize different social roles from childhood onward.

The notion of intersubjectivity is defined as the situation that allows subjects, characterized by subjective perceptions of the self, to form a society, community or common field that can be identified with the concept of "we" [COE 04], [ALL 13].

The perspective of intersubjectivity explores the idea that being a man, or a woman, goes far beyond a mere biological condition, and is a social construction that is defined in relation. It starts from a holistic perspective where gender must be analyzed in context, considering a set of cultural and historical variables that determine the role of individuals in society. It is from this perspective that we analyze gender in this study, as an intersubjective variable that is constructed from a set of relationships that are often power relationships, which develop in multiple social contexts that are interrelated and inter-influential. The gender perspective in the labor market cannot be dissociated from other contexts or from a set of variables that allow us to understand the position of individuals in society [SCO 16].

6.2.2. *Work, gender and generation*

Based on the assumption that it is possible to find common traits between people who share the same historical period, it is pertinent to approach the labor market from a generational perspective [CER 11], [SAN 24]. Three generations currently coexist in the job market: the baby boomers, who were born between 1944 and the early 1960s, generation X, who were born in the 1960s to the 1980s, and generation Y, who were born from the 1980s to the 1990s [SAN 23]. However, we must also consider the newest group entering the job market, generation Z, born in the 1990s, who have been familiar with technology since birth [CER 11]. In this study, we are working with generations X and Y from a gender perspective.

Women now dominate the job market, while we are witnessing a dominant presence of generation Y with the characteristics that are recognized in the literature related to immediacy, the use of technology, the search for meaning in work that relates to their purpose in life, inclusive values and solidarity, the search for pleasant environments and the importance of reconciling work and personal life, among other factors [EST 13].

The behaviors and values of generation Y influence organizations and force them to adapt. One of the factors that characterizes generation Y is precisely the appreciation of the role of women and their professional progression and salary situation on equal terms with men. According to [EST 13], most generation Y women in the workplace aspire less to having children and instead having more access to work and education.

However, despite generational changes, occupational and sectoral segregation by gender is remarkably persistent over space and time and is one of the main factors explaining the pay gap between men and women [BOR 20].

6.2.3. *Organizational attributes: gender and generation differences*

The aim of this section is to identify several work-related factors that are perceived differently by generations X and Y. It also demonstrates the differences and similarities in how work is viewed from a gender perspective. In this generational and gender perspective, the benefits, quality of work life, career progression, access to training and working relationships with colleagues and supervisors are addressed.

6.2.4. *Remuneration, rewards and benefits*

Remuneration, rewards and benefits are consistently identified as one of the main motivational factors by individuals from both generations X and Y [DEH 10], [TRI 17]. Benefits and incentives regarding differences between genders have also been studied in the context of motivational factors.

The formal benefits and incentives are stimuli to increase worker retention and enhance stability, security and flexibility in the workplace. Currently, benefits and incentives tend to be customizable or individualized [FUL 18], because employees react differently or perceive varying values in specific organizational remuneration or benefit offerings, depending on their personal characteristics [FUL 18]. This optimizes the incentive effects in a diverse workforce. [ERI 14] studied demographic variables, such as age/generation and gender, as factors that determine differences in preferences and reactions to benefits and total rewards. For example, personality traits such as risk aversion, more prevalent in women than in men [CRO 09], can influence the degree to which people prefer or are attracted to organizations offering certain types of health and well-being benefits. These traits may also influence candidates' preferences for specific configurations of total reward packages [ERI 14]. According to the needs and expectations of different generations, fringe benefits, from the perspective of human resource management, should be flexible and differentiated to value the various generations currently working in organizations.

6.2.5. *Social responsibility: quality of life at work/flexibility/work–family balance*

Business strategy involves making decisions that have economic as well as social consequences, which is why the subject of corporate social responsibility has been gaining weight in the reference literature [PES 09], [CRU 12].

The issue of environmental sustainability or minimizing damage is becoming increasingly important in companies' social responsibility strategies and, according to [PES 09], can generate competitive advantages. While most studies on social responsibility focus on its external dimension, it is important to recognize that effective human resource management and quality of life at work necessitate the development of an internal social responsibility strategy; the truth is that the issue of human resource management and quality of life at work implies the development of an internal social responsibility strategy, which can help increase individuals' commitment to organizations and contribute to improving quality of life at work [CRU 12].

Concerns about gender equality fall within the scope of internal social responsibility and, according to [PER 08], it comprises a set of dimensions related to the management of working time in order to facilitate the reconciliation of work and personal life, access to leadership positions for women, the construction of instruments and indicators to monitor the progress of measures promoting gender equality and action plans for gender equality in companies.

The perceptions of quality of work life also vary across generations, with flexibility and work–life balance practices being more highly valued by generation Y. Quality of work life, particularly the work environment and work–life balance issues, are especially appreciated by individuals from generation Y. Another factor studied is the concept of work–life balance, which means the "balance between professional and personal life", which involves the creation of a favorable/healthy work environment with flexible hours that simultaneously support the balance between individuals' personal and professional responsibilities [ALL 14].

Gender has been a variable of interest in numerous studies into the work–family relationship [SHO 17], particularly in studies on work–family conflict, which is considered bidirectional. However, in most societies, gender and work roles have changed over the past few decades. In Portugal, the proportion of women in the workforce has increased over time, and employed men continue to be more represented in leadership positions than employed women according to [POR 23].

Much research has focused on the work/family/life balance as a dilemma from the perspective of women who must reconcile professional obligations with personal/family responsibilities, and on the value and preference women have for corporate cultures where this flexibility is allowed. However, more recently, other studies, particularly meta-analysis, have not found clear and significant statistical support for this gender difference [SHO 17]. There are also generational studies that highlight specific preferences according to gender within generation Y: generation Y women prioritize job security and work–life balance, while men focus on career advancement, continuous learning and skill development [SHO 17]. Generation X women prioritize opportunities for recognition, relationships and flexibility to achieve work/life balance [FEY 05], while generation X men prioritize career development and work engagement [FER 09].

6.2.6. *Career development*

Another key factor highlighted in the literature is career progression. [TRI 17] state that the opportunity for professional growth is one of the determining factors for both men and women in generations X and Y. However, they identify generational differences, noting that generation X is more concerned with the benefits offered by the company, such as meal vouchers, health insurance or opportunities for professional development. These factors are not as significant to generation Y employees, who place a higher value on professional opportunities as a measure of job satisfaction [TRI 17].

In the analysis of the profiles of generations X and Y, some authors [DEH 10] support the position that personal career development is very important for generation Y. They note that generation Y is characterized by elevated expectations regarding job content, training, career development and financial rewards, suggesting that these expectations are deeply rooted in this generation. Similarly, [BEN 18] conclude that generation X employees have a strong need to develop their skills and experience and are more likely to take action to ensure their needs are met immediately. [BEN 18] also noted that generation Y places greater importance on the individualistic aspects of jobs, in line with – and reinforced by – the emphasis on individualism in contemporary human resource management. Furthermore, they assert that generation Y may have realistic expectations regarding their first job and salary but seek rapid career progression and the development of new skills.

[FAN 15] argues that the motivations and career perspectives of individuals within organizations go beyond age and generational factors, being influenced by various dimensions such as personal values and behavior patterns, among others. [FAN 15] also mentions that in organizations, it is common to see the peaceful coexistence of individuals from various generations, with employees of different age groups working together and sharing work experiences.

The fact that gender is a key factor in career development is well documented, for example, in [ONE 08], the literature review on women's career patterns since 1990 and, more specifically, in studies on women's career preferences [GAL 11]. Women's careers can be paradoxical and challenging [ONE 08]. In analyzing the barriers to women's career development at higher organizational levels, [SUL 07] noted that women face unique barriers to career progression and are often less satisfied with traditional male-oriented human resources management practices. A study conducted on UK generation Y graduates confirms that in career transition, men recorded more progress and advantages in career development than women [MAX 14].

6.2.7. *Professional recognition*

Professional recognition is one of the factors identified by [FEY 05] as being more valued by generation Y women compared to men of this generation, serving as a motivator and a factor for retention within the organization. The pursuit of professional recognition is associated with seeking opportunities to build successful careers and to receive constant feedback from supervisors and colleagues.

Recognition is also perceived differently when we talk about generation X and generation Y. For generation X, working on a project basis and identifying with the company and its values are motivating factors. For generation Y, extrinsic rewards are highly valued, not only in material terms but also in terms of recognition [SAN 24]. In this context, praise and professional appreciation hold much higher importance for generation Y compared to that for generation X. The absence of such recognition can be a source of demotivation and discouragement for generation Y [LIP 10], [SAN 24].

6.2.8. *Training*

As a fundamental human resources management practice, training and development refers to the degree to which organizations provide training to employees to foster their skills. Access to training is considered an important and valued factor by both generations X and Y. However, they differ in how they value training. Generation X tends to value expertise, a culture of quality and merit, whereas generation Y values access to training as part of a continuous learning and adaptation project [LEE 22]. [MAG 20] noted that generation Y tends to resist the overwhelming amount of information they encounter but values training and improvement in an unprecedented way compared to previous generations, often using new, more personalized and applied learning methods. Training is seen as a means of professional development, adaptation and improvement for individuals of various generations [PAR 11], [MAG 20] and [LEE 22].

6.2.9. *Work relations*

As we have seen, working relationships are valued by both generations, but women place a higher value on issues related to the work environment, especially if it allows for a greater possibility of reconciling work and family life [FEY 05]. In terms of generations, working relationships tend to be valued by individuals from both generations and are associated with the working environment. However, there is a difference in the way people experience work, with generation Y tending to develop more competitive relationships and reject imposed authority, which necessarily influences working relationships with colleagues and managers and the working environment itself [LIM 22].

6.2.10. *Gender and generations: the role of human resource management*

Some authors highlight that gender issues and social processes in the workplace are manifested in human resource management policies and practices [HUT 04], [STI 10].

[ACK 90] states that "the view of organisations as gender neutral facilitates an individualistic view of relative successes, influence, and power – the view that people succeed because of superior abilities, dedication, and performance".

From this perspective, human resources management policies and programs are based on developing tend to develop processes and practices centered around what Acker calls the "ideal worker" (which is implicitly understood as masculine), or "organisational man" [ACK 90]. [DEN 76], on the other hand, equated the organizational man with the administrative man, a concept based on outdated norms of masculinity that penalizes both fatherhood and motherhood. [LOU 08] highlights that "organisational structures, practices, and policies are shaped according to the male gender" and a dominant individualistic and masculine culture, forcing women to fit into this culture [CAR 02]. [DIC 88], from this perspective, refers to the existence of an implicit male model of the universal worker, directed at both men and women.

Despite considerable progress in regulation and the promotion of equal opportunities over the past decade, research continues to explore the subtle ways, within human resource management, in which gender practices shape workplace processes and practices, generating and perpetuating historical or existing inequalities [MAS 16].

According to these authors, these can be found in human resource management activities such as recruitment and hiring, performance evaluation and job descriptions, and rewarding attendance or presenteeism [MAS 16].

Human resource managers can no longer assume workplace equity based on the number of male and female workers in their organizations. Indeed, women's preferences regarding work, as well as the practices developed by organizations, may not be based on gender representation but rather on deeply rooted roles and expectations, particularly in matters of work–life balance, such as flexible working hours and the proximity of work to home.

Statistics on the growth of women in full-time employment or examples of women in top management are, from this perspective, insufficient as proxies for more equitable and humane workplaces [MAS 16].

In addition to examining gender-related human resource management policies, in recent years, scholars have been focusing on intergenerational differences as they present challenges, particularly for human resource managers, in effectively managing the workforce [BEN 11]. Although the research is limited, this perspective can be readily seen in the articles/books titled "Generational differences in work values in the workplace" [KIN 23], "Motivation at work for generation Z" [GOL 23] and "Managing a Multi-Generational Workforce" [SUR 20], [SAN 23, 24].

Despite the fact that, globally, there continue to be substantially unequal gender ratios in workplaces, both horizontally (i.e. across industries and sectors) and vertically (i.e. in leadership positions) [FIN 20], and despite the existence of arguments sometimes based on an incorrect view of women as fundamentally different from men in how they think, feel and behave [ANN 17], there are several studies spanning generations X and Y that identify differences and similarities in the motivating work factors for women and men.

Understanding gender and generational differences means, in the context of human resources management policies and practices, that these differences require tailored human resources management strategies to effectively address the unique needs of generation X or Y employees, ensuring their satisfaction and retention in the workforce. The research highlights that although there are some differences in how men and women experience and interact with human resource management practices, there are many similarities, particularly in leadership and decision-making [JON 23]; gender-focused HR policies can improve organizational performance by promoting gender diversity. Understanding these gender and generational differences is essential for creating more equitable, effective and inclusive human resources management systems, seeking to differentiate responses to the needs of each in favor of the company's success. [LOU 08] even refers to the need to "make structural adjustments so that women, under equal conditions, can prove their ability to perform roles of leadership, prestige, and high responsibility".

The HR practices that have the most significant impact on generations X and Y, particularly on the women and men of these generations, are flexible work arrangements. These specifically address the flexibility of working hours, promoting an increased balance between work and family life, enhancing motivation and meeting professional expectations without neglecting personal expectations. Developing an organizational culture that supports the implementation of work–life balance practices, inclusive of both women and men, requires rethinking the relationship between work and family. Organizations can make a difference and contribute to the well-being of their employees, which in turn enhances their engagement and effort, leading to greater competitiveness and organizational performance [SAN 20].

6.3. Methodology

The study adopts quantitative and exploratory methodology, using a questionnaire survey.

The survey was applied online via the link:

https://forms.gle/cnRXzXHax83JkYqB8

and was self-completed. The survey contains 14 questions, divided into five sections: sociodemographic characterization, which asks about the participants' age, gender and education and the area of study. Work situation; social origins; job satisfaction factors where participants are asked to give their opinion, based on a Likert scale, on the following items: job content, benefits and rewards, salary, working environment, working hours, autonomy, recognition, flexibility, career progression, access to training, teamwork, relationship with colleagues, relationship with management, work–life balance, quality of managers, quality of life at work, socially responsible company, feedback, and an environment that fosters innovation. The fifth item is professional experience.

421 responses were obtained.

6.3.1. *Ethical considerations*

This instrument was presented to the Institution's Ethics Committee of Setúbal University of Applied Sciences, which gave a positive opinion on it, regarding the presentation and analysis of the data and safeguarding the identity and sensitivity of the respondents.

6.3.2. *Objectives*

The general objective is to understand whether the organizational factors that contribute to individuals belonging to generations X and Y staying in or leaving organizations vary according to gender.

In terms of specific objectives, the aim is to:

– identify the factors most valued by individuals from generations X and Y in the job market;

– understand whether there are variations in the factors analyzed according to gender;

– present human resources practices that facilitate the attraction and retention of the population studied.

6.3.3. *Data collection procedure*

A comprehensive data collection process was employed to investigate the relationship between factors most valued by individuals from generations X and Y by gender.

The characterization of generations X and Y was based on the bibliographical research of reference literature and empirical information collected through questionnaire surveys applied to graduates of a Portuguese Public Higher Education Institution, with various academic areas.

The study aimed to address its objectives by contributing knowledge on what participants value in the job market and professional life, with a particular focus on gender and generational differences.

6.3.4. *Sample and participants*

The participants in this study were selected through a non-probabilistic sampling process, using the convenience sampling method. Therefore, although the data contribute a deeper understanding of generation Y, it cannot be extrapolated [HIL 09]. Efforts were made to include a balanced distribution of generational classes, with a focus on capturing responses from generations X and Y.

The sample consisted of 421 graduates from a Public Higher Education Institution in Portugal, located in the Lisbon Metropolitan Area.

6.3.5. *Data processing*

The data collected was entered into a database, standardized and processed, including the creation of some new variables through recoding demographic variable (age group, generation, etc.). Generation Cohort was categorized in such a way for participants to indicate the year of birth into the following categories 1960–1980 (generation X) and 1981–2000 (generation Y).

The data was subjected to a series of descriptive and correlational statistical operations, allowing frequencies, means, medians, modes, standard deviation, association and correlation relationships between variables to be calculated. The Statistical Package for Social Sciences (SPSS) version 29 software was performed for this purpose.

The procedures made it possible to identify the respondents' perspectives on the factors that contribute to them staying in or leaving organizations.

6.4. *Results and discussions*

The questionnaire analysis included an internal consistency test for the nineteen (19) questions related to job retention factors, using a six (6)-point Likert scale where 1 means not at all important and 6 means very important. The simple reliability, from the set of observable variables, was measured through Cronbach's alpha. A value greater than 0.9 (0.963) was obtained, indicating very high internal consistency [PES 05].

6.4.1. *Participants*

	Variable	N	%
Generational class	Generation X (1960–1980)	64	15.2%
	Generation Y (1981–2000)	357	84.8%
Age group	> 37	106.0	44.4%
	29–37	128.0	30.4%
	< 29	2	25.2%
Gender	Female	257	61.0%
	Male	164	39.0%
Educational level completed	Bachelor's degree	309	73.4%
	Master's degree	105	24.9%
	Doctoral degree	7	1.7%
Educational area	Education	22	15.2%
	Engineering	120	28.5%
	Health	47	11.2%
	Management	190	45.1%
Labor situation	Is studying	7	1.7%
	Is working and studying	45	10.7%
	Is working	354	84.0%
	Unemployed	14	3.6%

Table 6.1. *Summary of demographics (N=421)*

Table 6.1 provides a summary of the respondents' demographic characteristics. The results indicate that generation Y comprises a significantly larger proportion of the sample than generation X, accounting for 85% of the total. Generation X was aged 43 or over at the time the information was collected in 2023, so they were born up to 1980, while generation Y corresponds to those born from 1980 to 2000. Most participants are over 37 years old (44%), followed by those aged 29–37 (30%), and those under 29 years old (25%). The respondents in 2023 were aged between 23 and 61. Of the 421 valid responses, approximately 61% were female, highlighting the active participation of women in the Portuguese labor force. In terms of educational attainment, 73% of respondents hold a bachelor's degree, 25% have a master's degree and 17% possess a doctorate. Most respondents have a degree in management (45%), followed by engineering (29%), education (15%) and health (11%). Concerning their professional status, 84% were employed, 11% were both working and studying, 4% were unemployed and 2% were exclusively studying. Most respondents are either employed or both working and studying (96.4%), while 3.6% are unemployed.

An analysis of the generations by gender, as shown in Table 6.2, reveals that the percentage of generation Y is slightly higher among females than males (87.9% versus 79.9%).

	Relative frequencies (%)			
	Generation X	**%**	**Generation Y**	**%**
Female	31	12.1%	226	87.9%
Male	33	20.1%	131	79.9%

Table 6.2. *Generations by gender (N=421)*

Table 6.3 shows that the percentage of unemployed respondents is 2.7% among females and 4.2% among males, with no significant differences between genders situations.

	Relative frequencies (%)			
	Female	**%**	**Male**	**%**
Is studying	4	1.6%	3	1.8%
Is working and studying	31	12.2%	14	8.5%
Is working	213	83.5%	141	85.5%
Unemployed	7	2.7%	7	4.2%
Total	255	100%	165	100%

Table 6.3. *Employment situation by gender (N = 421)*

6.4.2. *Descriptive analysis by generation*

Factors	Generation X			Generation Y		
	Avg	SD	Var	Avg	SD	Var
Job contents	5.00	1.24	1.54	5.17	0.95	0.91
Benefits and incentives	4.86	1.29	1.66	4.82	1.31	1.71
Salary	4.91	1.31	1.71	4.87	1.33	1.76
Work time table	5.12	1.26	1.59	5.04	1.19	1.41
Work environment	5.20	1.19	1.42	5.13	1.06	1.12
Autonomy	5.21	1.18	1.40	5.11	0.97	0.93
Flexibility	5.17	1.21	1.46	5.17	1.10	1.20
Recognition	5.03	1.32	1.75	4.93	1.29	1.65
Career progression	4.89	1.37	1.88	4.74	1.44	2.08
Access to training	4.89	1.30	1.70	4.68	1.27	1.62
Teamwork	4.80	1.28	1.63	4.76	1.22	1.49
Relationship with colleagues	5.12	1.21	1.46	4.97	1.04	1.08
Relationship with superior	5.09	1.25	1.56	4.99	1.05	1.10
Work–life balance	5.33	1.28	1.64	5.19	1.19	1.42
Quality of managers	5.05	1.33	1.76	4.80	1.27	1.62
Life at work quality	5.15	1.19	1.42	5.03	1.16	1.34
Socially responsible company	4.70	1.28	1.63	4.40	1.38	1.90
Feedback	4.82	1.24	1.53	4.67	1.22	1.48
Company that encourages innovation	4.82	1.32	1.75	4.56	1.35	1.83

Table 6.4. *Descriptive analysis by generation (N = 421)*

Table 6.4 presents the factors identified by generations X and Y, highlighting those considered most important by both generations (based on an average score of 5 or higher). The differences between generations are minimal, with no factors showing significant variation between them. The factors with the highest scores are as follows:

i) Work–Life Balance (Generation X: 5.33; Generation Y: 5.19);

ii) Autonomy (Generation X: 5.21; Generation Y: 5.11);

iii) Work Environment (Generation X: 5.20; Generation Y: 5.13);

iv) Flexibility (Generation X: 5.17; Generation Y: 5.17);

v) Working Hours (Generation X: 5.12; Generation Y: 5.04);

vi) Work–Life Balance (Generation X: 5.15; Generation Y: 5.03);

vii) Job Content (Generation X: 5.00; Generation Y: 5.17).

Factors	N	Min.	Max.	Avg	SD	Var
Job contents	421	1	6	5.14	1.01	1.01
Benefits and incentives	421	1	6	4.83	1.30	1.70
Salary	421	1	6	4.87	1.32	1.75
Work time table	421	1	6	5.06	1.20	1.44
Work environment	421	1	6	5.14	1.09	1.19
Autonomy	421	1	6	5.12	1.01	1.01
Flexibility	418	1	6	5.16	1.12	1.26
Recognition	419	1	6	4.94	1.29	1.68
Career progression	413	1	6	4.76	1.43	2.06
Access to training	418	1	6	4.71	1.28	1.65
Teamwork	414	1	6	4.76	1.23	1.51
Relationship with colleagues	415	1	6	5.00	1.07	1.15
Relationship with superior	418	1	6	5.01	1.09	1.18
Life at work quality	419	1	6	5.21	1.21	1.45
Quality of managers	417	1	6	4.83	1.29	1.66
Life at work quality	414	1	6	5.04	1.17	1.38
Socially responsible company	416	1	6	4.45	1.37	1.87
Feedback	413	1	6	4.69	1.22	1.50
Company that encourages innovation	412	1	6	4.60	1.36	1.84

Table 6.5. *Descriptive analysis by generational factors (N = 421)*

The descriptive analysis by generational factors, as presented in Table 6.5, indicates that all items are valued. Notably, the following items stand out with average scores above 5 (out of 6): work–life balance, flexibility, work environment, job content, autonomy, working hours, quality of life at work and relationship with manager. Conversely, the type of company and feedback are the least valued items, with average scores below 4.70 (out of 6):

i) Work–life balance (5.21);

ii) Flexibility (5.16);

iii) Work environment (5.14);

iv) Job content (5.14);

v) Autonomy (5.12);

vi) Working hours (5.06);

vii) Quality of life at work (5.04);

viii) Relationship with superior (5.01);

ix) Relationship with colleagues (5.00).

Tables 6.6–6.8 present the results of the analysis disaggregated by gender, highlighting differences across several factors, including benefits and incentives, autonomy, flexibility and access to training.

Table 6.6 shows that women, across both generations X and Y, place greater importance on incentives and benefits compared to men, selecting higher values on the "considerable importance/very important" scale.

		Relative frequencies (%)							
		Generation X				Generation Y			
		Female	%	Male	%	Female	%	Male	%
Benefits and incentives	Not important	0	0.0%	1	3.0%	8	3.5%	2	1.5%
	Not very important	2	6.5%	2	6.1%	10	4.4%	6	4.6%
	Important	0	0.0%	6	18.2%	18	8.0%	11	8.4%
	Quite important	3	9.7%	2	6.1%	32	14.2%	26	19.8%
	Considerably important	13	41.9%	9	27.3%	64	28.3%	40	30.5%
	Very important	13	41.9%	13	39.4%	94	41.6%	46	35.1%

Table 6.6. *Generations by benefits and incentives analysis (N = 421)*

Table 6.7 indicates that the autonomy factor tends to have higher values for women in both generations X and Y compared to their male counterparts, with women in generation X showing slightly higher values.

Tables 6.8 and 6.9 indicate that both flexibility in the workplace and the preference for a socially responsible company are slightly more valued by women than by men within these generations.

		Relative frequencies (%)							
		Generation X				Generation Y			
		Female	%	Male	%	Female	%	Male	%
Autonomy	Not important	0	0.0%	2	6.1%	1	0.4%	0	0.0%
	Not very important	0	0.0%	1	3.0%	6	2.7%	1	0.8%
	Important	2	6.5%	1	3.0%	8	3.5%	5	3.8%
	Quite important	2	6.5%	3	9.1%	35	15.5%	25	19.1%
	Considerably important	11	35.5%	8	24.2%	81	35.8%	46	35.1%
	Very important	16	51.6%	18	54.5%	95	42.0%	54	41.2%

Table 6.7. *Generations by autonomy (N = 421)*

		Relative frequencies (%)							
		Generation X				Generation Y			
		Female	%	Male	%	Female	%	Male	%
Flexibility	Not important	0	0.0%	2	6.1%	3	1.3%	1	0.8%
	Not very important	1	3.2%	1	3.0%	8	3.6%	3	2.3%
	Important	1	3.2%	1	3.0%	11	4.9%	5	3.8%
	Quite important	2	6.5%	3	9.1%	21	9.4%	17	13.1%
	Considerably important	12	38.7%	9	27.3%	64	28.6%	44	33.8%
	Very important	15	48.4%	17	51.5%	117	52.2%	60	46.2%

Table 6.8. *Generations by flexibility analysis (N = 421)*

Table 6.9 shows that access to training, viewed as a human resource management practice from a career development perspective, is rated more positively by women than by men.

		Relative frequencies (%)							
		Generation X				Generation Y			
		Female	%	Male	%	Female	%	Male	%
Socially responsible company	Not important	0	0.0%	1	3.0%	8	3.6%	8	6.3%
	Not very important	0	0.0%	5	15.2%	9	4.0%	12	9.4%
	Important	1	3.2%	3	9.1%	23	10.2%	20	15.7%
	Quite important	4	12.9%	6	18.2%	61	27.1%	26	20.5%
	Considerably important	16	51.6%	8	24.2%	55	24.4%	41	32.3%
	Very important	10	32.3%	10	30.3%	69	30.7%	20	15.7%

Table 6.9. *Generations by socially responsible company analysis (N = 421)*

		Relative frequencies (%)							
		Generation X				Generation Y			
		Female	%	Male	%	Female	%	Male	%
Access to training	Not important	0	0.0%	2	6.1%	4	1.8%	5	3.8%
	Not very important	1	3.2%	1	3.0%	13	5.8%	5	3.8%
	Important	2	6.5%	3	9.1%	19	8.5%	12	9.2%
	Quite important	3	9.7%	8	24.2%	41	18.3%	33	25.4%
	Considerably important	11	35.5%	5	15.2%	68	30.4%	44	33.8%
	Very important	14	45.2%	14	42.4%	79	35.3%	31	23.8%

Table 6.10. *Generations by access training (N = 421)*

6.4.3. *Association/correlation analyses*

Analyses of association and correlation allow for the intensity of any existing associations, but this intensity is only credible if there is statistical significance. When we say that something is statistically significant, it does not necessarily imply that it holds clinical or educational significance. While an effect or difference may be statistically significant, it could be clinically irrelevant, and conversely, something that is not statistically significant might still have important practical implications [LOU 11].

6.4.4. *Association between gender and ordinal variables relating to perceptions*

The correlations analyzed include the age group variable and several perception-related variables. The association analyses are intended to examine how these perceptions vary according to gender.

Phi and Cramér's V are employed to assess the degree of association between two categorical variables, with Phi used when the variables have no more than two categories and Cramér's V applied in other cases [COM 20]. "There are indications to consider p-values ≤ 0.010 as significant." [COM 20].

	Value	**Approx. Sig.**
Cramér's V	0.212	0.002
N valid cases	416	

Table 6.11. *Symmetrical measures – socially responsible company*

Regarding the possible association between gender and the attributions across the 19 ordinal items, statistical significance was found only for the "Socially Responsible Company" item ($p < 0.05$). However, the degree of association was moderate, considering the six degrees of freedom in this case.

Cramér's V was calculated as 0.212, with an approximate significance level of 0.002, based on 416 valid cases.

6.4.5. *Spearman's correlations between ordinal variables*

Spearman's correlation assesses the strength and direction of the relationship between two variables, yielding values between -1 and 1, with these values indicating the correlation's direction and intensity. This method is particularly appropriate for ordinal variables [HEL 10]. Spearman's correlation analysis was conducted between the age group variable and all variables related to respondents' perceptions; however, no statistically significant correlations were found. The analysis was also performed for the generation group (Table 6.11) variable in relation to the respondents' perceptions, but again, no statistically significant correlations were observed.

Variable	JC	BI	SAL	WTT	WE	AU	FLE	REC	CP	AT	TEA	RC	RS	BBPPL	QM	WQ	SRC	FEE	CEI
Job content (JC)																			
Benefits and incentives (BI)	.537**																		
Salary (SAL)	.459**	.772**																	
Work time table (WTT)	.420**	.588**	.595**																
Work environment (WE)	.440**	.516**	.495**	.580**															
Autonomy (AUT)	.505**	.474**	.443**	.506**	.608**														
Flexibility (FLE)	.431**	.494**	.468**	.561**	.551**	.683**													
Recognition (REC)	.502**	.644**	.671**	.608**	.622**	.614**	.616**												
Career progression (CP)	.489**	.664**	.686**	.520**	.571**	.524**	.555**	.770**											
Access to training (AT)	.415**	.458**	.428**	.445**	.557**	.477**	.472**	.588**	.646**										
Teamwork (TEA)	.359**	.342**	.315**	.372**	.616**	.518**	.448**	.488**	.507**	.645**									
Relationship with colleagues (RC)	.303**	.322**	.269**	.392**	.648**	.463**	.415**	.455**	.418**	.526**	.714**								
Relationship with superior (RS)	.327**	.418**	.361**	.414**	.590**	.555**	.542**	.560**	.508**	.547**	.663**	.770**							
Balance pers/profes. life (BBPPL)	.441**	.610**	.632**	.739**	.583**	.525**	.592**	.671**	.633**	.510**	.424**	.408**	.444**						
Quality of managers (QMs)	.444**	.545**	.534**	.611**	.630**	.568**	.580**	.712**	.653**	.632**	.520**	.506**	.616**	.625**					
Life at work quality (LWQ)	.452**	.526**	.541**	.636**	.701**	.579**	.589**	.713**	.648**	.608**	.560**	.554**	.611**	.697**	.772**				
Socially responsible company (SRC)	.335**	.338**	.250**	.390**	.552**	.479**	.409**	.484**	.454**	.573**	.601**	.591**	.577**	.403**	.580**	.575**			
Feedback (FEE)	.421**	.508**	.446**	.513**	.601**	.531**	.527**	.639**	.639**	.647**	.639**	.546**	.602**	.563**	.709**	.711**	.718**		
Company encourages innovation (CEI)	.431**	.457**	.400**	.397**	.537**	.525**	.477**	.548**	.567**	.609**	.569**	.524**	.558**	.457**	.626**	.589**	.707**	.759**	
Generation (GEN)	0.033	-0.023	-0.004	-0.038	-0.051	-0.072	-0.004	-0.039	-0.032	-0.077	-0.033	-.103*	-0.091	- 0.065	-0.092	-0.056	-0.095	-0.062	-0.089

N = 421, correlation is significant at **0.01 and *0.05 level of significance.

Table 6.12. *Spearman's correlation coefficient (rho) matrix*

6.4.6. *Limitations and implications*

In an expanded future version of the study, the intent will be to increase sample size and ensure more balanced demographic representation, particularly regarding gender and generational cohort. Given the exploratory nature of this study, further research should involve developing a more comprehensive set of hypotheses to enrich the study and provide a deeper understanding of the underlying dynamics.

In examining the findings of a study that identified no statistically significant differences between generational groups and gender in their perception of retention factors, it is crucial to consider the broader implications for organizational practices and workplace dynamics.

6.5. Conclusion

This study of graduates from a University of Applied Sciences found no significant differences in the aspects most valued at work based on gender, generation or age group. These results are consistent with existing evidence, which indicates that empirical differences between generations are extremely limited compared to the differences observed within groups [SHO 17].

Through our empirical findings of examining the expectation and preferences of retaining generations X and Y employee retention factors with an emphasis on the gender, it was concluded that there are no significant differences between generations for the factors studied and for our sample. Individuals from the two generations analyzed a socially responsible company as a factor that is highly valued at work. The results also led to the conclusion that a socially responsible company, a supportive working work–life balance, autonomy, flexibility and life at work quality, are more important for female employees, which corresponds to the reference literature.

From a theoretical point of view, this study contributes to the enrichment of the scientific literature regarding the relationships between generations X and Y and the organizational factors to retain employees from this generation, offering a perspective about gender differences, in the actual society.

Based on the empirical findings, Human Resources managers can develop effective intervention strategies to enhance the retention of generations X and Y employees by fostering a supportive work environment. These insights are valuable for creating tailored strategies and human resource practices that are centered around employee needs. Organizations that are committed to gender equality should develop human resource management practices that contribute to the process of becoming aware of the nature of gender in organizations, seeking to adapt their policies, values, norms and formal and informal structures to this reality, rather than adopting gender-neutral practices that reflect the attributes, needs and standards of the socially male worker.

We must also think in terms of functional contents that they are also the ones who will work side-by-side with robots (5.0) and who will feel the transformations in the organization of work resulting from the growing use of artificial intelligence (6.0).

6.6. References

[ABR 11] ABRANTES P., "Para uma teoria da socialização", *Sociologia Revista da Faculdade de Letras da Universidade do Porto*, vol. XXI, pp. 121–139, 2011.

[ACK 90] ACKER J., "Hierarchies, jobs, bodies: A theory of gendered organizations", *Gender & Society*, vol. 4, no. 2, pp. 139–158, 1990.

[ACK 00] ACKER J., *Revisiting Class: Thinking from Gender, Race, and Organizations*, Oxford University Press, Oxford, 2000.

[ALL 13] ALLEGRANTI B., "The politics of becoming bodies: Sex, gender and intersubjectivity in motion", *The Arts in Psychotherapy*, vol. 40, no. 4, pp. 394–403, 2013.

[ALL 14] ALLEN T., FINKELSTEIN L., "Work-family conflict among members of full-time dual-earner couples: An examination of family life stage, gender, and age", *Journal of Occupational Health Psychology*, vol. 19, no. 3, pp. 376–384, 2014.

[ANN 17] ANNIS B., NESBITT R., *Results at the Top: Using Gender Intelligence to Create Breakthrough Growth*, Wiley, Hoboken, 2017.

[ARR 02] ARRUDA A., "Teoria das representações sociais e teorias de gênero", *Cadernos de pesquisa*, vol. 117, pp. 127–147, 2002.

[BAN 03] BANERJEE P., SRIVASTAVA M., KRISHNAMOORTHY B., "A gender-based comparative analysis of generation X and Y on emotional contagion", *The Qualitative Perspective. Business Perspectives and Research*, vol. 11, no. 3, pp. 336–354, 2003.

[BAR 18] BARRIOPEDRO E., RIPOLL R., TELLO A., "Análisis de la situación actual de la felicidad corporativa de los millennials en España, desde el barómetro del centro de investigaciones sociológicas", *Revista Académica Eco*, vol. 18, pp. 1–22, 2018.

[BEN 11] BENSON J., BROWN M., "Generations at work: Are there differences and do they matter?", *The International Journal of Human Resource Management*, vol. 22, no. 9, pp. 1843–1865, 2011.

[BEN 18] BENSON J., BROWN M., GLENNIE M. et al., "The generational "exchange" rate: How generations convert career development satisfaction into organisational commitment or neglect of work", *Human Resource Management Journal*, vol. 28, no. 4, pp. 524–539, 2018.

[BER 23] BERNIER J., *L'intelligence artificielle et les mondes du travail. Perspectives sociojuridiques et enjeux éthiques*, Presses de l'Université Laval, Quebec, 2023.

[BOR 20] BORROWMAN M., KLASEN S., "Drivers of gendered sectoral and occupational segregation in developing countries", *Feminist Economics*, vol. 26, no. 2, pp. 62–94, 2020.

[BOW 13] BOWLEG L., "'Once you've blended the cake, you can't take the parts back to the main ingredients': Black gay and bisexual men's descriptions and experiences of intersectionality", *Sex Roles*, vol. 68, nos 11–12, pp. 754–767, 2013.

[CAN 18] CANNON L., KENDIG H.,"'Millennials': Perceived generational opportunities and intergenerational conflict in Australia", *Australasian Journal on Ageing*, vol. 37, no. 4, pp. 127–132, 2018.

[CAP 15] CAPPI N., ARAUJO B. "Satisfação no trabalho, comprometimento organizacional e intenção de sair: um estudo entre as gerações XEY", *REAd. Revista Eletrônica de Administração (Porto Alegre)*, vol. 21, pp. 576–600, 2015.

[CAR 02] CARVALHO T., CABRAL-CARDOSO C., "Women in human resource management: The Portuguese case", in *2nd EURAM Conference*, Stockholm, 2002.

[CER 11] CERETTA S., FROEMMING L., "Geração Z: compreendendo os hábitos de consumo da geração emergente", *RAUnP*, vol. 3, no. 2, pp. 15–24, 2011.

[CIN 02] CINAMON R., RICH Y., "Gender differences in the importance of work and family roles: Implications for work–family conflict", *Sex Roles*, vol. 47, pp. 531–554, 2002.

[COE 04] COELHO N., FIGUEIREDO L., "Figuras da intersubjetividade na constituição subjetiva: dimensões da alteridade", *Interações*, vol. 9, no. 17, pp. 9–28, 2004.

[COM 20] COMERLATO P., SANTOS S., ZIEGLER N. et al., "Métodos estatísticos para desfechos qualitativos", in CAPP E., NIENOV O. (eds), *Bioestatística quantitativa aplicada*, UFRGS, Porto Alegre, pp. 101–124, 2020.

[CRO 09] CROSON R., URI G., "Gender differences in preferences", *Journal of Economic Literature*, vol. 47, no. 2, pp. 448–74, 2009.

[CRU 12] CRUZ F., DE AQUINO CABRAL A., PESSOA M. et al., "Relações entre responsabilidade social interna e comprometimento organizacional: um estudo em empresas prestadoras de serviços", *Revista Brasileira de Administração Científica*, vol. 3, no. 3, pp. 163–182, 2012.

[DEH 10] DE HAUW S., DE VOS A., "Millennials' career perspective and psychological contract expectations: Does the recession lead to lowered expectations?", *Journal of Business and Psychology*, vol. 25, no. 2, pp. 293–302, 2010.

[DEN 76] DENHARDT R., PERKINS J., "The coming of death of administrative man", *Public Administration Review*, vol. 36, no. 4, pp. 379–384, 1976.

[DIC 88] DICKENS L., "What HRM means for gender equality", *Human Resource Management Journal*, vol. 8, pp. 23–40, 1988.

[ERI 14] ERIKSSON T., KRISTENSEN N., "Wages or fringes? Some evidence on trade-offs and sorting", *Journal of Labor Economics*, vol. 32, no. 4, pp. 899–923, 2014.

[EST 13] ESTEVES S., "Como elas serão no futuro?", *GV-EXECUTIVO*, vol. 12, no. 1, pp. 58–61, 2013.

[FAN 15] FANTINI C., SOUZA N., "Analysis of motivational factors of baby boomers, X, Y and Z generations and their expectations about professional career", *Piracicaba*, nos 3/4, pp. 126–145, 2015.

[FED 05] FEDATO M., Responsabilidade social corporativa: benefício social ou vantagem competitiva? Um estudo das estratégias de atuação social empresarial e sua avaliação de resultados, PhD Thesis, Universidade de São Paulo, 2005.

[FER 09] FERNANDEZ S., Comparing generation X to generation Y on work-related beliefs, Master's Thesis, San José State University, CA, doi: 10.31979/etd.fen9-fcyh, 2009.

[FEY 05] FEYERHERM A., VICK Y., "Generation X women in high technology career", *Development International*, vol. 10, pp. 216–227, 2005.

[FIN 20] FINE C., SOJO V., LAWFORD-SMITH H., "Why does workplace gender diversity matter? Justice, organizational benefits, and policy", *Social Issues and Policy Review*, vol. 14, no. 81, pp. 36–72, 2020.

[FUL 18] FULMER I., SHAW J., "Person-based differences in pay reactions: A compensation-activation theory and integrative conceptual review", *Journal of Applied Psychology*, vol. 103, no. 9, pp. 939–958, 2018.

[GAL 11] GALLHOFER S., PAISEY C., ROBERTS C. et al., "Preferences, constraints, and work-lifestyle choices: The case of female Scottish chartered accountants", *Accounting, Auditing and Accountability Journal*, vol. 24, no. 4, pp. 440–470, 2011.

[GOL 23] GOLDBACH D., TAYAUOVA G., ZHUMABEKOVNA T., "Motivation at work for generation Z", *FAIMA Business & Management Journal*, vol. 11, no. 3, pp. 68–74, 2023.

[GUZ 07] GUZMÁN M., PÉREZ A., "La teoría de género y su principio de demarcación científica", *Cinta de Moebio*, no. 30, pp. 283–295, 2007.

[HEL 10] HELD U., "Pièges des corrélations : les coefficients de corrélation de Pearson et de Spearman", *Forum Med Suisse*, vol. 10, no. 38, pp. 652–653, 2010.

[HIL 09] HILL M., HILL A., *Investigação por Questionário*, 2nd edition, Edições Sílabo, Lisbon, 2009.

[HUT 04] HUTCHINSON J., MANN H., "Feminist praxis: Administering for a multicultural, multigendered public", *Administrative Theory & Praxis*, vol. 26, no. 1, pp. 79–95, 2004.

[JON 23] JONG J., "The role of leader and subordinate gender in the leadership–empowerment relationship", *Public Personnel Management*, vol. 52, no. 3, pp. 317–343, 2023.

[KIN 23] KINGER N., KUMAR S., "Generational differences in work values in the workplace", *Folia Oeconomica Stetinensia*, vol. 23, no. 2, pp. 204–221, 2023.

[KOR 23] KORTE A., TSCHUSCHKE Q., "Media's stranglehold on storm and stress – The sorrows of Generation Z about sex and gender", *Zeitschrift fur Kinder- und Jugendpsychiatrie und Psychotherapie*, vol. 51, no. 5, pp. 351–365, 2023.

[LAN 02] LANCASTER L., STILLMAN D., *When Generations Collide: Who They Are, Why They Clash, How to Solve the Generational Puzzle at Work*, Collins Business, New York, 2002.

[LAS 23] LASSLEBEN H., HOFMANN L., "Attracting Gen Z talents: Do expectations towards employers vary by gender?", *Gender in Management: An International Journal*, vol. 38, no. 4, pp. 545–560, 2023.

[LEE 22] LEE C.C., LIM H.S., DONGHWI S.J. et al., "Examining employee retention and motivation: The moderating effect of employee generation", *Evidence-based HRM*, vol. 10, no. 4, pp. 385–402, 2022.

[LIM 22] LIMPINHO M., A relação entre o Turnover e o grau de satisfação no trabalho da Geração Y (Millennials) – O caso dos trabalhadores portugueses, Master's Dissertation, Universidade de Évora, 2022.

[LIP 10] LIPKIN N., PERRYMORE A., *A Geração Y no trabalho: como lidar com a força de trabalho que influenciará definitivamente a cultura de sua empresa*, Elsevier, Rio de Janeiro, 2010.

[LOU 08] LOUREIRO P., CABRAL-CARDOSO C., "O género e os estereótipos na gestão", *Revista de Estudos Politécnicos*, vol. 6, no. 10, pp. 221–238, 2008.

[LOU 11] LOUREIRO M., GAMEIRO M., "Interpretação crítica dos resultados estatísticos: para lá da significância estatística", *Revista de Enfermagem Referência*, vol. III, no. 3, pp. 151–162, 2011.

[LOW 09] LOWEN A., SICILIAN P., ""Family-Friendly" fringe benefits and the gender wage gap", *Journal of Labor Research*, vol. 30, pp. 101–119, 2009.

[MAG 20] MAGNI F., MANZONI B., "Generational differences in workers' expectations: Millennials want more of the same things", *European Management Review*, vol. 17, no. 4, pp. 901–914, 2020.

[MAS 16] MASTRACCI S., ARREOLA V., "Gendered organizations: How human resource management practices produce and reproduce administrative man", *Administrative Theory & Praxis*, vol. 38, pp. 137–149, 2016.

[MAX 14] MAXWELL G., BROADBRIDGE A., (2014). "Generation Y graduates and career transition: Perspectives by gender", *European Management Journal*, vol. 32, pp. 547–553, 2014.

[MOS 61] MOSCOVICI S., "La représentation sociale de la psychanalyse", *Bulletin de psychologie*, vol. 14, no. 194, pp. 807–810, 1961.

[MUS 19] MUSKAT B., REITSAMER B.F., "Quality of work life and Generation Y: How gender and organizational type moderate job satisfaction", *Personnel Review*, vol. 49, no. 1, pp. 265–283, 2019.

[NIC 94] NICHOLSON L., "Interpreting gender. Signs", *Journal of Women in Culture and Society*, vol. 20, no. 1, pp. 79–105, 1994.

[ONE 08] O'NEIL D., HOPKINS M., BILIMORIA, D., "Women's careers at the start of the 21st century", *Journal of Business Ethics*, vol. 80, no. 4, pp. 727–743, 2008.

[PAN 15] PANT J., VIJAYA V., "Management of Gen Y employees through psychological contract – An exploratory study", *IT/ITES Companies*, no. 6, pp. 24–39, 2015.

[PAR 11] PARRY E., URWIN P., "Diferenças geracionais nos valores do trabalho: uma revisão da teoria e da evidência", *Revista Internacional de Revisões Gerenciais*, vol. 13, no. 1, 79–96, 2011.

[PAZ 87] PAZY A., "Sex differences in responsiveness to organizational career management", *Human Resource Management*, vol. 26, pp. 243–256, 1987.

[PER 08] PERISTA H., GUERREIRO M.D., JESUS C. et al., "A igualdade de género no quadro da responsabilidade social: O projecto equal diálogo social e igualdade nas empresas", *Ex æquo*, no. 18, pp. 103–120, 2008.

[PES 05] PESTANA M., GAGEIRO J., *Análise de dados para as ciências sociais, a complementaridade do SPSS*, 4th edition, Sílabo, Lisbon, 2005.

[PES 09] PESSOA R.W.A., DO NASCIMENTO L.F., NEVES J.A.D. et al., "Estratégia e vantagem competitiva da responsabilidade social empresarial", *Gestão & Regionalidade*, vol. 25, no. 74–30, pp. 79–93, 2009.

[POR 23] PORDATA, "Disparidade no emprego entre homens e mulheres", *Fundação Francisco Manuel dos Santos*, available at: https://www.pordata.pt/pt/estatisticas/emprego/populacao-empregada/disparidade-no-emprego-entre-homens-e-mulheres, 2023.

[SAN 20] SANTOS J., PEDRO L., "Reinventing human resource management to increase organizational efficacy", in MACHADO C., DAVIM J. (eds), *Entrepreneurship and Organizational Innovation*, Springer, Cham, pp. 23–36, 2020.

[SAN 23] SANTOS J., PEDRO L., QUINTAS C., "Gestão de talentos — Desafios da Geração Y para o mercado de trabalho", in MACHADO C., DAVIM J. (eds), *Indústria 5.0. Pessoas, Tecnologia e Sustentabilidade*, Actual Editora, Lisbon, pp. 281–312, 2023.

[SAN 24] SANTOS J., PEDRO L., QUINTAS C., "Managing and engaging a multigenerational workforce in Portugal", in MACHADO C., DAVIM J. (eds), *Smart Engineering Management*, Springer, Cham, pp. 123–146, 2024.

[SCA 21] SCARBOROUGH W., FESSENDEN D., SIN R., "Convergence or divergence? The generational gap in gender attitudes, 1977–2018", in DEMOS V., SEGAL M.T. (eds), *Gender and Generations: Continuity and Change*, Emerald Publishing Limited, Bingley, pp. 73–94, 2021.

[SCO 16] SCORSOLINI-COMIN F., SANTOS M., "Construir, organizar, transformar: considerações teóricas sobre a transmissão psíquica entre gerações", *Psicologia Clínica*, vol. 28, no. 1, pp. 141–159, 2016.

[SHA 16] SHARON M., VERONICA I.A., "Gendered organizations: How human resource management practices produce and reproduce administrative man", *Administrative Theory & Praxis*, vol. 38, no. 2, pp. 137–149, 2016.

[SHO 17] SHOCKLEY K., SHEN W., DENUNZIO M. et al., "Disentangling the relationship between gender and work–family conflict: An integration of theoretical perspectives using meta-analytic methods", *Journal of Applied Psychology*, vol. 102, no. 12, p. 1601, 2017.

[STI 10] STIVERS C., "A tactless question", *Administrative Theory & Praxis*, vol. 32, no. 2, pp. 220–224, 2010.

[SUÁ 22] SUÁREZ J., PAREDES S., "El factor humano y su rol en la transición a Industria 5.0: una revisión sistemática y perspectivas futuras", *Entreciencias: Diálogos en la Sociedad del Conocimiento*, vol. 24, no. 10, pp. 1–18, 2022.

[SUL 07] SULLIVAN S., MAINIERO L., "The changing nature of gender roles, alpha/beta careers, and work–life issues: Theory-driven implications for human resource management", *Career Development International*, vol. 12, pp. 38–263, 2007.

[SUR 20] SURBHI J., *Managing a Multi-Generational Workforce*, Society Publishing, Burlington, available at: https://search.ebscohost.com/login.aspx?direct=true&AuthType=ip,shib&db=edsebk&AN=2725361&lang=pt-pt&site=eds-live&scope=site, 2020.

[TER 07] TERJESEN S., VINNICOMBE S., FREEMAN C., "Attracting Generation Y graduates: Organisational attributes, likelihood to apply and sex differences", *Career Development International*, vol. 12, pp. 504–522, 2007.

[TRI 17] TRIDAPALLI J., DA SILVA E., ZIEDE M. et al., "Os conflitos da geração X e Y no mercado de trabalho", *Core*, vol. 7, no. 20, pp. 132–156, 2017.

[YIL 23] YILDIZ Y., BILGIN R., "How do young generations perceive gender differences? A case study", *Amazonia Investiga*, vol. 12, no. 64, pp. 28–44, 2023.

7

Add More Marketing to Marketing Doctoral Programs – Answering Hunt and Yadav's Calls

In a situation based on ambiguity and disruption, this research seeks to expose the new frontier for marketing education within doctoral programs as a meaning and design search space for the discipline's raison d'être, wherein the locus of innovation is based on answering the contemporary open calls by [HUN 20b] and [YAD 20], within the marketing discipline to reinvigorate marketing's cognitive identity within the emerging cyber-physical realities from the lenses of the STEAM education.

Based on two different AI-driven softwares, Knime and Python, the author has analyzed close to 30,000 articles to account for a histography of marketing science. The author performed an extensive literature review wherein 28,253 articles from 17 top journals in marketing science were analyzed. The task encompassed the creation of an Excel list including all relevant information (authors, title, journal, keywords, abstract, citations, citations per year). Additional tasks were to code the keywords to manually generate 50/30 preliminary topics, and then the tokenization of information about title, authors and keywords, punctuation of erasure, filtering out numbers, deleting words with less than three characters and the creation of a stop list to exclude specific common words. Single-parallel distributed implementation of LDA in KNIME was used for detecting the most relevant topics. The node uses the MALLET topic modeling library. Generation of 50/30 topics was performed, based on sensitivity analysis of five terms each and default beta of 0 and asymmetrical alphas, which were individually determined for each topic. Furthermore, grouping of topic terms and weighting with qualitative evaluation and pivoting of topics over years and also MCA was applied to identify the relationship between topic terms and keywords. MCA performance was also applied on the relevant journals for this research project and generation of the time series plots with trendlines show the assigned articles per topic over the years. The additional task was CA analysis to present the interrelation of assigned topics to grouped years and journals, and designated box plots.

The author concludes that marketing is facing many challenges and finds itself in a fractured state of evolution. Marketing must take a holistic approach to overcome the losses of the fields

Chapter written by Qeis KAMRAN.

of SCM and CB within the sphere of the discipline. Furthermore, marketing must proclaim an ambidextrous and transformative raison d'être; hence, to re-institutionalize the field according to Hunt's call, the author argues that it is required avant la lettre to establish the question of whether a reinstitutionalization is the adequate path. The focus of this research is to prepare the marketing discipline with a designerly lens pertaining the major shifts that are occurring within the discipline and to the discipline marketing. The author's findings align with Yadav's call, which states that toward reclaiming a discipline's cognitive identity, the doctoral programs of that field must reproduce the essence of its collective history and the origin, so that the field can reproduce itself and maintain the ability of indigenous theory-building capacity.

The research is possibly the first to study all of the marketing science literature based on the complete content of the 17 top journals in marketing science. Furthermore, it answers the most essential calls within the top journals of the discipline. The study paves the way for a solid rapprochement for the fields of marketing science and the STEAM education to merge in a manner that is beneficial for the viability of the marketing discipline and the Ph.D. students' capacities of scientifically issuing a holistic critique toward the viability of the field.

7.1. Introduction

This research seeks to expose the new frontier of innovation in post-graduate qualifications, particularly to establish a foundation for doctoral schools of marketing, wherein the locus of innovation lies within the development of indigenous theory and conceptual foundations within the field to face the emerging cyber-physical realities of the contemporary epoch.

7.1.1. *The research problem*

Marketing science as a discipline is in a state of crisis [REI 09]; [YAD 14, 20]; [ELA 18]; [HUN 20b]; [KAM 21]; [KEY 21]; [HEL 22]; [VAR 22]. [LEV 18] doubts the existence of the crisis and calls it a point of danger (from the Greek crisis), which requires a decision or a cri de coeur for a change coming, thus being eventually beneficial for the discipline. [PAR 21] postulated a design for an integrative theory. However, the evidence identifies a fractured and identity-lacking scientific domain with a borrowing attitude from other sciences. While interdisciplinary studies are necessary to advance a field's relevance toward practice and address the evolving societal challenges, the relationship of marketing toward other diverse scientific domains needs to be based on more reciprocal co-creative grounds, hence the theory exporting foundations within marketing science toward other fields has been profoundly absent [BAR 74]; [REI 09]; [HUN 20a, 20b] and the theory-loaning department has had its bubble burst.

The contemporary state of marketing is based on the field's raison d'être to have capitulated by becoming a "borrowing only domain" within managerial social sciences. Marketing's intellectual health needs indigenous theory development capability [HUN 20a].

The main controversy identified and addressed here is to deliver an answer to Yadav and Hunt's calls [HUN 20a]; [YAD 10, 20]:

1) Do the challenges within the domain of marketing science portend a de-institutionalization of the discipline in marketing's Era IV (1980–2020) and its potential reinstitutionalization in Era V (2020-?)?[1]

2) The marketing discipline needs to integrate more marketing into marketing doctoral programs; thus, for a discipline to be able to reproduce itself, it must have the capacity for reproduction avant la lettre by what is taught in the doctoral schools [YAD 20].

Based on the in-depth research conducted, the author goes one step further than [HUN 20a]; hence, the evidence within the literature identified proclaim that a reinstitutionalization of the field may not cure the nature of the crises. Therefore, for the marketing domain to address the challenges of the contemporary era and that of the future, the field needs a trans-disciplinary foundation. This means designing a new home within the realm of a new scientific domain and standing on more robust scientific, practical, educational and teaching pillars for the next generations of doctoral scholars in marketing science [REI 09]; [YAD 14, 20] [ELA 18].

7.1.2. *Research objectives*

The research fulfills the following objectives:

1) to establish a solid historical account of the marketing discipline since the inception of the field;

2) to identify the roots of the crises within the marketing discipline within the diverse eras of the field's evolution;

3) to establish a clear foundation wherein the next generations of marketing scholars need to be trained;

1 For a more profound understanding of the eras in marketing science, reference is made to the following articles: "Marketing's identity crisis: Insights from the history of marketing thought" [ELA 18]; "For re-institutionalizing the marketing discipline in Era V" [HUN 20a]; "Re-institutionalizing marketing" [KEY 21], and "Fostering scholarly discussion in marketing" [VAR 22].

4) to develop a foundational curriculum to strengthen the doctoral schools in marketing science, which embraces the cognitive identity of the field and thus represents a microcosm of the discipline in an evolutionary, sustaining manner.

The doctoral schools represent the collective aspirations and expectations for the future of a discipline, but moreover, they must cope with the imperfections and flaws of a discipline. A borrowing-only scientific domain lacks disciplinary, collective and evolutionary capacity and tends to disintegrate…"therefore, it is imperative that doctoral programs place theory at the heart of the curriculum that shapes the thinking, priorities and research efforts of our new scholars" [YAD 20].

7.1.3. *Actuality/novelty of the research*

This research is novel because it gathers the essence of traditional marketing, its history and possible futures.

The fabric and nature of traditional marketing are changing, and it is pivoting toward the holistic discipline of digital humanities. New lines of service(s) [VAR 08] are emerging, which need(s) a new lens to decipher their design, production and management. To address the shifting needs of the marketing discipline, a new domain of "trans-institutionalization"[2] has been identified.

7.1.4. *Research questions*

The following research questions are postulated:

1) What are the essential topics in marketing literature based on all published articles within 17 top journals since their inception?

2) What are the most impactful articles in the field of marketing?

3) Which authors have had the most impact on the history of marketing discipline?

4) How have the topics shifted in importance and scholarly attention within the span of 80 years of research history?

2 Trans-institutionalization based on the Oxford Reference means: *"A process whereby individuals, supposedly deinstitutionalized as a result of community care policies, in practice end up in different institutions, rather than their own homes."* Applied to the marketing discipline and its contemporary disillusioned state and within this research, it means finding a new domestic home, where the discipline could flourish by nurturing ingenious theory building and theory enhancing capabilities. Here, the development of viable curriculums for the doctoral schools plays a major role to sustain the field's raison d'être.

5) Based on the historical context developed, how fit is the field to address the challenges facing marketing science and practice today?

6) What are the most foundational pillars of science and technology, whereupon marketing as a discipline could construct a viable future?

7) What essential foundations of science, theory, methodology, empirical knowledge and practice need to be taught at the doctoral schools of marketing?

7.2. Literature review

This section will establish a holistic and in-depth history of marketing, covering the last 80 years of the field's evolution within the most relevant journals.

Marketing as a foundational field of management science has been challenged from many sides, and this has affected its raison d'être. These challenges embrace the rise of technicity, a global movement for sustainability, a lack of indigenous theoretical foundation and cognitive identity, and a lack of development of a solid framework for organizational structures. As no other field in management has had such a long and prosperous history, 80 years of research is sufficient time for a field to reflect upon itself, account for a solid history and thus propose a future direction for the field, while understanding the many challenges affecting its foundation and simultaneously making the field fit for the challenges ahead.

To awaken the interest of the scholarly researcher/practitioner, the modus of introduction to the themes of marketing deriving from this research is based on [GRA 11].

Who cares?

Marketing researchers and scholars are interested in having a holistic and in-depth analysis that encompasses the history of the field seen from the published research within the top journals available; and, at the same time, based on the three essential jobs which define marketing as a field of management. Based on the latest definition of marketing [NEL 17], marketing has three foundational jobs: (1) to create; (2) to communicate; and (3) to deliver value to a target at a profit. Therefore, the choice of journals was based on the most essential publications that could deliver answers to all three jobs that marketing must accomplish concurrently.

Which research questions are postulated?

The essence of this research is based on giving a holistic and in-depth history of marketing as a field of managerial science. Our research was guided by the following main questions:

1) What are the essential topics in marketing literature based on all published articles within 17 top journals since their inception?

2) What are the most impactful articles in the field of marketing?

3) How have the topics shifted in importance and scholarly attention within the span of 80 years?

4) Based on the history provided, how fit is the field to address the challenges facing marketing science and practice?

5) Which authors have had the most impact on the history of marketing?

Why is it exciting and important in theory and practice?

Based on the challenges facing the cognitive identity of the discipline, the lack of an indigenous theoretical foundation and the grand strategy of putting back marketing within marketing postgraduate programs, this research gives a single, actualized and novel analysis which enhances the scholarly debate and delivers a source of historical account; it also delivers a foundation in marketing, whereupon the fragmented state of the field could be brought (back) together in concert. Furthermore, the research paves the way for a possible future of marketing science in cyber-physical realities.

What do we know so far?

[WEB 13] describes the contemporary situation wherein marketing finds itself in the following manner: "...the marketing discipline faces an urgent need for a rethinking of its fundamental purpose, premises, and implicit models that have defined marketing for at least the past 50 years". Furthermore, additional voices within marketing describe the state of maturity of the discipline not going forward within this context: "...the increase in [marketing] knowledge [presently] occurs at a decreasing rate, and marketing knowledge has reached a stage of maturity.... The more mature a research field is, the less groundbreaking its new findings are, which therefore leads to a lower increase in knowledge..." [EIS 15]. The author sees the lack of indigenous development within marketing based on the field's overreliance on consumer behavior (CB) for such a long time and that the indigenous research and theory-building capacity of the field was largely overwritten by the successes of consumer research (CR); this was due to the quantified method of empiricism they provided and the models of behavioral research they produced. Until the CB-laden marketing research was conducted, marketing had a cognitive identity. Then, when CB research left marketing by becoming a field under its own domain, the separation left marketing to its own demise, and thus, so too did the whole field of supply chain management (SCM) by creating their own journals and putting marketing under the umbrella of SCM. The next frontier of disruption has already hit marketing, as illustrated in Figure 7.1. Thus, [MOO 19] established the following scenario:

> The discipline of marketing has made great strides in developing and testing models related to markets, brands, channels, and customer behavior. However, the discipline's continued relevance and academic value proposition founded on theoretical and conceptual research is more important than ever as marketing managers (and scholars) face unprecedented marketplace and technological disruption.

The essence of this chapter is to contribute toward solving the challenges the discipline faces, especially in making it reflect on marketing itself for seeking a cognitive raison d'être [KAM 21].

What do we not know, and so what?

Marketing is the most foundational field in management science, and contemporary technological developments need to be derived from adequate marketing philosophies to decipher what the new cyber-physical environment is and the nature of exchange, co-creation and co-evolution. How does the sustainability movement challenge marketing?[3] And what is the established bridge between marketing and organizational structures? Furthermore, how can marketing overcome its shortcomings by establishing its own cognitive identity?

Figure 7.1. *The emerging trends affecting contemporary food marketing. Source: [KAM 21]*

3 See Denmark's recent decision of adding a carbon tax: according to Reuters 24.06.2022: *"Danish lawmakers on Friday agreed a new corporate carbon tax, the highest in Europe, which will target companies both in and outside the EU's carbon quota system, the government said. A high carbon tax is seen as crucial to help reach Denmark's ambitious 2030 target to cut greenhouse gas emissions by 70% from 1990 levels."*
(It is) the biggest single contribution so far to cut emissions by 2030 "said tax minister Jeppe Bruus in a statement." See Reuters 24.06.2022.

What key theoretical perspectives and empirical findings have already informed the topic or question?

To reflect on the possible future developments of marketing research, a brief historical account is necessary. The author produces a broad but non-exhaustive historical account avant la lettre to stimulate the debate on the history of the field. While many solid historical accounts have illustrated and guided the intellectual development of the field [BAR 62, 67]; [KOT 69, 72, 05]; [LUC 69]; [LEV 79]; [WIL 03]; [SHA 05, 09]; [VAR 05]; [ALD 06a, 06b]; [TAM 11]; [POW 12]; [ELA 18], [HUN 20a, 20b], the historical account here is constructed on contributions and developments that complement these works. This brief but preliminary general overview of marketing history was inspired by [HUN 20a]: "For-re-institutionalizing the marketing discipline in Era V"; [WIL 03]: "Scholarly research in marketing: Exploring the "4 eras" of thought development"; and [KOT 05]: "The role played by the broadening of marketing movement in the history of marketing thought" and [ELA 18]: "Marketing's identity crisis: Insights from the history of marketing thought." While research identifies [KEY 21] analysis on the state of unsettledness of marketing, [YAD 10] on the state of conceptual papers in marketing journals, and [YAD 20], to be highly relevant, this paper provides a contemporary view of the state of top doctoral programs in marketing in the United States. It concentrates mainly on establishing a historical structure of the field based on a content analysis of top marketing journals.

The author's essential challenge, however, is not the general consent on the historical narrative but rather proposing a future direction for marketing science while reviewing the past.

Furthermore, the author believes that a reversal of the non-confirmed divorces of marketing with fields such as consumer behavior (CB) and supply chain management (SCM), which were once under the umbrella of marketing science, is only possible if the field makes the research and the embracing capacity of the field sufficiently attractive so that the lost loved ones' return. At this contemporary historic juncture for the discipline, this research, based on the emerging new domain of the field of marketing going from a de-institutionalized state toward a re-institutionalized and a trans-institutionalized state, delivers a solid platform for innovation of educating future marketing scholars.

Research has also identified that the challenges that the field is facing contemporarily and in the future are highly disruptive; thus, we need to act and prepare for them today while still coping with the fractured state of marketing in the present [KAM 21]. The last five decades were pivotal in developing marketing theory and practice, with significant advances in marketing science and the development of rigorous methodologies for the field [SHE 21]. However, marketing has become a methodology and innovation borrowing.

Marketing as a discipline, with a clear agenda of an autopoietic body of knowledge, was founded based on three dimensions that took shape historically [SHE 90]:

1) Developments of channels were ignored by economics, leading to the institutional school of marketing;

2) Product developments leads to the commodity school of marketing;

3) Concentration on functions or processes leads to the functional school of marketing.

We could technically add the service school of marketing as the fourth dimension deriving from a combination of research streams, where a broader horizon of foresight for the field was needed beyond the packaged goods industry, where marketing originated. Furthermore, the service school delivered a rich ground for the reproduction of knowledge within marketing theory and its practice: developments of service thought of marketing [SHO 77]; [GUM 79, 08]; [GRÖ 82]; [VAR 08]; [SKÅ 15].

The idea of service is not new. [LEV 84]'s HBR publication "Marketing Myopia" was a milestone in developing service thought, as were the developments within the Nordic school of marketing. In addition, Christensen's major theories include the "The innovator's dilemma. When new technologies cause great firms to fail" [CHR 08], "Competing Against Luck. The Story of Innovation and Customer Choice" [CHR 16a] and "Know your customers' 'jobs to be done'". [CHR 16b] contributed much to the notion of service in terms of how disruption in an industry takes place and how not knowing your customers and the jobs that are contributing to their self-fulfillment for which they would hire the product that a firm produces, would contribute to the demise of the firm and the industry. This dimension could also be observed within the spectrum of B2B marketing as tightening the social systems between the provider, the customer and the customer's customer would contribute to a more rapid response to the changing nature of the environment, thus contributing to the resilience of the individual industry.

Thinking with the end in mind was discussed, with examples, in [LEV 84]. Essential streams in service were developed in Europe. According to [BER 93] and [SKÅ 15], these schools were established in the 1970s, which draws back mainly on [GRÖ 82]; [GUM 08] and other researchers also in France [EIG 76].

However, the introduction of the "Service-Dominant Logic" in 2004 started an oeuvre of publications. It contributed much to the development of the field and made the term "*service*" the major revolution in recent marketing thought [VAR 04], thus contributing much toward the broadening of the marketing movement from a historical perspective [KOT 69]; [WIL 03] and its relevance for further research [KAM 21].

Era IV of marketing [HUN 20a] had to cope with many changes, such as the globalization of markets [LEV 83] and many technological changes in Boorstinian's term of "the Republic of Technology" [BOO 78], which Neill Postman referred to as "Technopoly" [POS 93], while the foundational developments took place during Era III. Therefore, the dimensions that marketing developed in that era are still valid today. These foundations are: brand image [GAR 55], market segmentation [SMI 56], marketing management concept [MCK 57], the 4 Ps of marketing [MCC 78], marketing myopia [LEV 84], the marketing mix [BOR 64] and marketing management [KOT 67].

[SHA 05] identified many schools of thought in marketing, starting from Plato and Aristotle [SHA 05] to [GRE 76]. The schools of thought that were identified are: marketing functions school, marketing commodities, marketing institutions, marketing management, marketing systems, consumer behavior, macro marketing, exchange and marketing history.

[HUN 15] gave a solid overview of the literature and intellectual structure of marketing in JAMS (2010–2015), clustering the essential themes and topics and illustrating the essential authors. [YAD 10] also established a timely analysis of 30 years (1978–2007) of publications in major marketing journals on the state of conceptual articles within the field. He concluded by stating that the decline of conceptual articles weakens the theoretical core of the discipline and that more balanced, sustained and multipronged efforts are necessary to reinvigorate the field. While [YAD 20] illustrated the state of knowledge reproduction or field's homeostasis and its autopoiesis in terms of providing a list of the content and structure of 11 marketing doctoral programs in the United States, he illustrated the obvious but much-ignored fact that strengthening marketing means strengthening the doctoral programs in marketing; or, to which [HUN 20a] also referred to as putting marketing back in our marketing doctoral programs.

[KEY 20] revealed the myopic approach to rigor and relevance which marketing scholars have widely adopted, thus resulting in a discipline-wide rift and majoring in minors. [BEL 20] summed up the two end runs around marketing, which are disruptive for marketing. These dimensions include the shift in control of brands from marketers to consumers, which has largely been due to the Internet and its affordances. The second seismic shift, however, has equally contributed to the rise of the Internet and the rise of the tech sector.

[ELA 18] built on [BAR 62] by establishing a solid historical analysis of the three paradigms in marketing, while adding the fourth dimension of market systems into the marketing field:

– the traditional paradigm (1910s to late 1950s);

– the macro-to-micro paradigm shift management paradigm (late 1950s to late 1960s);

– the broadening/genetic boundary expansion paradigm (late 1960s to present);

– market systems [WIE 48]; [ASH 56]; [BEE 59, 95]; [BER 69].

The author agrees with [ELA 18]. We see the systems view in marketing as having great implications for in-depth research capabilities and the possible integration of essential subfields. Marketing can be an attractive and fruitful ground, thereby providing a robust foundation for all three essential jobs of marketing [NEL 17]: 1) create, 2) communicate and 3) deliver value that could be aligned.

What major, unaddressed puzzle, controversy or paradox does this study address?

While many histories of marketing have been established and led the way, a holistic, actualized and comprehensive history, especially by applying contemporary technological advances deriving from the evolution of AI, wherein a complete history of a field could be put under scrutiny and analysis, has so far not yet been published.

Why does it need to be addressed?

Therefore, a construct of history built by the published research within the 17 most prominent journals and the entirety of their respective publications, which amount to over 28,000 articles over the course of more than 80 years of research, seems to be necessary for the contemporary research to address the shifting milieu within the field seen through the lenses of science and practice.

What will we learn?

A comprehensive and holistic history of the marketing discipline was established based on the top journals related to the field, which have not been put into an integrative approach. How do the topics relate to each other, and how do they develop over time? What are the most influential articles and authors in the marketing field?

How does the study fundamentally change, challenge or advance scholars' understanding?

These research contributions identify the key issues for rethinking the current state of marketing based on a comprehensive study that establishes a solid longitudinal reflection of the field; this is for scholars to self-assess how the current disruptions within the state of practical and scholarly discourses affect the contemporary path of indigenous knowledge development within the field of marketing. Based on a foundational analysis, as this research provides, a holistic historical foundation could be visualized, thus to establish a flourishing frontier for innovation in the future and thus fostering a designerly collective imagination among doctorial scholars.

7.2.1. *Limitations*

This section's spectrum of contributions is limited to the 17 journals in marketing. Additional relevant journals, such as the Journal of Marketing (JM), were included in this study. At the same time, the author has paved the way for including diverse journals to establish a more adequate and holistic literature and content analysis based on the most relevant marketing-related journals. The journals relating to consumer behavior (CB) and consumer research based on the Association for Consumer Research (ACR), such as the Journal of Consumer Research (JCR), were also included. While building on [WIL 03], [HUN 20a] and [SHE 82], and as the field of CB has divorced itself from marketing, a foundational premise of this research is, however, to lay a possible future direction for marketing to develop indigenous theory-building capacity. Hence, the first 900 issues of the JCR mention marketing in the title only three times, which is a clear indication that another rich and vibrant field has evolved as CB, which was initially embedded as a major part of marketing science and also within the domain of marketing. It is currently a long and established domain outside of the marketing spectrum. The historical account here is founded on the shoulders of the giants of the marketing field. The author has tried to establish the research analysis only on observable and variable themes deriving from the historical accounts, which were integrated from a conceptual lens and based on the large data analyzed in this research, thus embracing the complete content of the 17 top journals.

7.2.2. *Method*

The proposed methodology of the research embraces many essential facets and necessary dimensions to establish itself as a contribution for future researchers, who either must cope with a large amount of data or would like to deliver solid contributions to establish an exhaustive account of the histories of the different fields and disciplines.

The author's essential focus was to avoid bias in conducting this research. Therefore, all of the topics analyzed within the scope of this paper went through machine learning algorithms (MLA) analysis to establish the historical account of the field of marketing and in-depth CA. Usually, conceptual articles are not free from intellectual bias, which goes with these types of papers. Above all, the author has analyzed all of the textual data from 17 journals with the help of unsupervised MLA to establish an objective account of the historical narrative. Thus, the author has combined the scholarly analysis of the *"lay of the land"* in marketing with the techniques and tools of AI for more precision and objectivity in establishing unbiased labeling of the *"topic-keyword relations."*

The LDA assumes that each topic is associated with multiple terms and vice versa; terms can belong to multiple topics but have a different impact on the overall topic distribution.

Based on the contemporary developments in ML capacity, this research's methodology delivers a robust foundation not only for the paper but also for additional research pursuits, wherein a vast amount of textual data could be analyzed for more holistic and grounded theory design possibilities and results. This research delivers a dynamic approach to highlight the evaluation of topics over time.

A CA was first performed on all papers published in the Academy of Management Journal, AMS Review, International Journal of Research in Marketing, Journal of Supply Chain Management, Journal of the Academy of Marketing Science, Journal of Business Logistics, Journal of Consumer Psychology, Journal of Consumer Research, Journal of Marketing, Journal of Marketing Research, Journal of Operations Management, Journal of Retailing, Journal of Service Research, Manufacturing & Service Operations Management, Marketing Science, Operations Research and Production and Operations Management.

Editorial notes, errata or commentaries were excluded to examine the content of the marketing literature, trace its evolution and identify its main streams or subfields. This analysis is used to objectively, systematically and quantitatively consider the published articles while allowing for an interpretation of the shifting priorities of editors, reviewers and authors, which shaped the evolution of the field.

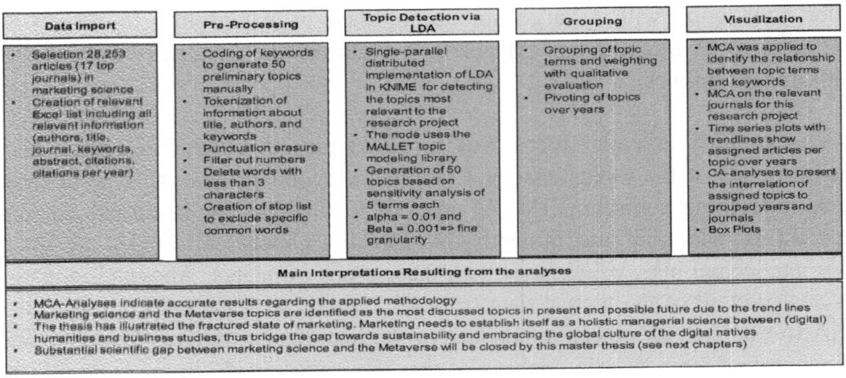

Figure 7.2. *Methodology of the analysis. For a color version of this figure, see www.iste.co.uk/machado/industry.zip*

As a first step, the articles to be analyzed were selected. The author included every article published in the respective journals from their foundation until our analysis began in the spring of 2022. A total of 28,253 articles were selected and analyzed. The number of articles per journal and per year is shown in Table 7.1. The journals with the most articles in this analysis are the Journal of Operations Research, with a share of 18.9%; the Journal of Marketing, with 12.4%; and the Journal of Marketing Research, with a share of 10.2%. For the analysis, the author collected all keywords. Where there were no keywords (especially in the early issues of longer established journals, where no keywords were indicated), the author used the titles of the papers to analyze their content.

Figure 7.3. *Illustration of word cloud term weight. For a color version of this figure, see www.iste.co.uk/machado/industry.zip*

Years of Publishing	Academy of Management Journal	AMS Review	International Journal of Research in Marketing	Journal of Supply Chain Management	Journal of the Academy of Marketing Science	Journal of Business Logistics	Journal of Marketing Research	Journal of Operations Management	Journal of Retailing	Journal of Service Research	Journal of Consumer Psychology	Journal of Consumer Research	Journal of Marketing	Manufacturing & Service Operations Management	Marketing Science	Operations Research	Production and Operations Management
1936													23				
1937													19				
1938													46				
1939													50				
1940													53				
1941													48				
1942													70				
1943													51				
1944													41				
1945													33				
1946													26				
1947													40				
1948													34				
1949													62				
1950													39				
1951													33				
1952													23				
1953													26				
1954													34				
1955													19				
1956									25				48			40	
1957									23				28			53	
1958	25								25				27			54	
1959									24				42			52	
1960									32				39			57	
1961	24								25				57			62	
1962	20								26				54			57	
1963	22								23				56			67	
1964	23						47		23				58			63	
1965	28						48		24				45			66	
1966	24						42		26				49			78	
1967	31						45		25				52			73	
1968	31						47		25				41			89	
1969	34						47		36				42			66	
1970	27						80		24				46			72	
1971	29						72		31				39			104	
1972	27						74		23				43			77	
1973	48				19		56		29				36			96	
1974	40				43		57		39			26	36			89	

1975	57				36		52		19			32	48			70	
1976	34				37		65		29			23	45			67	
1977	38				40		56		26			25	54			67	
1978	40				30		43		37			25	55			65	
1979	47				33		55		24			32	35			68	
1980	38				33		53		22			37	39			85	
1981	47				34		38		25			45	51			70	
1982	52				33		44		22			36	40		16	61	
1983	42				34		37		21			32	42		16	54	
1984	43				40		39		22			34	33		22	84	
1985	39				52		37		15			37	48		20	84	
1986	30				31		37		17			34	30		36	82	
1987	29				38		32		16			36	33		37	73	
1988	25				35		30		16			37	29		31	75	
1989	31				34		38		21			40	27		33	77	
1990	34				34		33		19			41	29		19	95	
1991	26				37		42		16			45	22		22	78	
1992	22				39		34		13		19	48	26		24	115	
1993	36				36		36		15		19	48	29		23	83	
1994	39				31		42		20		8	51	34		23	85	
1995	50		32		31		37		18		13	24	29		46	82	
1996	61		29		26		34		21		16	27	29		22	82	
1997	51		28				36		22		13	33	25		22	85	
1998	41		27				38		22		15	26	29		28	79	
1999	42		17		12		39	6	23		19	24	49	11	35	84	25
2000	67		23		37		37	26	23		20	28	26	24	23	81	28
2001	67		18		21		39	32	23		27	35	25	25	27	76	31
2002	69		22		33		36	37	23		32	35	29	16	29	96	32
2003	49		22		25		35	25	21		36	38	31	16	31	74	28
2004	54		23		27		37	30	24	12	36	67	39	20	49	69	28
2005	56		23		38		53	32	24	25	28	56	48	21	52	68	31
2006	55		29		48		57	41	31	27	25	57	43	23	71	82	36
2007	56		25		46		56	66	31	24	15	61	50	33	69	82	46
2008	44		22	18	43	23	53	32	31	25	14	77	54	42	84	107	47
2009	38		26	16	38	23	60	21	27	29	49	78	61	42	87	112	43
2010	63		26	16	50	29	90	35	33	30	34	72	49	38	72	126	48
2011	53	13	27	31	51	28	91	53	45	33	40	74	57	35	77	115	63
2012	60	8	32	24	48	16	70	37	37	29	26	84	48	46	69	106	68
2013	74	20	35	27	40	21	50	40	32	36	20	77	48	47	59	96	104
2014	72	6	33	20	38	22	51	34	37	31	31	90	48	41	52	92	148
2015	72	11	33	18	43	21	56	48	34	31	32	55	37	41	59	98	125
2016	88	14	61	20	43	24	66	41	35	29	25	76	37	37	55	95	120
2017	90	14	47	20	44	17	62	17	35	29	22	49	47	42	56	101	121
2018	87	19	39	20	56	15	59	31	28	28	28	67	47	47	53	106	125
2019	73	25	38	12	58	15	59	34	25	28	17	64	46	47	54	97	158
2020	71	27	41	13	60	18	60	38	36	31	18	36	45	95	61	104	140
2021	69	0	55	16	20	21	62	37	31	36	25	53	70	76	62	87	211
2022	0	0	0	0	0	0	8	4	0	0	0	0	8	0	4	0	0
total	**2854**	**157**	**833**	**271**	**1755**	**293**	**2889**	**797**	**1695**	**513**	**722**	**2227**	**3511**	**865**	**1730**	**5335**	**1806**

Table 7.1. *Output per journal per year*

Pre-processing

For conducting the analyses, the author used the machine learning software "KNIME," which is an API commercial software with an AI drag-and-drop solution. Fifty major topics were identified by coding the keywords of the articles. The keywords were retrieved from the databases *Web of Science* and *CrossRef*. The information about the title, author and keywords was then transformed into a document data type and adjusted using a word tokenizer (*OpenNLP English WordTokenizer*). Tokenization is the process of chopping the given sentence into smaller parts (tokens), which is used in tasks such as spell-checking, processing searches, identifying parts of speech, sentence detection and document classification (see TutorialsPoint 2021). The raw text is tokenized based on a set of delimiters, for example, whitespaces and punctuation. Numbers are filtered out, and words with two or fewer letters are deleted. Next, a built-in "stop list," which is integrated into KNIME and contains specific words that are very common in the English language, and a manual one with additional very common words in marketing that the author detected, were excluded in the analysis because their inclusion would distort the results with their weight and impact.

Topic detection via LDA

A simple parallel distributed implementation of LDA follows [NEW 09] with the sparse LDA sampling scheme and data structure from Yao et al. (2009). It was used to detect the topics most relevant to the marketing field during the almost ten decades of literature analyzed. Topic relevance was first assessed by the topic weightings given as an output measure by the LDA algorithm, and the subsequent quantitative assessment was judged by proximity to the originally published article keywords and the topic-terms distributions (see the following MCA) analysis. The author further embedded the topics into a dynamic context in the sense of a per-decade and topic assessment induced by an additional CA analysis. The LDA algorithm uses the "*MALLET: A Machine Learning for Language Toolkit*" topic modeling library. MALLET is a Java-based package for statistical natural language processing, document classification, clustering, topic modeling, information extraction and other machine learning applications to text [MCC 02].

Probabilistic topic modeling is a branch of unsupervised machine learning algorithms that aim to annotate large archives of documents with thematic information [BLE 12]. As the model is unsupervised, it does not require any prior labeling of the to-be-analyzed documents. The topics emerge from the analysis of the original texts in the form of a dimension reduction similar to the numeric counterpart method of a PCA. The only difference is that LDA aims to reduce the information of textual (therefore non-metric) data using Bayesian statistics.

With respect to previous topic modeling algorithms, such as the previously developed probabilistic latent thematic analysis (pLSI), LDA assumes that all documents in the collection share the same set of topics. Still, each document exhibits those topics in different proportions [BLE 12]. From a model-theoretic-based approach, a topic is distributed over a fixed number of words (also called vocabulary). It is assumed that the topics are to exist before any data has been generated.

LDA is a three-level hierarchical Bayesian model in which each word of a collection is modeled as a finite mixture over a pre-existing set of topics. Each topic is, in turn, modeled as an infinite mixture over a set of topic probabilities [BLE 03]. The three-level process can be characterized as follows: each document exhibits all topics in different proportions (step 1). A randomly selected topic is chosen from the per-document distribution over topics (step 2). Then, each word in each document is drawn from the previously chosen topic (step 3). To base our model on solid methodological ground, the author would like to closely describe the probabilistic model in more detail based on the preceding literature, such as [BLE 12]. The underlying problem traces back to calculating the so-called posterior distribution, i.e. the conditional distribution of the hidden variables for the given documents. To highlight these hidden variables, let $\beta_{1:K}$ denote the topics, where each β_k is a distribution over words. The topic proportions for the d-th document are θ_d, where $\theta_{d,k}$ is the topic proportion for topic k in document d. The topic annotations for the d-th documents are z_d, where $z_{d,n}$ is the topic annotation for the n-th word in document d. The observed words for document d are w_d, where $w_{d,n}$ is the n-th word in document d, which is an item from the fixed vocabulary that stems from the automatic read-out bag-of-words representation. With these notations, the author can describe the generative nature of the LDA process by using basic stochastic calculus on conditional probability. With this, the joined distribution of hidden and observed variables can be represented by

Equation 1: Joint Distribution of Hidden and Observed Variables

Source: [BLE 12]

$$p(\beta_{1:K}, \theta_{1:D}, z_{1:D}, w_{1:D}) = \prod_{i=1}^{K} p(\beta_i) \prod_{d=1}^{D} p(\theta_d) \left(\prod_{n=1}^{K} p(z_{d,n} \mid \theta_d)\, p(w_{d,n} \mid \beta_{1:K}, z_{d,n}) \right)$$

The equation above uncovers various dependencies. First, the topic assignment $z_{d,n}$ depends on the per-document topic proportions θ_d. Second, the observed word $w_{d,n}$ depends on the topic assignment $z_{d,n}$ as well as on *all* topics $\beta_{1:K}$.

These dependencies define LDA which are further summarized in the graphical model of Figure 7.4. In this figure, each node displays a random variable according to its role in the generative process. Shaded nodes characterize the words of the documents. The rectangles denote replication. The N plate denotes the collection words within documents; the D plate denotes the collection of documents within the collection [BLE 03].

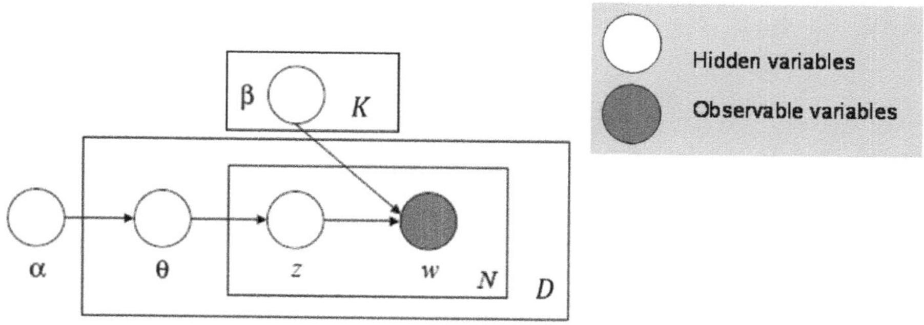

Figure 7.4. *The graphical model for LDA. Rectangular boxes indicate multiple iterations. Source: [BLE 03]*

Finally, the posterior probabilities can be calculated as follows:

Equation 2: Posterior Probabilities

Source: [BLE 03]

$$p(\beta_{1:K}, \theta_{1:D}, z_{1:D} \mid w_{1:D}) = \frac{p(\beta_{1:K}, \theta_{1:D}, z_{1:D}, w_{1:D})}{p(w_{1:D})}$$

The numerator describes the joint variable of all of the random variables (topic distributions as well as document-specific topic proportion, topic assignment and word collection). The denominator is the marginal probability of the words, which is *theoretically* computed by aggregating the joint distribution over every possible instance of the hidden topic structure. *Practically,* this distribution is hard to compute [DIC 83]. Therefore, [ATT 00] proposed a variational Bayesian framework as an approximation method for the posterior probabilities assuming a Dirichlet distribution for each topic β_k. Further, [BLE 03] proposes to use another Dirichlet approximation to estimate the per-document topic proportions θ_d based on a fixed parameter called α.

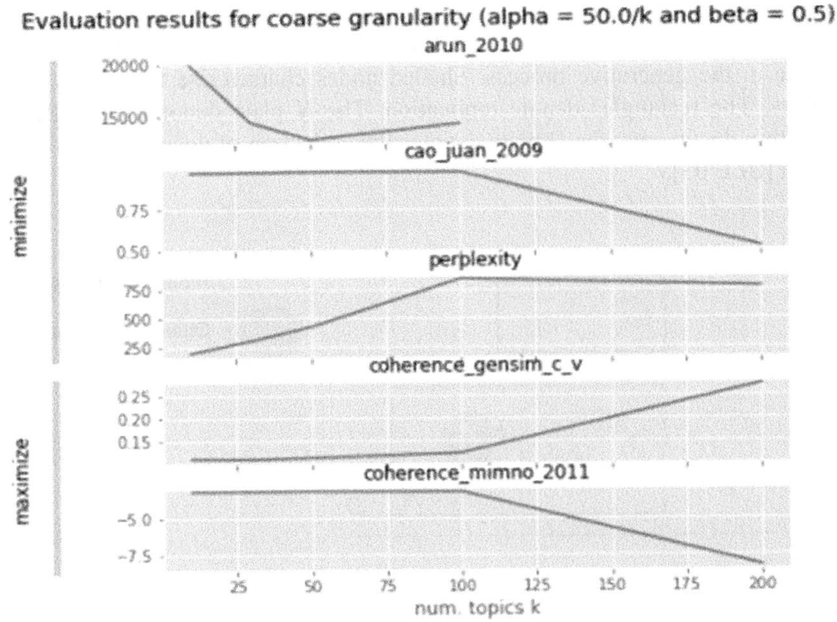
Evaluation results for coarse granularity (alpha = 50.0/k and beta = 0.5)
arun_2010
cao_juan_2009
perplexity
coherence_gensim_c_v
coherence_mimno_2011
minimize
maximize
num. topics k

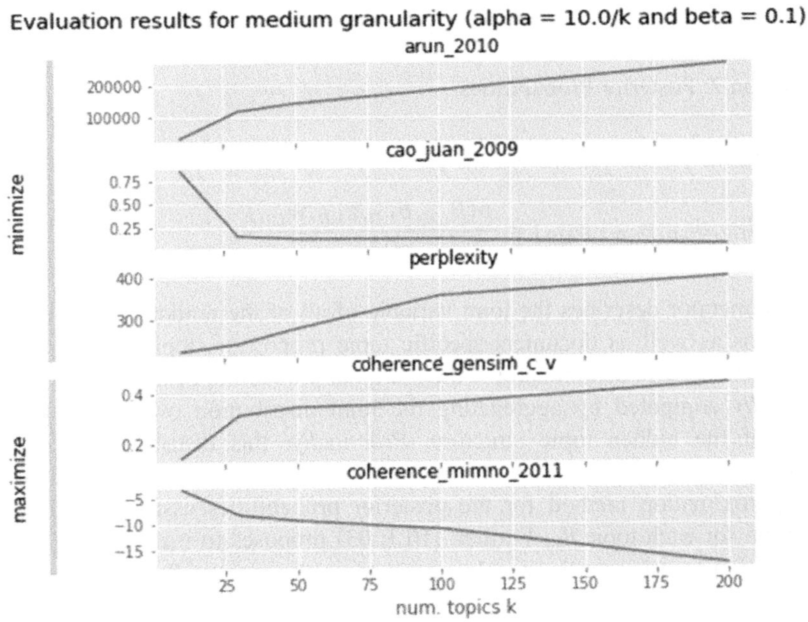
Evaluation results for medium granularity (alpha = 10.0/k and beta = 0.1)
arun_2010
cao_juan_2009
perplexity
coherence_gensim_c_v
coherence_mimno_2011
minimize
maximize
num. topics k

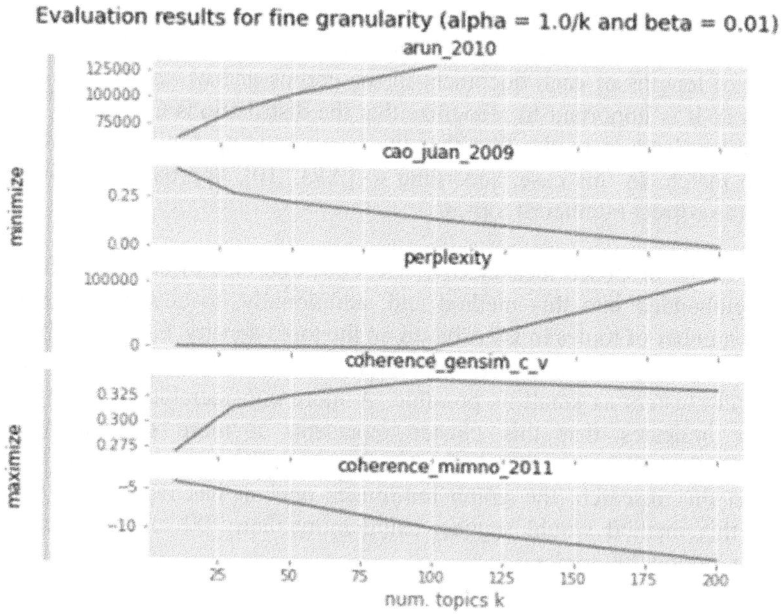

Figure 7.5. *Evaluation results for various granularities (coarse, medium and fine). For a color version of this figure, see www.iste.co.uk/machado/industry.zip*

To determine which number of topics best represents the results for this paper, evaluation results for coarse granularity, fine granularity and medium granularity regarding the analysis were performed. To explain the choice of 50 topics, the evaluation results are shown in Figure 7.5. According to [ARU 10], the identification of the "correct" number of topics in mechanisms like Latent Dirichlet Allocation (LDA) is of extraordinary importance as they determine the quality.

The first divergence measure in the scope of this analysis was proposed to find the correct number of topics [ARU 10]. The measure is based on a given Corpus C and a given topic K, resulting in two stochastic matrices as the LDA output. The measure for detecting the correct number of topics is the following:

Equation 3: Divergence Measure Proposed [ARU 10]

$$Proposed\ Measure\,(M1, M2) = KL(C_{M1}|C_{M2}) + KL(C_{M2}|C_{M1})$$

C_{M1} is the distribution of singular values of Topic-Word matrix M1. At the same time, C_{M2} is the distribution obtained by normalizing the vector L $*$ M2 (where L is 1$*$D vector of lengths of each document in the corpus and M2 is the Document – Topic matrix). It is important to recognize that the distributions CM1 and CM2 are in sorted order, leading to the fact that the corresponding topic components are expected to match. In this case, according to [ARU 10], approximately 50 topics seem to be the correct number of topics.

Moving to the proposed measure by [CAO 09], the idea of clustering based on density is embedded into this method and, additionally, to adaptively select the appropriate number of topics in LDA based on the topic density. Clustering based on density aims to enlarge the similarity in the intra-cluster while keeping the similarity between inter-clusters as small as possible. A more considerable similarity in the intra-cluster indicates that this cluster represents a more explicit meaning. Additionally, a smaller one between intra-cluster indicates a more robust topic structure. In this research, the graph minimizes once it has reached 100 topics. Therefore, this method would suggest using more than 100 topics, reaching its lowest point around 200 topics.

Additionally, the perplexity needs to be minimized to optimize the number of topics. A lower perplexity score indicates better generalization performance. The perplexity of a test set D_{test}, including M documents, is given by

Equation 4: Perplexity [CAO 09]

$$perplexity(D_{test}) = exp\left\{-\frac{\sum_{d=1}^{M} p(d_d)}{\sum_{d=1}^{M} N_d}\right\}$$

with N_d being the length of document d and $p(d_d)$ being the probability of the document d generated by the mode [CAO 09]. Once the graph reaches 100 topics, it starts decreasing. Therefore, the perplexity method suggests using as many or as few topics as possible.

Moving to the measures that need to be maximized, the coherence gensim according to [RÖD 15] combines the indirect cosine measure with the normalized pointwise mutual information (NPMI) and the Boolean sliding window. Here, the graph is being maximized as more topics are selected and increases once it reaches 100 topics.

Lastly, the coherence "mimno" measure was applied to evaluate the best number of topics. This measure calculates as follows: $D(v)$ is the *document frequency* of word type v (i.e. the number of documents with least one token of type v), whereas $D(v, v')$ is the co-document frequency of word types v and v' (i.e. the number of documents containing one or more tokens of type v and at least one token of type v'). Therefore, the topic coherence can be defined as follows:

Equation 5: Topic Coherence [MIM 11]

$$C\left(t; V^{(t)}\right) = \sum_{m=2}^{M} \sum_{l=1}^{m-1} \log \frac{D\left(v_m^{(t)}, v_l^{(t)}\right) + 1}{D\left(v_l^{(t)}\right)}$$

with $V^{(t)} = (v_1^{(t)}, \dots, v_M^{(t)})$ being a list of the M most probable words in topic t (see [MIM 11]). The last graph in Figure 7.5 clearly indicates that the topic quality after 100 topics decreases. Hence, 50 topics are still a solid number of topics whereupon judgments regarding the nature of evolution within a field could be made.

Considering all five measures evaluated in the sensitivity analyses, no specific optimal number of topics can be identified at first glance and no granularity model indicates the best fit. The different methods do not all calculate the same number of topics as optimal. At this point, a certain amount of human judgment is now required to determine the number of topics. Due to the amount of data, it seems reasonable to use a larger number of topics here, so 49 topics have been chosen. However, to use more than 49 topics would also blow up the model's interpretability. Further sensitivity analyses also shows that the results would not significantly improve. In the context of this paper, with 49 topics, $\alpha = 0.5$ and $\beta = 0.001$ is worked, even if this does not follow the practitioners' approach for one of the three sensitivity analyses previously performed because the number of topics and amount of collected articles seem to be represented well by using these parameters. Additional robustness checks have been performed. Even if an increasing number of topics had led to even higher accuracy, the increase would not be significant due to the following reasons. A large increase in topic numbers would not lead to better topic distributions from the qualitative perspective. Some topics might be redundant concerning most keywords associated with that topic, and topic labels would be even harder to identify. Furthermore, the visualization of such a variety of topics would be impracticable in subsequent analyses. This results in the decision to use the parameters as described ($\alpha = 0.5$, $\beta = 0{,}001$, $K = 49$ and $T = 3$). The $T = 3$ is set for the following reason: using three keywords indicating the highest probability that terms are associated with that specific topic, the author ensured to label the topics as accurately as possible. Nevertheless, using more keywords would not deliver additional value as the keywords would become less important for the respective topics, and the precision of the analysis would decrease.

In our analysis, the author chose the constant parameters $\alpha = 0.1$ and $\beta = \beta_k = 0.001$ as well as $K = 50$. Each topic is associated with $T = 3$ major terms, in the sense of keywords indicating the highest probability that terms are associated with that topic. As introduced above, the alpha parameter defines the Dirichlet prior on the per-document topic distributions and the prior weight of the topics in a document. The Knime library proposes to use α of normally less than 1 for all topics to prefer sparse topic distributions, i.e. few topics per document. Per default, the parameter is set to 0.1 but could be changed accordingly. The β parameter defines the prior on per-topic multinomial distribution over words. The KNIME library suggests using a number *much* smaller than 1 to strongly prefer sparse word distributions. The author decided to use the pre-set values in KNIME ($\alpha = 0.5$, $\beta = 0.001$, $K = 50$ and $T = 3$) which the author expects to have been approved in prior practices.

Finally, topic terms and weights were then interpreted, and a qualitative assessment from the author named each topic before they were pivoted over the years. Various robustness checks have been performed to assess the quality of the resulting topic-term distributions. The author analyzed the association of terms to keywords and topics qualitatively for an independently chosen subset of articles, which suggests that the LDA clusters were economically sound. In the following section, the author presents a multiple correspondence analysis to quantitatively validate concordance among real published article keywords and topic-term associations. The author used a two-component-based MCA, which, similarly to the metric PCA equivalent, aims to project the underlying variation of the data into two main factors. Enlarging the number of components does not lead to any increases in explained variation in terms of maximum Eigenvalue criteria that are depicted in the sensitivity analysis. The author, therefore, expects that this validation analysis with two main components is sufficient to test the suitability of the LDA projections.

Limitations of the LDA methodology and comments

One assumption that LDA makes is the "bag of words" representation, i.e. the order of the words in the document does not matter. It is problematic from a semantic point of view, especially when it comes to language generation. In uncovering the sheer structure of all of the documents, this shortcoming does not pose a major limitation to our analysis.

Another assumption is that the order of documents does not matter. As we will see in the time series analysis below, interest in certain topics may vary over time, so it may be reasonable to estimate multiple LDAs over time. In this approach, a topic would be a sequence of distributions over words. Similarly, the author conducted a CA (see Figure 7.9) to assess the topic's relevance over decades.

A third assumption about LDA is that the number of topics is assumed to be known and fixed. The Bayesian non-parametric topic model provides an elegant solution [TEH 06]. In this setting, the number of topics is determined by the collection during posterior inference analysis, and new documents can exhibit previously unseen topics.

	1936-40		1941-50		1951-60		1961-70		1971-80		1981-90		1991-00		2001-2010		2011-2020		2021-2022		Total
	abs.	in %	abs.	in %	abs.	in %	abs.	in %	abs.	in %	abs.	in %	abs.	in %	abs.	in %	Abs.	in %	Abs.	in %	
channel management	6	3%	14	3%	12	2%	29	1%	48	2%	57	2%	78	2%	128	2%	160	2%	12	1%	536
consumer behavior	9	5%	23	5%	41	6%	142	7%	320	11%	242	8%	202	6%	220	4%	289	3%	28	3%	1516
international markets	8	4%	16	4%	18	2%	50	2%	43	1%	57	2%	72	2%	96	2%	128	2%	5	1%	493
macromarketing	7	4%	8	2%	6	1%	15	1%	28	1%	18	1%	11	0%	41	1%	59	1%	10	1%	203
marketing science	60	31%	94	21%	113	16%	266	13%	281	10%	204	7%	175	5%	183	3%	221	3%	31	3%	1628
organizational performance	4	2%	3	1%	12	2%	63	3%	198	7%	233	8%	234	7%	282	5%	327	4%	35	4%	1391
relationship management	1	1%	2	0%	6	1%	21	1%	19	1%	31	1%	94	3%	173	3%	216	3%	24	3%	587
retailing	21	11%	71	16%	92	13%	159	8%	144	5%	105	3%	62	2%	95	2%	113	1%	14	1%	876
advertising	4	2%	18	4%	23	3%	34	2%	90	3%	86	3%	68	2%	60	1%	110	1%	15	2%	508
experiment design	1	1%	1	0%	5	1%	17	1%	38	1%	56	2%	78	2%	382	6%	131	2%	13	1%	714
food management	3	2%	11	2%	1	0%	10	0%	14	0%	20	1%	32	1%	136	2%	190	2%	27	3%	444
pricing	13	7%	23	5%	20	3%	35	2%	54	2%	68	2%	106	3%	168	3%	248	3%	13	1%	748
rational behavior	3	2%	8	2%	14	2%	74	4%	99	3%	99	3%	86	2%	188	2%	212	3%	24	3%	727
structural equation modeling	1	1%	1	0%	3	0%	18	1%	36	1%	35	1%	33	1%	45	1%	46	1%	0	0%	218
sustainability management	13	7%	9	2%	5	1%	5	0%	14	0%	6	0%	14	0%	34	1%	192	2%	32	3%	324
firm performance	1	1%	3	1%	3	0%	28	1%	32	1%	73	2%	126	4%	219	4%	244	3%	17	2%	746
market research	4	2%	15	3%	32	4%	58	3%	153	5%	112	4%	80	2%	59	1%	48	1%	8	1%	569
model-based management	3	2%	13	3%	12	2%	45	2%	59	2%	114	4%	120	3%	129	2%	99	1%	10	1%	604
optimal systems	4	2%	11	2%	57	8%	118	6%	113	4%	75	2%	39	1%	29	0%	41	0%	3	0%	490
risk management	2	1%	0	0%	2	0%	13	1%	13	0%	20	1%	35	1%	131	2%	279	3%	40	4%	535
sales performance	2	1%	9	2%	13	2%	30	1%	56	2%	41	1%	36	1%	37	1%	57	1%	9	1%	299
value-based management	3	2%	0	0%	7	1%	15	1%	15	1%	24	1%	30	1%	117	2%	202	2%	18	2%	431
consumer identity	2	1%	1	0%	2	0%	5	0%	7	0%	11	0%	32	1%	49	1%	250	3%	18	2%	377
dynamic competition	4	2%	13	3%	14	2%	24	1%	37	1%	81	2%	70	2%	253	4%	453	5%	57	6%	976
modeling	2	1%	4	1%	12	2%	59	3%	86	3%	130	4%	126	4%	179	3%	198	2%	19	2%	814
product design	1	1%	4	1%	3	0%	22	1%	34	1%	46	1%	53	2%	195	2%	173	2%	11	1%	451
production planning	2	1%	0	0%	21	3%	46	2%	51	2%	89	3%	143	4%	90	2%	73	1%	7	1%	522
segmentation	2	1%	3	1%	5	1%	13	1%	24	1%	32	1%	23	1%	26	0%	59	1%	17	2%	204
compensation	1	1%	5	1%	8	1%	13	1%	22	1%	33	1%	44	1%	63	1%	98	1%	11	1%	298
customer relationship management	1	1%	4	1%	6	1%	13	1%	11	0%	23	1%	41	1%	213	4%	227	3%	14	1%	553
forecasting	2	1%	10	2%	22	3%	55	3%	69	2%	77	3%	44	1%	38	1%	56	1%	3	0%	378
programming optimization	1	1%	3	1%	29	4%	185	9%	229	8%	157	5%	226	7%	235	4%	266	3%	22	2%	1353
performance management	0	0%	4	1%	6	1%	14	1%	20	1%	18	1%	52	2%	252	4%	180	2%	25	3%	571
service management	0	0%	2	0%	3	0%	11	1%	13	0%	25	1%	58	2%	186	3%	276	3%	37	4%	612
social media	0	0%	11	2%	6	1%	17	1%	21	1%	11	0%	18	1%	106	2%	359	4%	63	7%	612
branding	0	0%	2	0%	5	1%	18	1%	26	1%	27	1%	64	2%	140	2%	152	2%	20	2%	454
consumer response	0	0%	3	1%	7	1%	24	1%	49	2%	117	4%	183	5%	218	4%	268	3%	27	3%	897
decision theory	0	0%	1	0%	6	1%	12	1%	19	1%	11	0%	22	1%	22	0%	46	1%	5	1%	144
logistics	0	0%	4	1%	5	1%	14	1%	6	0%	7	0%	10	0%	45	1%	118	1%	15	2%	224
service optimization	0	0%	8	2%	17	2%	44	2%	59	2%	55	2%	111	3%	273	5%	284	3%	23	2%	874
knowledge management	0	0%	1	0%	3	0%	6	0%	14	0%	13	0%	22	1%	68	1%	100	1%	10	1%	237
consumer choice	0	0%	3	1%	4	1%	24	1%	63	2%	90	3%	89	3%	111	2%	137	2%	12	1%	533
information technology	0	0%	2	0%	3	0%	15	1%	18	1%	14	0%	8	0%	41	1%	113	1%	27	3%	241
queuing	0	0%	1	0%	34	5%	133	6%	102	3%	110	4%	95	3%	97	2%	124	1%	15	2%	711
field experiments	0	0%	2	0%	5	1%	12	1%	14	0%	11	0%	7	0%	19	0%	113	1%	29	3%	203
healthcare	0	0%	0	0%	1	0%	7	0%	12	0%	11	0%	11	0%	43	1%	196	2%	21	2%	302
product management	0	0%	0	0%	3	0%	10	0%	15	1%	11	0%	25	1%	79	1%	154	2%	18	2%	315
product innovation	0	0%	0	0%	2	0%	24	1%	47	2%	37	1%	49	1%	129	2%	192	2%	12	1%	492
corporate social responsibility	0	0%	0	0%	0	0%	6	0%	14	0%	23	1%	22	1%	56	1%	178	2%	29	3%	327
total	191	1	444	1	729	1	2070	1	2949	1	3015	1	3444	1	5998	1	8458	1	955	1	28253

Table 7.2. *Topic labels pivoted over the years*

7.2.3. *Analyses*

A multiple correspondence analysis (MCA) was used to see the relationship between the topic terms and keywords. MCA is an exploratory data analysis technique for the graphical display of multivariate categorical data [BEN 82];

[LEB 84]; [HOF 86] which aims to explain the interdependence among a set of categorical variables (keywords and topics) that is like a principal component analysis [HOF 94]. It allows the researcher to explore and analyze multi-way tables to detect a structure in the relationships between nominal variables [FUR 08]. The analysis shows which rows and columns of a frequency table have similar patterns of counts.

As depicted in Figure 7.8, collected keywords from the articles (blue points) and associated terms (green points) and subordinate topics (orange points) overlap mostly, thus underpinning the good fit of the LDA topic extractor. In this analysis, the author chose a two-dimensional factor, which is the most suitable representation for a graphical outline. The corresponding eigenvalues were both around 1, with a total inertia of about 655 and explained inertia of 0.15% for each component. As noted by [KAC 90], the proportion of the total variance explained by the dimensions is often very small. This is closely connected to the (inevitably) binary nature of the transformed nominal data [LEB 84]. Additional sensitivity analyses for higher orders of components (n = 2–30) have been performed, but they did not improve the maximum explained inertia per component.

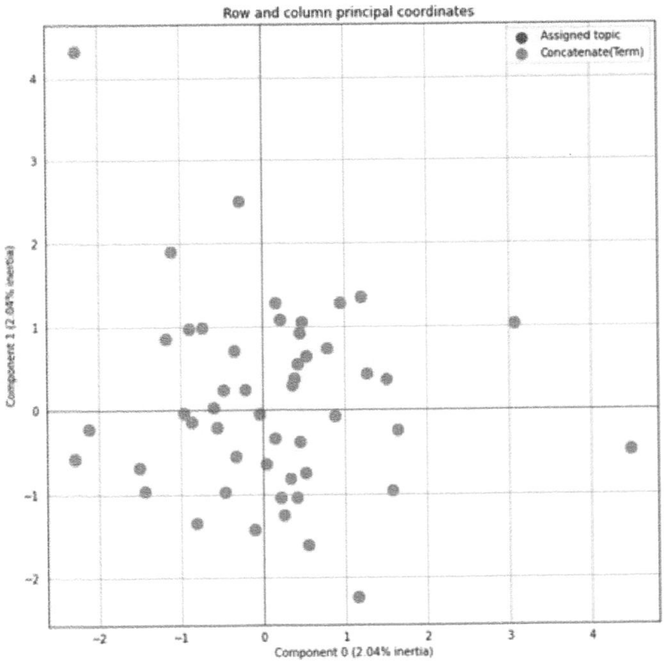

Figure 7.6. *MCA analysis showing the proximity of topic assignment (blue dots) and (concatenated) terms assignment (orange dots). The dot size indicates the number of articles affiliated with the categories. For a color version of this figure, see www.iste.co.uk/machado/industry.zip*

Based on the observations within the research and the experience of a marketing community, the author would like to propose another possible solution to [HUN 20a] call; thus, instead of focusing on a reinstitutionalizational framework for marketing, that marketing science does not only look back at what has been lost but simultaneously for what is to come in the future.

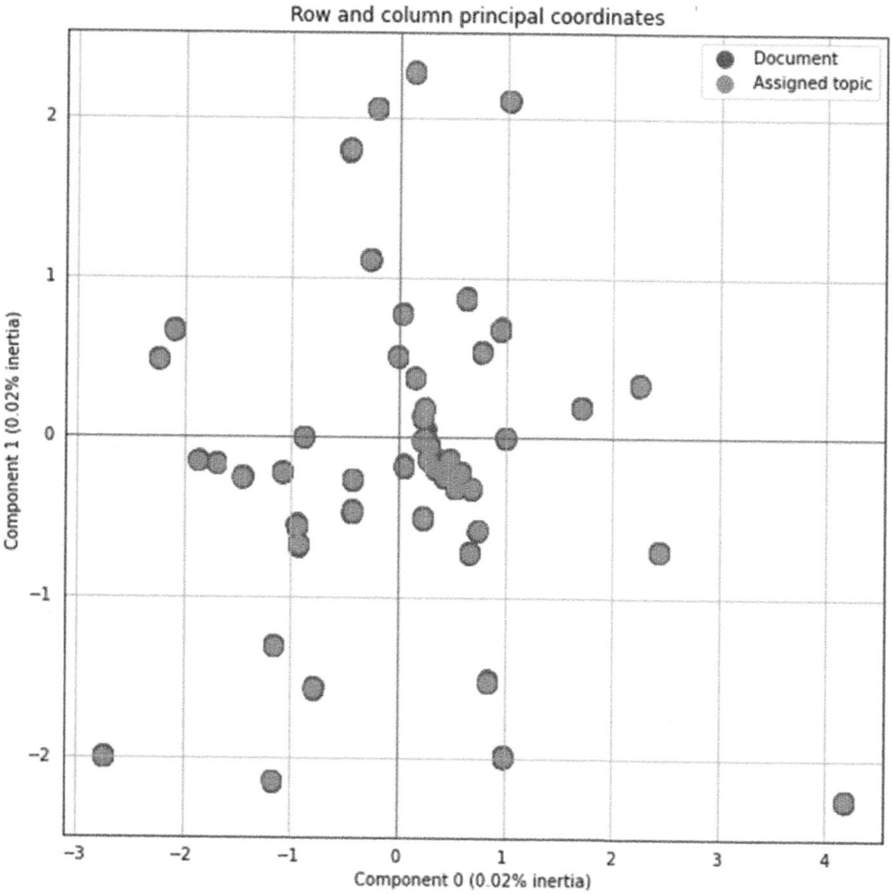

Figure 7.7. *MCA analysis showing the proximity of observable research article keywords (blue dots) and topic assignment (orange dots). For a color version of this figure, see www.iste.co.uk/machado/industry.zip*

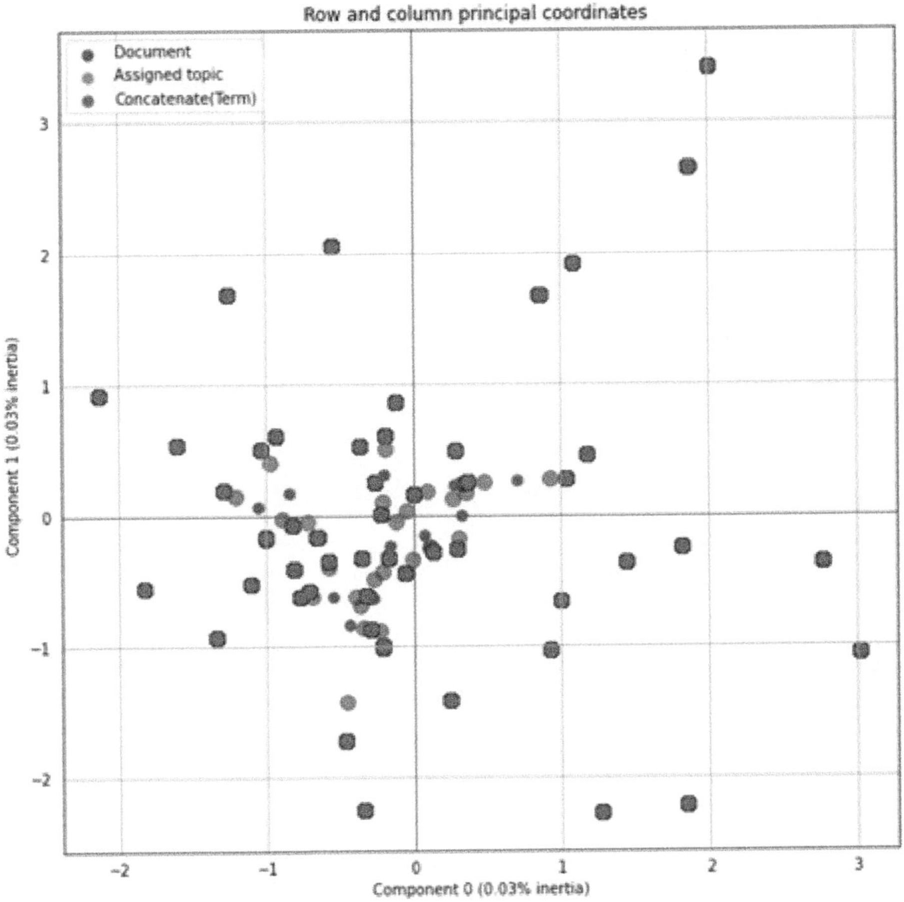

Figure 7.8. *MCA analysis showing the proximity of observable research article keywords (blue dots), topic assignment (gray dots) and (concatenated) terms assignment (green dots). For a color version of this figure, see www.iste.co.uk/machado/industry.zip*

In principle, the above MCA results (Figures 7.6–7.8) indicate a high match between the observed literature articles and the topic-term distributions. In successive analysis, researchers can assess advanced visualization methods as presented in the Python library LDA developed by [SIE 14]. Moreover, the author defines a new *relevance* score to assess the affiliation of a term to a topic by a mixture of the log-probability of n, end subscriptions approximated.

Variation Bayesian distribution (as introduced in the theory part above) is over the empirical term frequency. This measure overcomes the shortcomings of simply ranking terms purely by their probability under a topic, as suggested by prior studies by [BIS 12]. A visualization method comparable to [SIE 14] was performed using an MCA among topic-term distributions, as described in the previous section. To also highlight the dynamic changes of topics over time, the author further performed a CA between topic and years association.

7.2.4. *CA analysis/results*

A CA analysis was carried out to show the interrelation of the assigned topics to grouped years (see Figure 7.9). The frequency of published articles within a topic was based on a pivoting table drawn from the assigned topics over the years after LDA extraction. Again, a two-factor model to simplify the graphical representation was chosen. Topics (blue bubbles) that are closely related to a time period (orange bubbles) are also geometrically (in terms of the Euclidean distance) close together. The bigger the topic bubble, the more articles were published on the topic; and the bigger the years' bubble, the more articles were published in the respective time span. The analysis provides a pair of coordinates in the two-dimensional space for each of the 28,253 articles included in the dataset. As such, a large number of dots (for each article) on the map would become uninterpretable, so the author decided to represent only the position of the topics. However, as the previously discussed MCA analysis suggests, article keywords and topics are closely correlated to each other (see Figures 7.6–7.8).

The two dimensions of the map in Figure 7.9 can be interpreted as follows. The horizontal dimension separates older topics (on the left) from newer ones (on the right), with a cut around the themes "Macromarketing", "Segmentation", "Rational Behavior" and "International Markets," which are historically topics that have defined marketing and thus the history of globalization. The themes "Pricing", "Channel Management" and "Agency Theory" have also contributed to this historical evolution and the themes that have driven traditional marketing, whereas the vertical axis separates topics focusing from a focal point of marketing with a foundation-building narrative on "Retailing", "Marketing Science" and "Optimal System" aspects, relating to the efficiency movement that also guided marketing to enhance the global value chains and value streams of produced goods (at the top-right) and "Consumer Behavior"- and "Marketing Research"-related aspects (at the bottom-right). These dimensions reflect "poles" of topical orientation within marketing. The dimensions that cover the upper left, as the spiral flow of the themes illustrates, are based on the newer topics that are driving marketing within the contemporary era. These topics are "Field Experiments", "Sustainability Management", "Social Media", "Healthcare" and "Information Technology", especially the re-concentration of marketing on "Logistics" due to the Covid-19 global pandemic, which created much disruption within the supply chains of the world, "Consumer Identity" and "CSR", "Risk Management" and at the lower-left part, the dimension of "Dynamic Competition and "Consumer Motivation." Combining the two dimensions provides insights into the relationships between the topics.

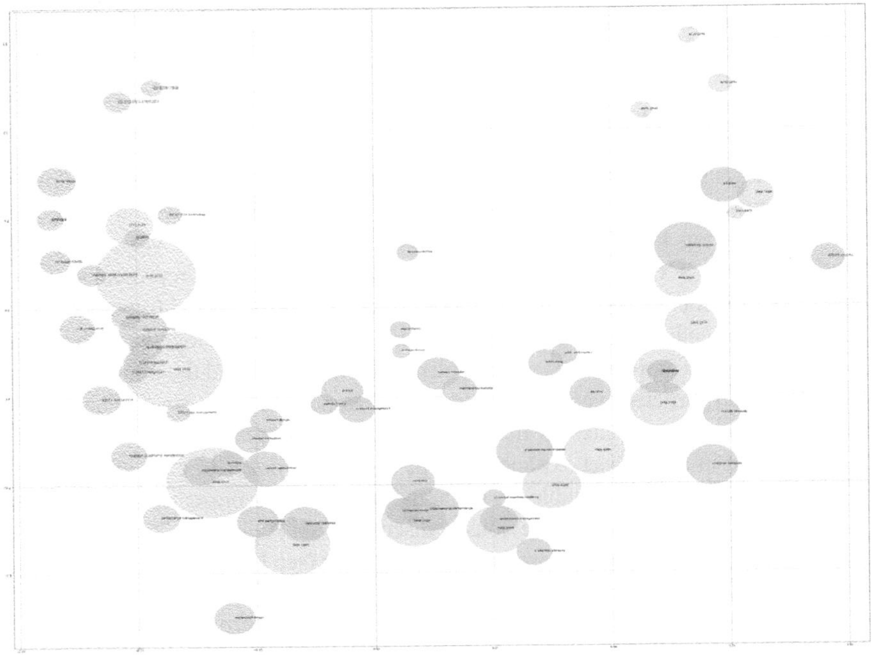

Figure 7.9. *CA analysis indicating the association of topics (blue bubbles) with time decades (orange bubbles). The size of the bubbles indicates the number of articles associated with either the topic or year category, respectively. Bubbles closer together indicate thematic proximity induced by the underlying LDA classifier. For a color version of this figure, see www.iste.co.uk/machado/industry.zip*

The timeline follows a spiral form. On the top left, we can see the newer articles with topics such as those discussed above. At the top are articles from the recent themes within the period 2021–2022, with topics such as "Logistics" as discussed, while in the time frame 2016–2020, the essence of the theme "CSR" displays dominance, and the period 2011–2015 was occupied by themes such as "Value Based Management", Food Management", "Product Management", "Service Management" and "Knowledge Management."

While topics such as "Branding", "Relationship Management", "Product Design" and "Service Optimization" were important during the period 2006–2010, the essence of "Firm Performance" and "Consumer Response" also became essential themes during the time frame between 2001 and 2005. The turn of the millennium (1996–2000) is represented in the bottom middle quadrant when especially topics such as "Modeling", "Consumer Choice" and "Organizational Performance" derived by the period 1991–1995, where theses such as "Model-Based Management", "Structural Equation Modeling" and "Product Planning" became more important. This trend continues to this day, as the most essential and highly cited papers in marketing science are based on "Structural Equation Modeling" themes.

At the top right, the reader finds older articles and themes illustrated. There was a clear trend toward "Retailing in the 1950s" (top right), but also other traditional marketing science dominating themes within the other periods 1960s–1980s.

To assess the evolution of topics over time, the author also performed a time series analysis to delve into the themes more in-depth. A positive (negative) slope of the trendline (dashed red line) illustrates an increase (decrease) in the average number of articles associated with the respective topic over the years 1930–2022. It is essential here to establish that the applied topic modeling methodology relates to the category of unsupervised learning algorithms (USLA) that were not fed by a priori labeled themes or terminologies as in prior literature studies such as [FUR 08]. The algorithm combined thematic close keywords into groups of related themes, which, as the prior robustness checks illustrated, seem to be meaningful from a marketing professional/scholar perspective. It, therefore, improves subjective topic labeling by scholars embedded in a specific research domain, which might yield biased themes' assessments and topology concentrations. The only qualitative assessment that the author has committed to here was the sub-summation of terms to topics which, in many settings, are similar to terms such as "Corporate", "Social" and "Responsibility" to CSR or the whole terms relating to the efficiency movement, etc., were self-explanatory. The source of Figure 7.11 to Figure 7.60 is the author's own illustration.

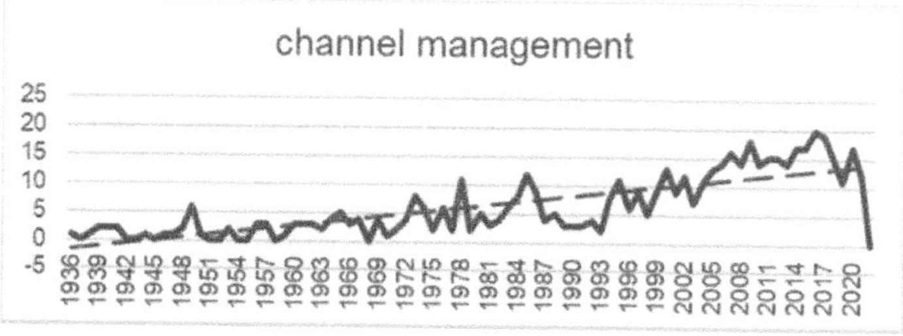

Figure 7.10. *Time plot channel management*

The importance of channel management as a founding pillar of marketing has been ubiquitous in the development of marketing throughout its history.

Contrary to the general assumption of many in the research community, digital marketing and technologically driven e-commerce activities have emerged as a new type of marketing; our collected history illustrates that the rise of the Internet has had no significant effect on the constant rise of the importance of channels in marketing.

Themes such as franchising, retailing and distribution have been essential throughout the evolution of channel management.

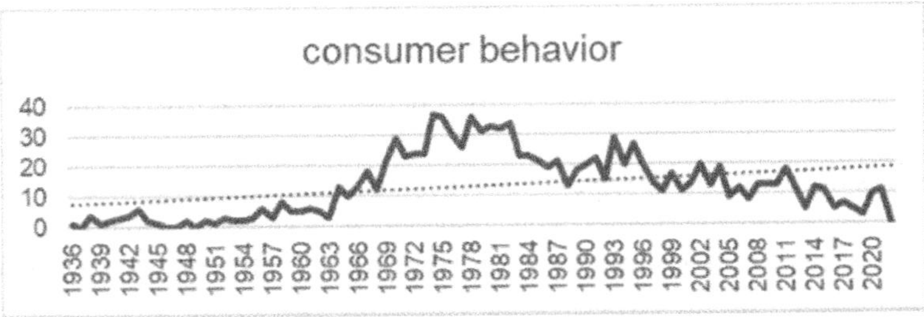

Figure 7.11. *Time plot consumer behavior*

This figure underpins the essential dimension of consumer behavior for marketing and the field's many solid contributions to marketing science.

However, the separation of the field from marketing science to evolve toward consumer behavior in the foundation of the Journal of Consumer Research in 1974, as illustrated in this figure, describes the evolution of the field. While a large bump can be observed from 1974 until the early 1980s, the number of publications associated with consumer behavior has declined until today.

Figure 7.12. *Time plot international markets*

International markets are constantly volatile. However, the topic's importance can be observed in the constant rise of academic publications within the field.

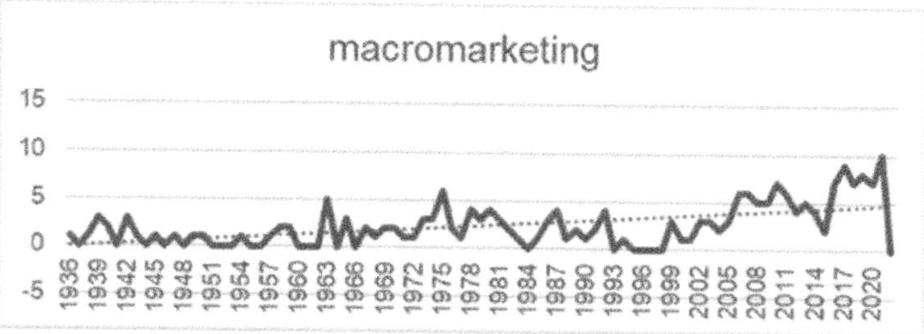

Figure 7.13. *Time plot macromarketing*

While the topic of macromarketing has constantly risen, the number of publications has been comparatively low until today. This could be due to the field of economics, which is foundationally concentrated on the macro frameworks and structures of the markets.

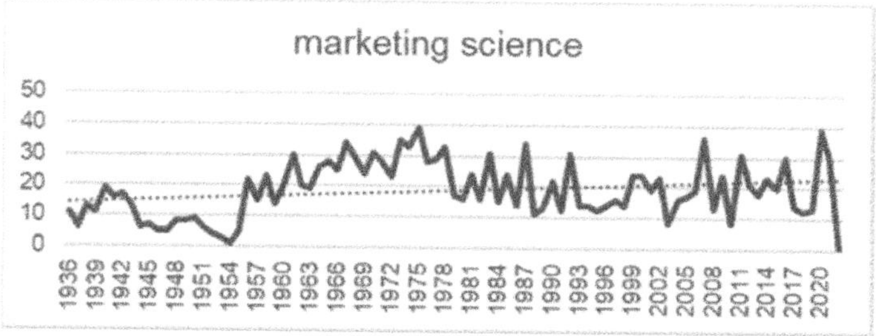

Figure 7.14. *Time plot marketing science*

The topic of marketing science has been important throughout marketing history, especially after 1973, when the Academy of Marketing Science journal was founded. Additional peaks were reached in 1987, 1992, 2007 and 2020.

The top articles are old; no significant indigenous theory development in the past 20 years can be observed.

The most cited articles were published within the period of 1981–2021; however, as the measure of most cited papers per year reveals the importance of publication over the years in terms of the papers' history and their impact on the scholarly landscape, they favored concurrently recent publications within the field, which could also mean that these papers could go both ways, respectively, maintain their impact or vanish. Therefore, the most significance is given to articles such as [FOR 81], [BAG 88], [MOR 94] and [HEN 15], who have maintained solid importance over the years.

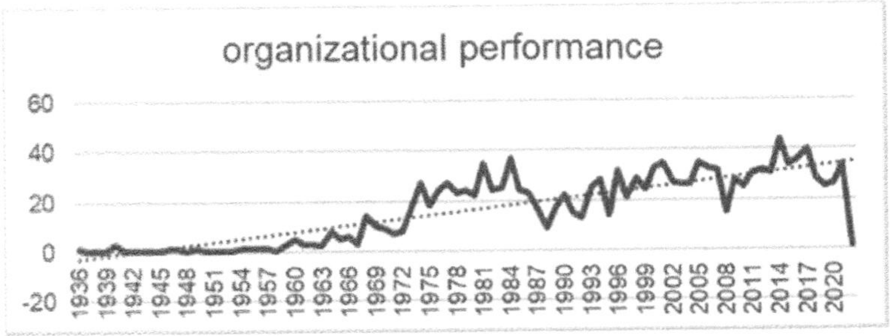

Figure 7.15. *Time plot organizational performance*

The essence of marketing is to connect the firm with its wider environment. This history can be observed within marketing, where the notion of organizational performance is essentially connected to the evolution of the field.

Figure 7.16. *Time plot relationship management*

The notion of relationship management has had a slower start than other topics observed so far, and yet the years 2010 and 2011 illustrate an increased interest in the topic in recent years.

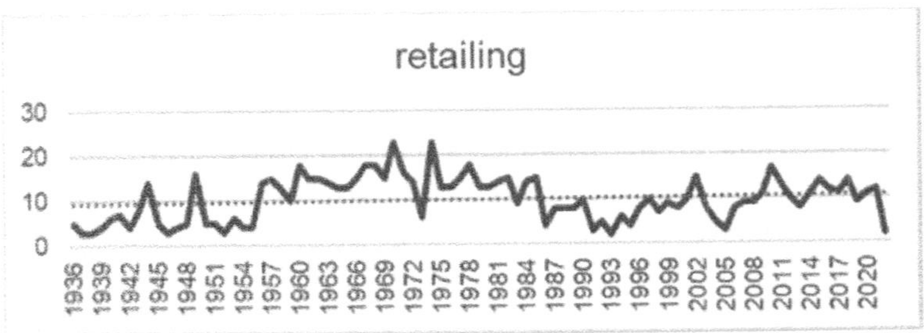

Figure 7.17. *Time plot retailing*

Retailing has been a founding pillar of marketing and, therefore, has maintained a steady level of publications throughout most of its history. The most interesting topic was observed between 1955 and 1985.

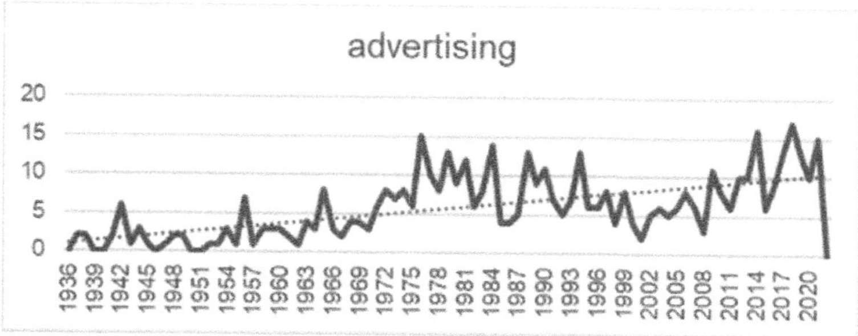

Figure 7.18. *Time plot advertising*

Based on the Google Ngram viewer search history, theme distribution has had a huge impact on the evolution of marketing, followed by the notion of advertisement, which has been highly influential on marketing as well. However, within the course of academic publishing, the topic of advertising has increased with the introduction of mass media such as television or social media, which explains the higher number of publications after 1970 and 2008.

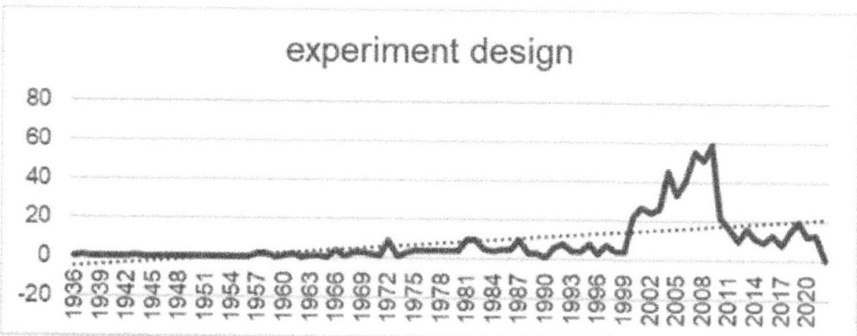

Figure 7.19. *Time plot experiment design*

The theme of expert design has gained momentum only since 1999, with a sharp fall between 2009 and 2011. Thus, it did not really recover from the downward turn but only had a minor peak in 2020. This could also be related to the Covid-19 pandemic, as no real experiments in terms of physical participation of the participants could be designed.

Figure 7.20. *Time plot food management*

The importance of food in marketing was sensed in the early 2000s, peaking in 2008. This trend has remained constant until today due to a rise in the general population's health consciousness.

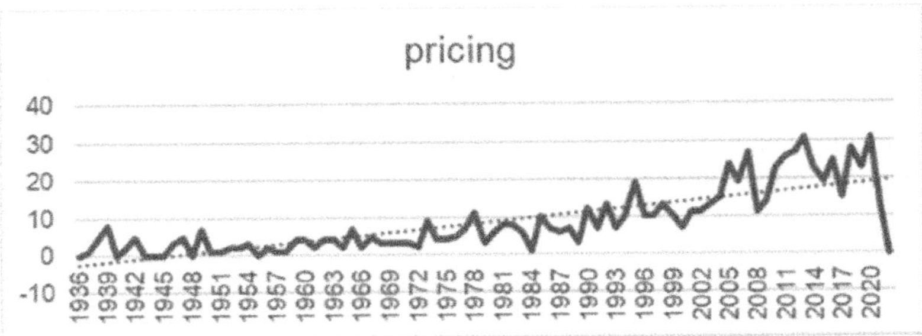

Figure 7.21. *Time plot pricing*

Pricing, based on the Google Ngram Viewer, has had the lowest number of book publications compared to advertising, distribution, promotion, product management and marketing. Regarding academic publications, however, the topic has steadily increased in importance, with peaks in 1995, 2007, 2014 and 2020. As pricing decides what type of marketing we are conducting, there is no discussion about the importance of pricing within the marketing context. Thus, our derived history of marketing reflects this foundation, and the generic strategy of marketing starts with deciding what pricing is pursued within the organization's marketing context.

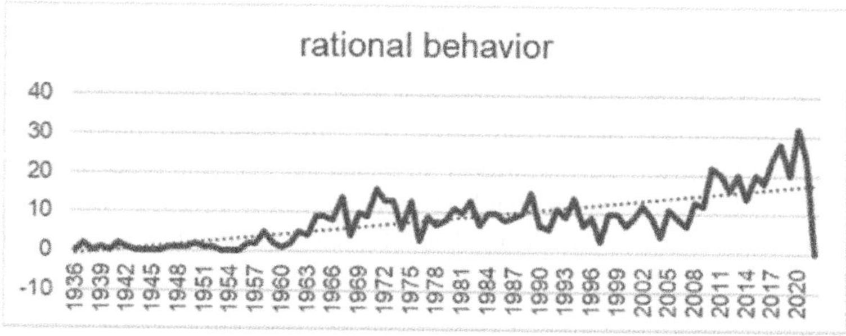

Figure 7.22. *Time plot rational behavior*

Throughout history, there has been broader interest in academic publishing on the notion of rational behavior. While multiple peaks can be observed throughout the years, 2010, 2018 and 2020 indicate the most publications in recent years.

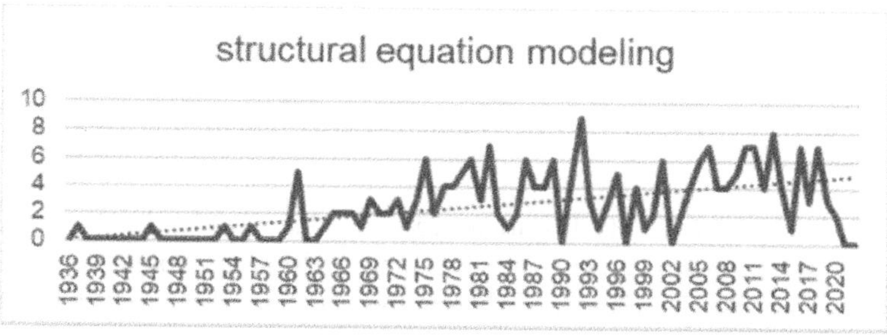

Figure 7.23. *Time plot structural equation modeling*

While structural equation modeling is represented in the top and most cited articles of our analysis, it has not had many publications throughout the years. However, there has been constant interest in the topic after the 1960s.

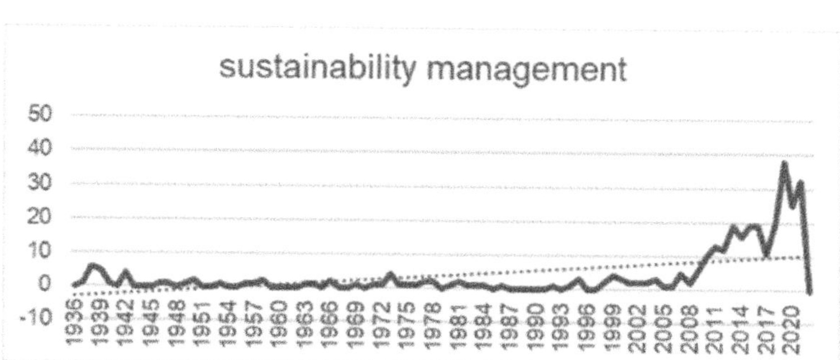

Figure 7.24. *Time plot sustainability management*

The essence of sustainability in marketing, especially in academic publishing, has been discussed mainly since 2010, when most Western countries introduced environmental protection acts and companies started to focus on their sustainability goals. Even though the Brundtland report of 1987 has been influential on the collective observation of the essentiality of sustainability, this essential dimension has not been solved, nor has a standard model on how to cope with this global challenge been developed, which has been founded on indigenous research from marketing.

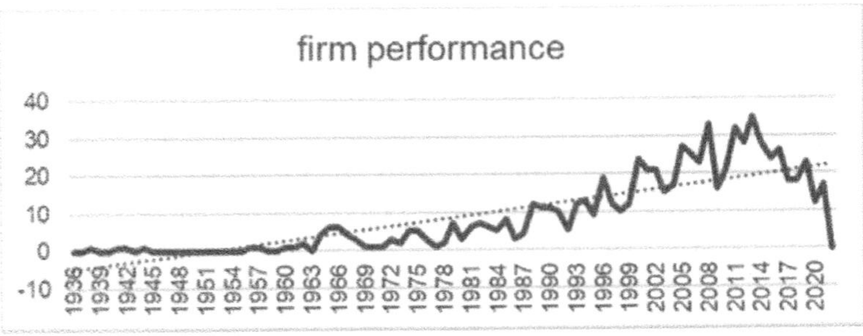

Figure 7.25. *Time plot firm performance*

The first major peak of this topic can be observed in 1996, while most publications were observed between 2000 and 2020. This could be due to the importance of industrial economics, wherein the structure of the industry was more important than the individual firm performance. However, the evolution of the theme dynamic capabilities of the firm within the strategic management field could have influenced scholarly attention toward firm performance in marketing.

Figure 7.26. *Time plot market research*

The first peak of the topic of market research can be observed in 1964 when the Journal of Marketing Research was founded. Much attention was paid to the theme, with large peaks in 1977 and 1979 and another peak in 1984. Afterward, a steady decline can be observed. Due to the rise of the Internet and creativity-based entrepreneurial pursuits, a data-based world does not require mailed-out surveys or market research activities, which were of vital necessity in the 1970s and 1980s. Therefore, especially after 2000, market research has not been of much interest in the marketing field.

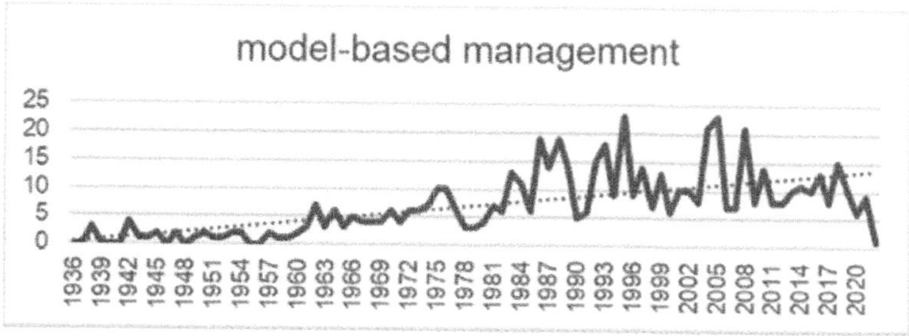

Figure 7.27. *Time plot model-based management*

Data management has contributed to more competitive dynamics for firms, wherein modeling, customer choice and consumer behavior were the essential themes. The author has organized the themes deriving from the topics under the umbrella of management.

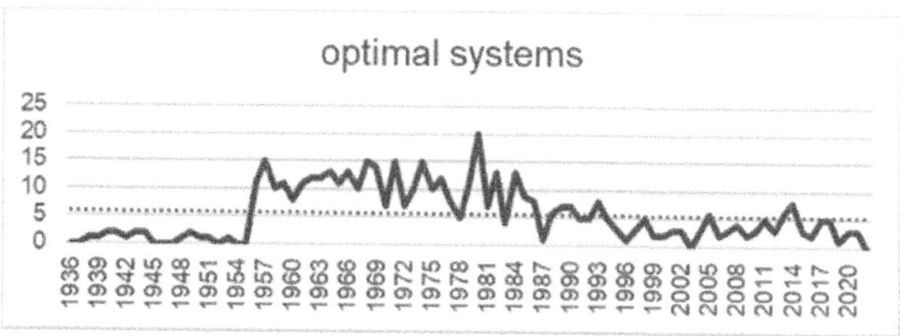

Figure 7.28. *Time plot optimal systems*

The notion of optimal systems has been found based on the evolution of statistical and operational research, wherein diverse performance measurements were developed to create an optimal model of efficiency management.

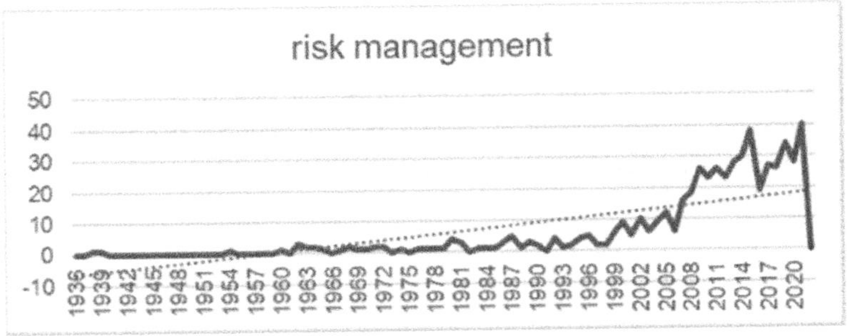

Figure 7.29. *Time plot risk management*

Risk management within the holistic marketing literature reached its peak in 2007. This development was due to the global financial crisis, which peaked again in 2015. Additional upward movements (peaks) could be observed due to the global Covid-19 pandemic, which started in late 2019 in China and then spread throughout the globe.

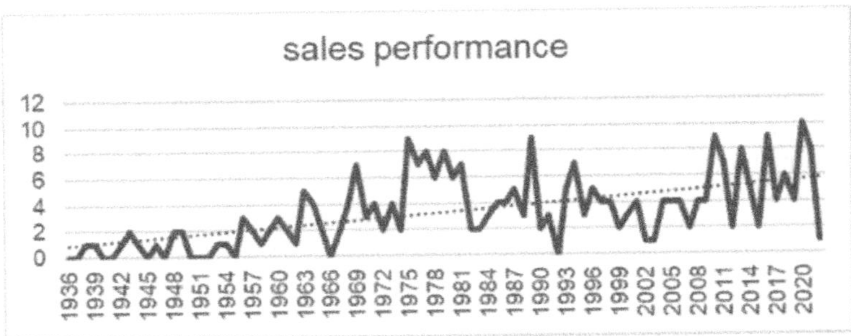

Figure 7.30. *Time plot sales performance*

The notion of sales has been essential for the raison d'être of marketing, and it has been highly volatile throughout the evolution of the marketing literature.

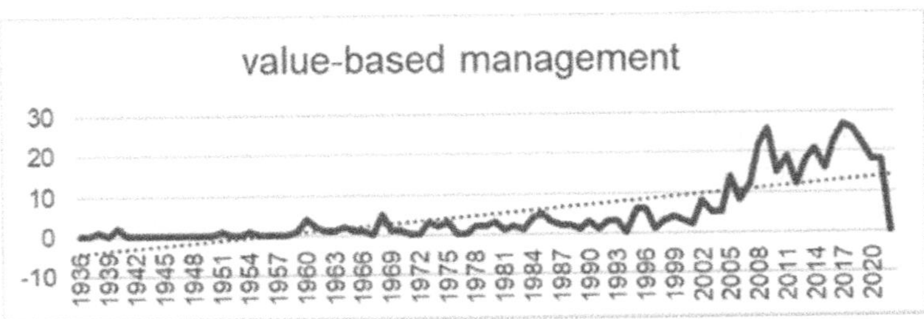

Figure 7.31. *Time plot value-based management*

Value-based management is the contrary approach to price-based and competition-based marketing, wherein the perception of value to the customer is the locus of innovation. This approach has contributed to creating vast amounts of growth and productivity and, hence, to the evolution of branding within marketing. Many firms like Apple (iPhone) and traditional companies such as Steinway & Sons have profited from this approach. These companies could market products at a much higher price than their competitors by increasing the value perception of their customers.

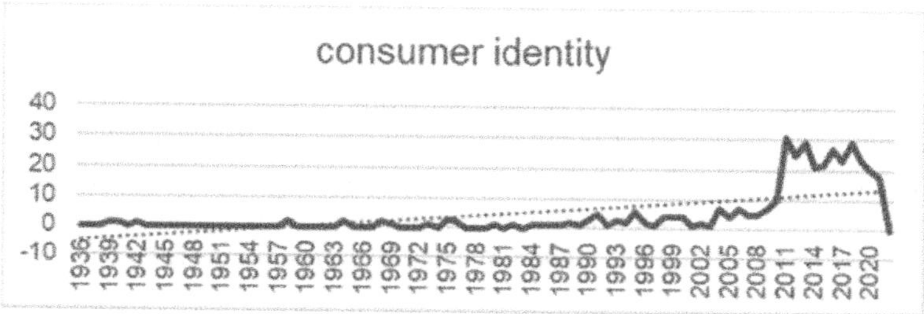

Figure 7.32. *Time plot consumer identity*

While many diverse publications have embedded consumer identity throughout the years, a huge peak in interest has been observed since 2011. This rise in interest has been stable and could stem from the rise of Internet-based technologies, which can thoroughly observe consumer behavior.

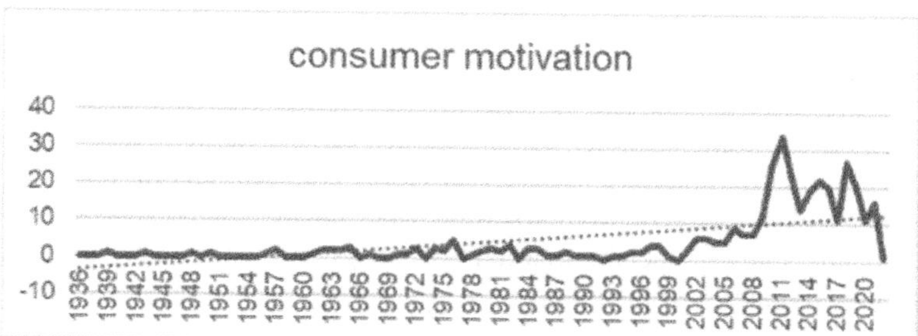

Figure 7.33. *Time plot consumer motivation*

Consumer motivation has been the primary driver of marketing. While many publications have been observed throughout the years, a huge peak was observed in 2011, and the interest in this topic seems to be fruitful for the field and good customer relationship management.

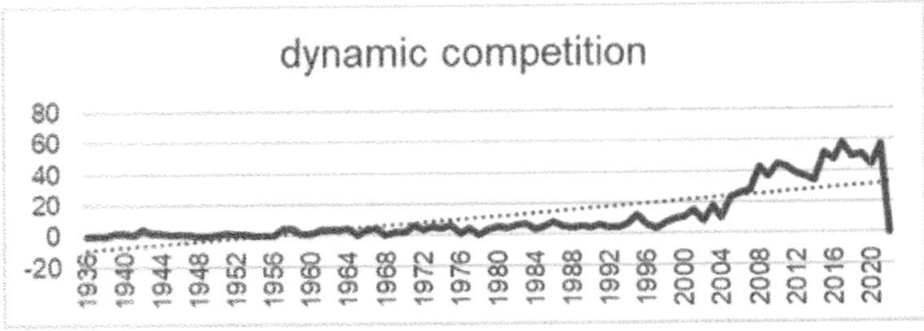

Figure 7.34. *Time plot dynamic competition*

Many strategic games, for example, pricing, focus and differentiation, are foundational to the evolution of the theme of dynamic competition, wherein multiple games of strategy have been played to achieve optimal results for the firm. The peaks in this topic are parallel to the evolution of technology and post-PC devices (for example, mobile phones), whereby much competition for the attention of consumers has been pursued.

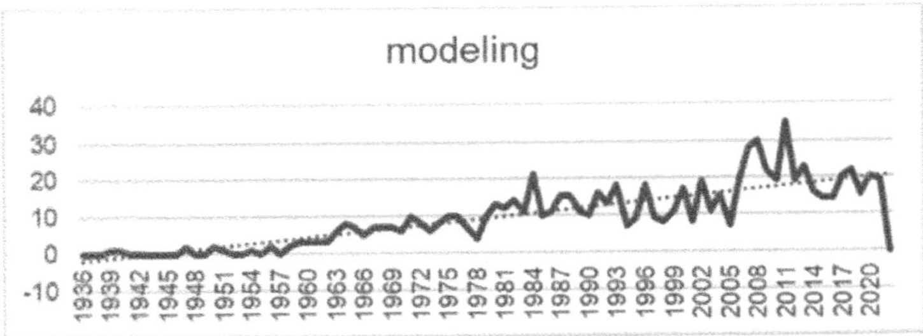

Figure 7.35. *Time pot modeling*

The evolution of modeling within marketing can be derived from the many statistics that were integrated into marketing. Thus, adequate data was translated into information and intelligence for marketing managers to make better decisions or understand the dynamics of the markets.

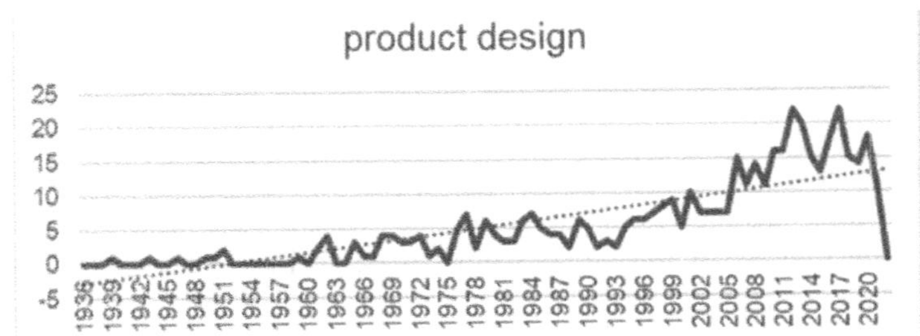

Figure 7.36. *Time plot product design*

The essence of product design has always been foundational in customer choice. However, based on the evolution of digital gadgets, mainly deriving from the introduction of the first iPhone in 2004 and additional design-based production strategies, the notion of design science research, information sciences and designing has become a core capability of diverse technology-based companies. This evolution substituted the focus on value-chain activities, which were based on inbound and outbound logistics and production to be outsourced. At the same time, channel relationships and direct consumer strategies within the cyber-physical duality have been considered as the core activity of marketers.

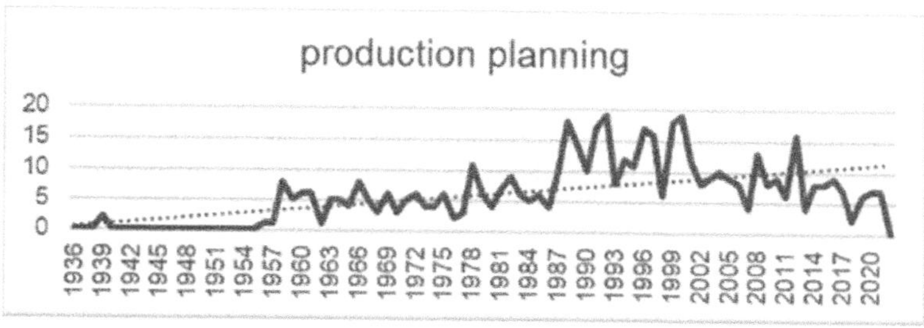

Figure 7.37. *Time plot production planning*

This evolution has been based on the foundations of total quality management and efficiency research, where much attention was put toward scheduling, service management, production, quality and production facility design.

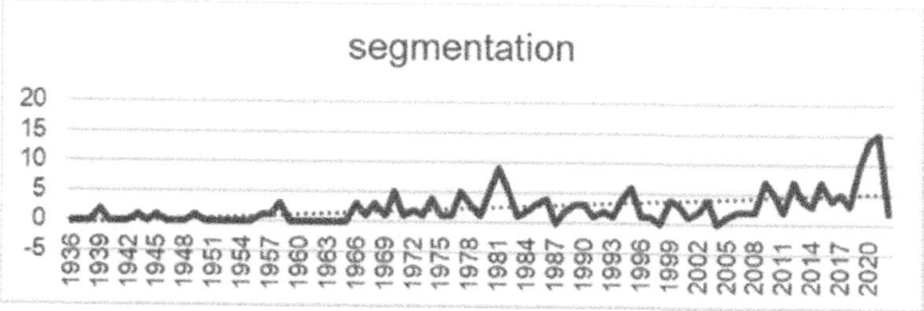

Figure 7.38. *Time plot segmentation*

As segmentation has been the foundational pillar on which marketing has nurtured, the evolution of segmentation has been ubiquitous and important. This essentiality has not decreased and, on the contrary, has increased within the times of hyper-digitality.

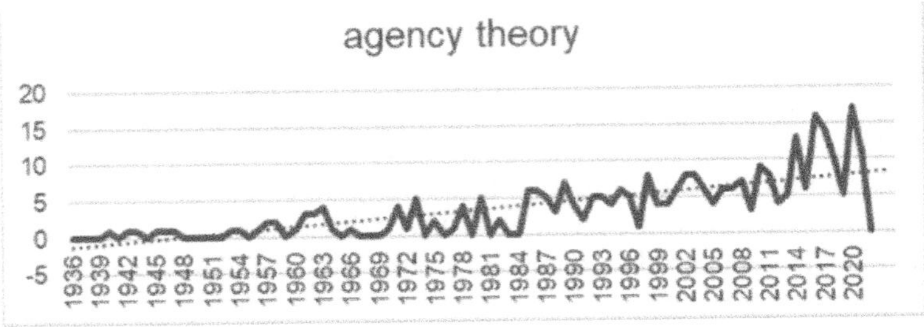

Figure 7.39. *Time plot agency theory*

The notion of agency theory steadily rose during the mid-1980s, with some drops in the late 1990s, possibly due to the Asian financial crisis and the late 2000s due to the 2007 and 2008 financial crises. However, a great rise in academic interest was observed between 2016 and 2020.

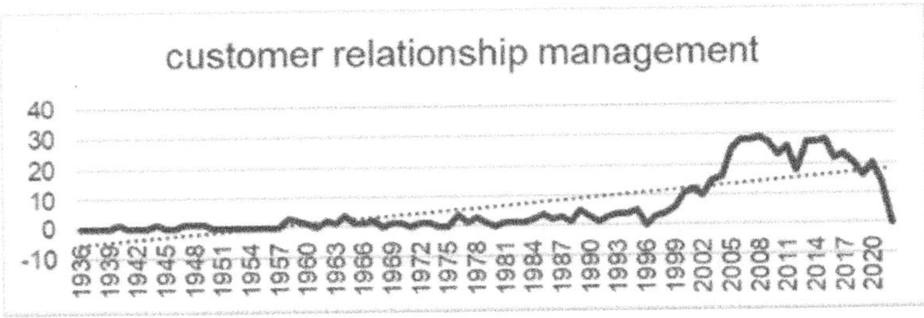

Figure 7.40. *Time plot customer relationship management*

Customer relationship management (CRM) has been a foundational topic for branding and marketing in general. While the theme was discussed in many marketing publications, a strong rise in the topic was observed after 1996. This is likely due to the rise of e-commerce and the Internet. The topic has not lost its importance. Based on the diverse digital tools and frameworks developed, much more is expected to be possible, especially with the rise of AI-driven organizational capabilities.

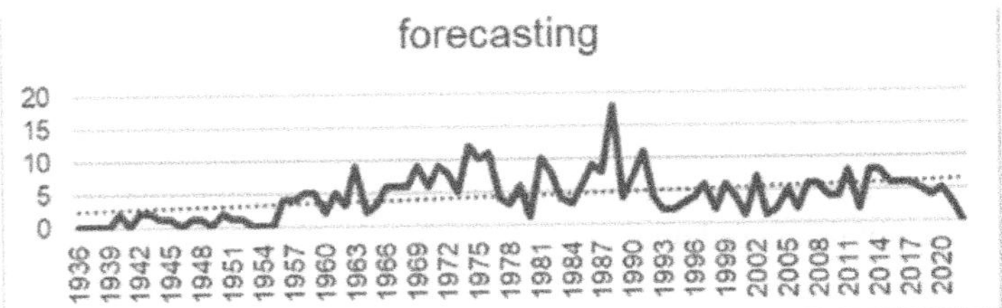

Figure 7.41. *Time plot forecasting*

Forecasting has been a major topic in SCM. As the author sees SCM to be the third job of marketing (deliver value to the customer), the theme, regardless of the artificial boundaries created by the academic reductionist approaches for SCM to be separated from marketing, the author sees this differently. While seen from the operations and efficiency management side, the importance of forecasting cannot be underestimated; the main peaks were observed during the late 1980s, with some declining interest after that.

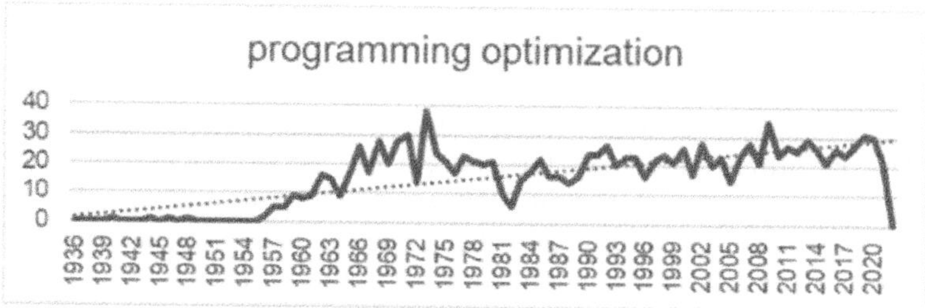

Figure 7.42. *Time plot programming optimization*

The essence of programming optimization greatly rose in academic interest in the 1960s, peaking in 1973 and 2009. This is due to the huge interest in AI and algorithmic modeling of stochastic models and networks.

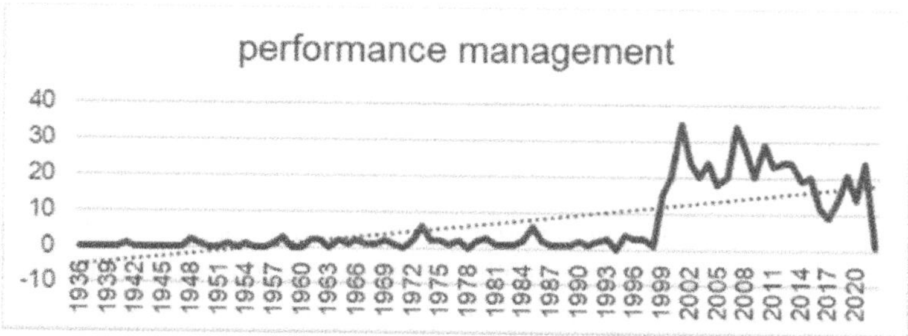

Figure 7.43. *Time plot performance management*

The theme of performance management had a huge peak in academic interest between 2001 and 2007. While some decline was sensed during the early 2000s, additional interest was sensed in 2021, when many themes such as performance, quality, technology and manufacturing were of interest.

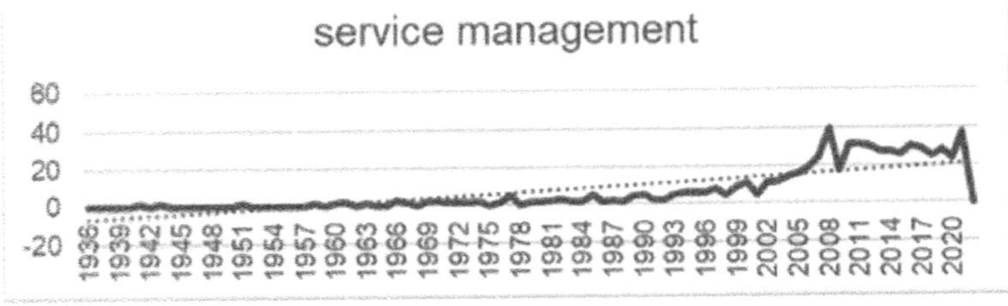

Figure 7.44. *Time plot service management*

The essential interest in the theme "service" was launched by the publication of [VAR 04]. The article introduces the concept of "Service-Dominant Logic." An update in continuing the evolution of SDL was introduced in 2008, which brought the broader dimension of service into the academic marketing discourse. So far, SDL could be regarded as a major theoretical framework adding to the body of knowledge of marketing and, thus, its cognitive identity.

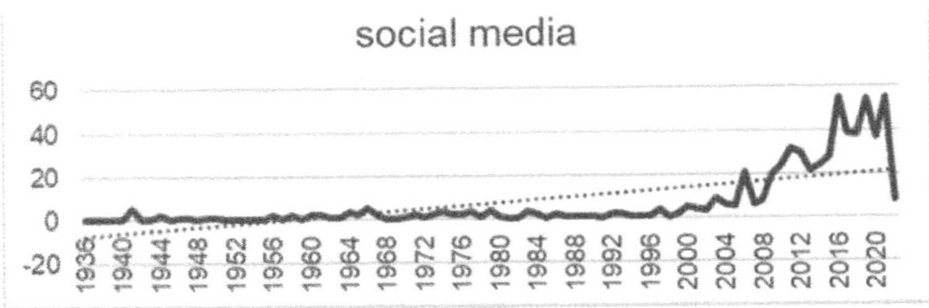

Figure 7.45. *Time plot social media*

The birth of social media marketing can be traced to February 4, 2004, when the firm went online. The first peaks in publications can be sensed in 2006, 2012, 2019 and 2021. Social media marketing has been essential to the growth and birth of many companies, entrepreneurial endeavors and jobs, and it could be considered one of the revolutionary moments in channel management of marketing thought.

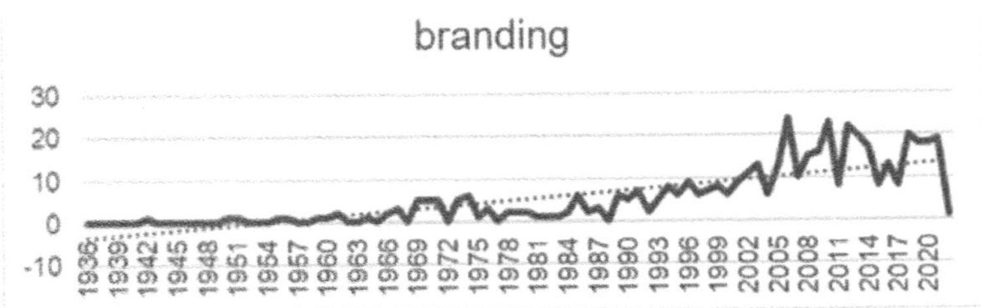

Figure 7.46. *Time plot branding*

Branding is a major foundation of marketing, wherein the ubiquity of quality, trust and continuity of service in loyalty is ubiquitous. Branding has had many smaller peaks. However, diverse research contributions peaked in the early 2000s, with a brief decline in 2007 and additional peaks in 2010, 2018 and 2021.

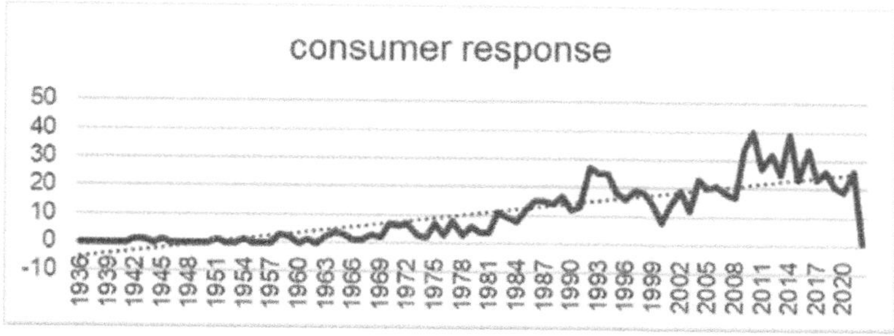

Figure 7.47. *Time plot consumer response*

The essence of good marketing planning and campaigns is to collect solid feedback based on the dimension of consumer response. The effects of advertisement, the processing of the information collected and redistributing the corrected iterative actions toward the marketing objectives are the keys to firms' success. A brief peak in publications could be sensed in 1992, with a decade of decline and a rise again in 2010 and 2014. Since then, some decline has been noted.

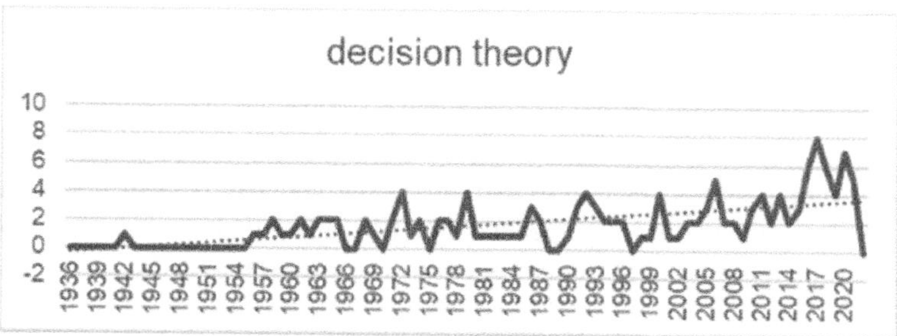

Figure 7.48. *Time plot decision theory*

Within the notion of decision theory, we observe decision-making, bias, location and facility. Academic interest in this topic has been constant, although a peak could be sensed in 2017 and 2020.

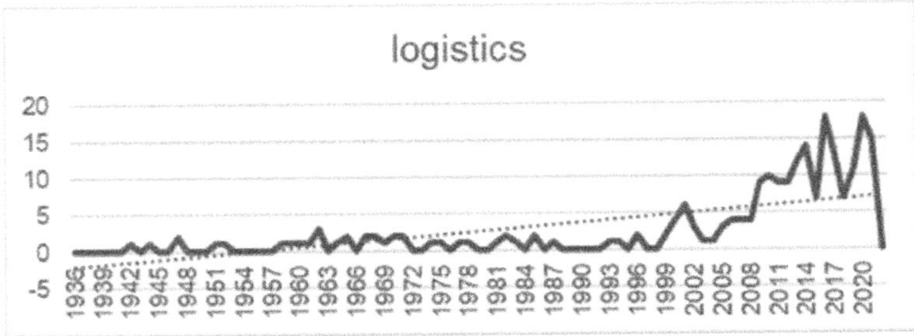

Figure 7.49. *Time plot logistics*

As SCM has been the third pillar of marketing, logistics has been the foundation of the rise of marketing and, thus, management science. A peak was observed in 2001 but also a constant surge in 2010, 2014, 2016 and 2020.

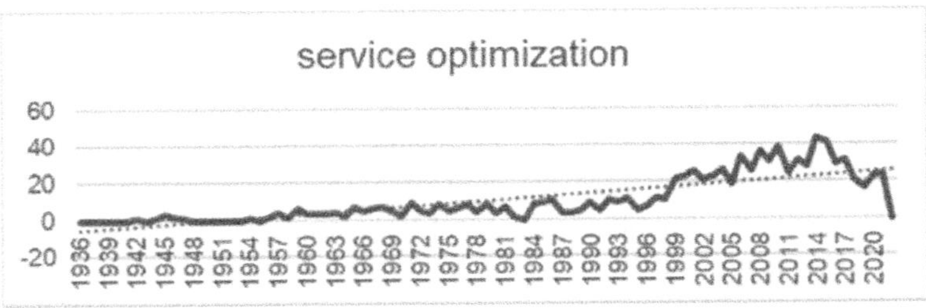

Figure 7.50. *Time plot service optimization*

The topic of service optimization could be attributed to the general efficiency movement, which started in the 1970s and 1980s and has peaked since the early 2000s. While academic interest peaked in 2010 and 2014, some decline can be observed within the marketing-related literature we have analyzed.

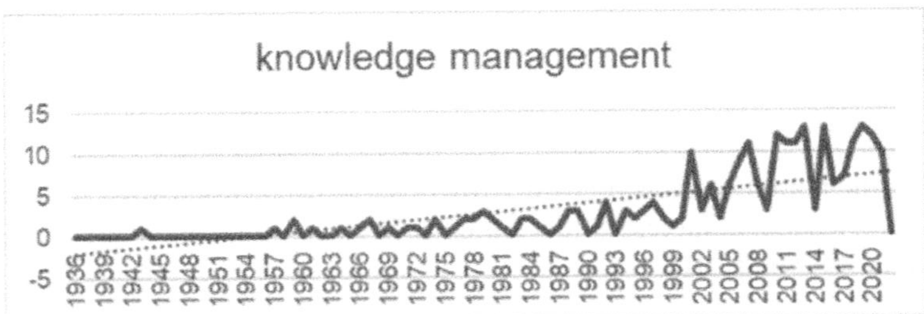

Figure 7.51. *Time plot knowledge management*

Knowledge management has been present throughout the marketing literature. Solid peaks in academic interest could be seen since 2001. In later years, this interest has not declined but, to the contrary, has peaked.

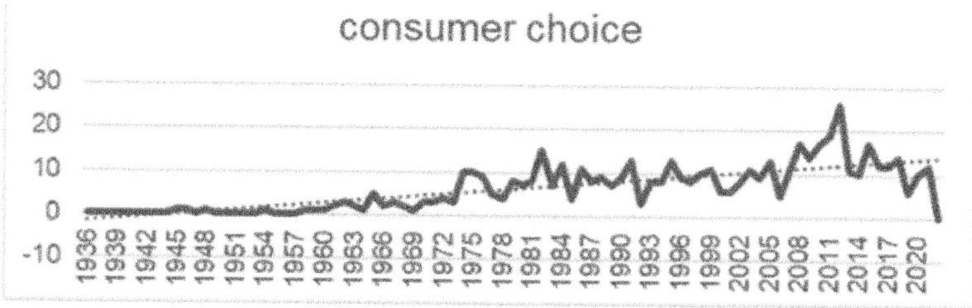

Figure 7.52. *Time plot consumer choice*

Consumer choice as a foundational theme within marketing literature has steadily gained academic interest. We can observe many peaks, especially in 1975, 1982 and 2012.

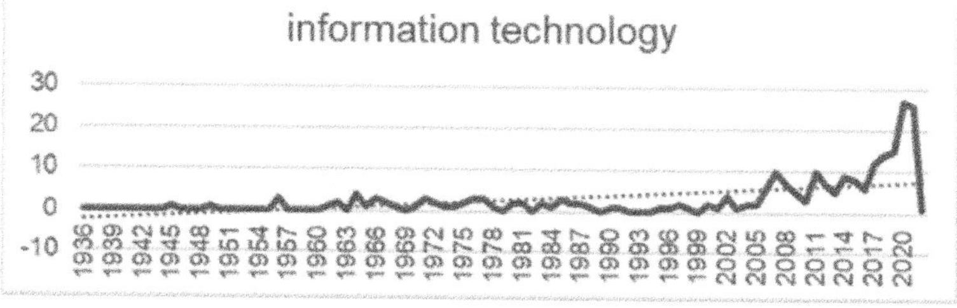

Figure 7.53. *Time plot information technology*

Contrary to general belief, the notion of information technology has had a rocky adventure throughout the marketing literature. The first marginal peaks were observed in 2007, 2011 and 2020, when much attention was paid to the theme.

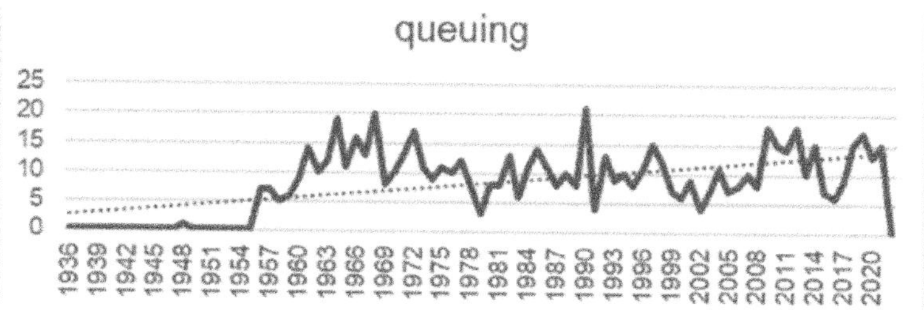

Figure 7.54. *Time plot queuing*

Service management systems and queuing have been influential throughout the broader marketing literature. Major peaks could be observed in the mid- and late 1960s, early 1990s and throughout the 2000s.

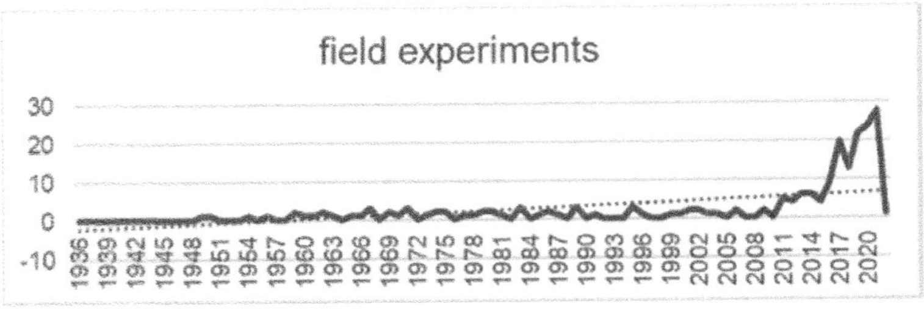

Figure 7.55. *Time plot field experiments*

The notion of field experiments observed here derives from the development of mobile technologies, where academic interest mainly began in the early 2020s.

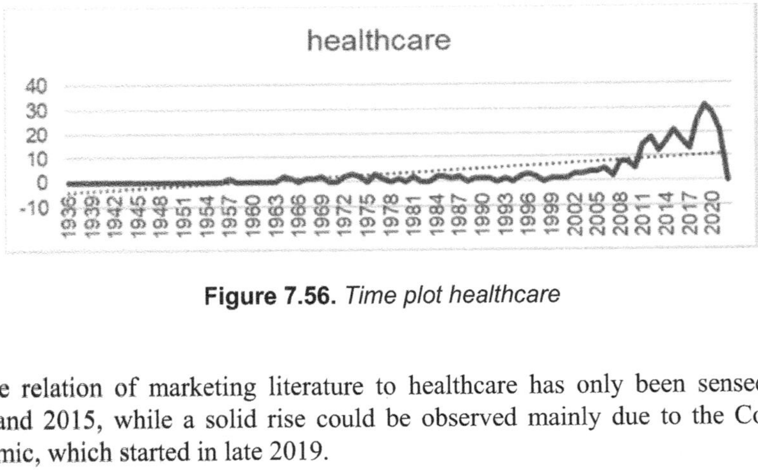

Figure 7.56. *Time plot healthcare*

The relation of marketing literature to healthcare has only been sensed since 2012 and 2015, while a solid rise could be observed mainly due to the Covid-19 pandemic, which started in late 2019.

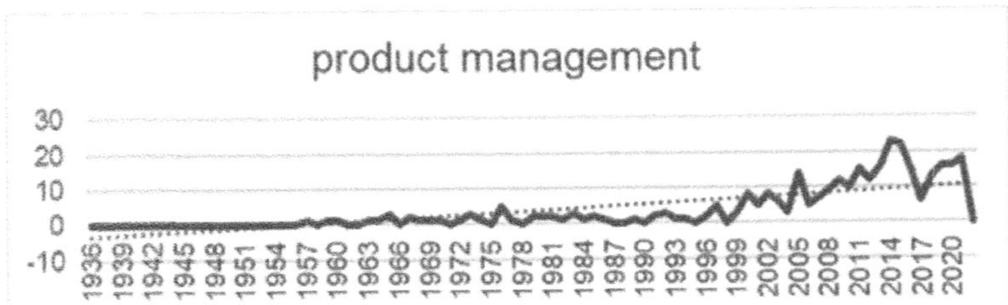

Figure 7.57. *Time plot product management*

Product management has always been vital to the evolution of marketing. However, as the themes of management, risk, design and project discussed within this topic illustrate, much attention has been paid to them in 2000, 2011, 2014 and again in 2021.

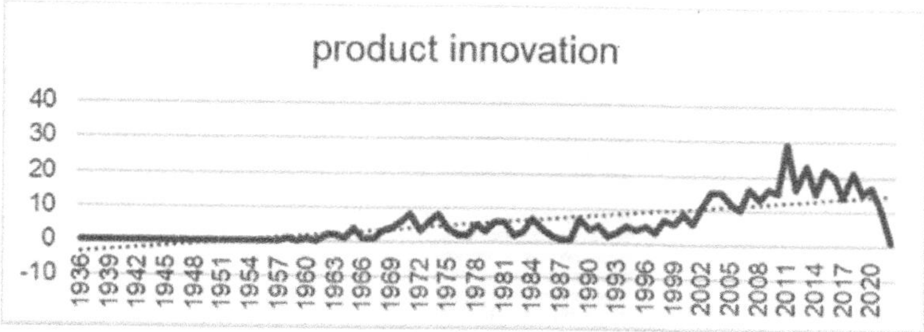

Figure 7.58. *Time plot product innovation*

Attention to product innovation was first observed in the early 1970s and has been on a constant rise since the 2000s. It peaked in 2011 and has declined slightly since then, but it has maintained constant academic interest in these discussed themes.

The time series illustrations account for the holistic history of marketing and the multiple streams of topics that were (and have been) relevant within the field.

We see that most of the topics are based on "Marketing Science", "Consumer Behavior", "Organizational Performance", "Programming and Optimization", "Dynamic Competition", "Consumer Response", "Retailing", etc.

While additional themes, as illustrated in a single figure below, demonstrate that many topics are relevant, we still see the dependence of marketing to CB as the second most essential theme to which 1,516 articles were assigned, while marketing science captures just a few more with 1,628 articles.

Here, marketing is possibly better served to put some of the essential intellectual output into the themes of the Metaverse as the essence of this work. Hence, marketing must cope with the digital and not leave the developments within the digital field to chance.

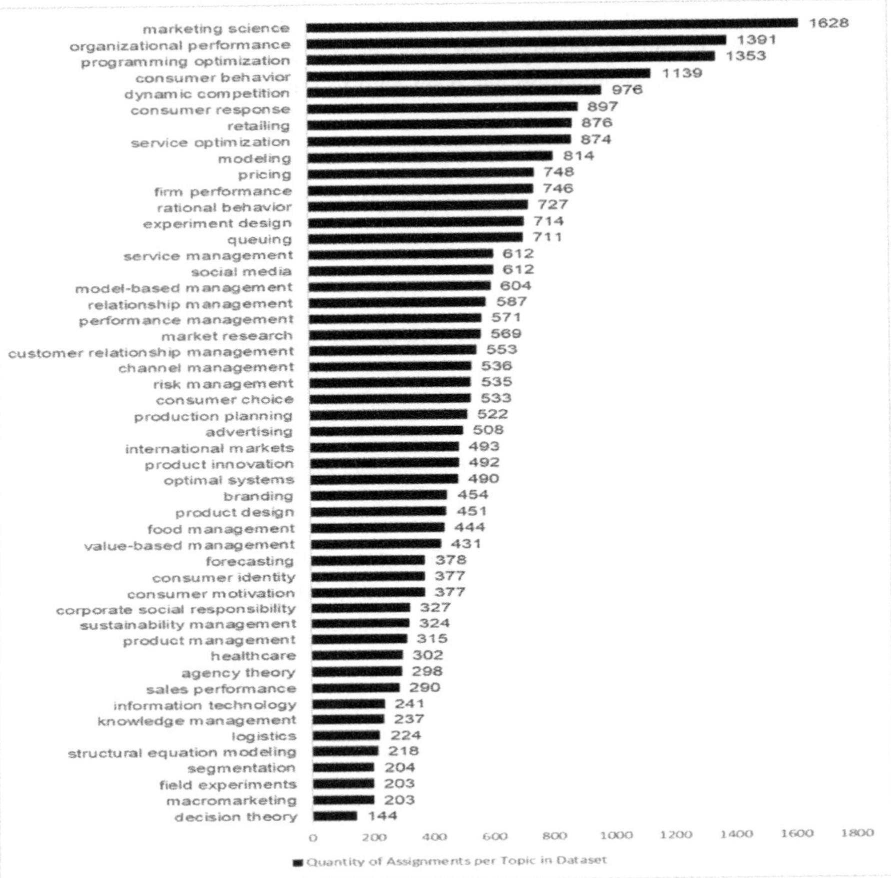

Figure 7.59. *Number of articles assigned to topic*

Investigating the evolution of the total marketing literature in all 17 top journals over time, we see a clear increase from the millennium to the 2020s. In contrast, the drop in the last period stems from the fact that they collected the data at the beginning of 2022, resulting in a significantly shorter period for the last time bucket compared to two years in total for the remaining time periods.

In conclusion, research in marketing is emerging from its relatively low level at the end of the millennium to steadily increasing publication activity over the years.

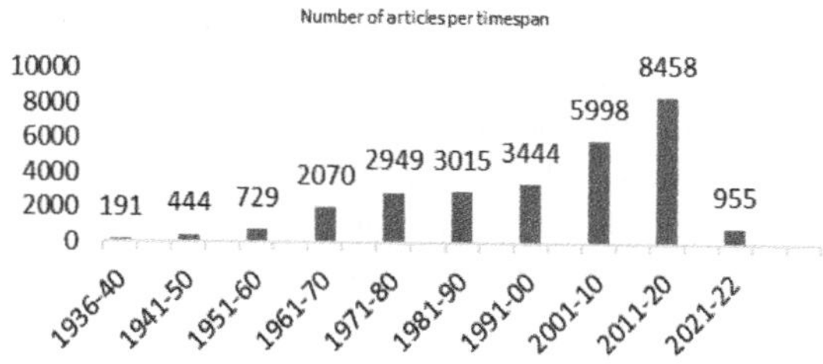

Figure 7.60. *Number of articles per time span*

7.2.5. *Contributing authors*

The study by [BER 06] found that the author's characteristics have the most explanatory power on an article's impact. They concluded that researchers who published the most in a period had the most substantial impact on the themes studied during the following period. Therefore, we analyzed the most published authors in marketing to better understand the field's past evolution and possible future directions.

The author identified and ranked those authors whose work has appeared most frequently in the 17 Journals. More than 32,199 different authors have written the 28,253 articles. Among these authors, 2,607 (8.1%) published five articles or more, 1,143 (3.5%) published four articles, 2,068 (6.4%) published three articles, 4,819 (15.0%) published two articles and 21,562 (67.0%) published only one article. Thus, only 33.8% of all authors have published more than one paper in all of the 17 journals. These proportions are comparable to those found by [FUR 08] in strategic management and [HEC 88] in the financial literature.

Table 7.3 shows a ranking of the top 40 authors based on both the total number of appearances and adjusted appearances, which reflect multiple-authored articles. If two writers collaborated on an article, each earned half of the credit; if three writers collaborated, each earned one-third of the credit, etc.; 7,036 (24.9%) papers were written by one author, 10,442 (37.0%) by two authors, 7,728 (27.4%) by three authors, 2,289 (8.1%) by four authors, 521 (1.8%) by five authors and 237 (0.8%) by six authors or more.

The five most prolific authors were Viswanathan Kumar, with 27.39 adjusted appearances and 73 total appearances; Dhruv Grewal, with 17.01 adjusted appearances and 62 total appearances; Christopher S. Tang, with 20.35 adjusted appearances and 53 total appearances; Christian Homburg, with 18.42 adjusted appearances and 48 total appearances; and Donald R. Lehmann, with 18.44 adjusted appearances and 44 total appearances.

This analysis emphasizes the contributions of researchers over an 86-year period by distinguishing between long-established and new generations of scholars. This latter group will undoubtedly and increasingly be tasked with defining the marketing field's future directions. As an author's number of published papers is proportional to the duration of their career, the writers at the top of our list also have the longest careers; as a result, they have had the most effect on the structure and evolution of the marketing field.

Rank	Author	Appearances	Adjusted Appearances
1	Kumar, Viswanathan	73	27.39
2	Grewal, Dhruv	62	17.01
3	Tang, Christopher S.	53	20.35
4	Homburg, Christian	48	18.42
5	Lehmann, Donald R.	44	18.44
6	van Wassenhove, Luk N.	44	15.11
7	Mahajan, Vijay	40	15.79
8	Hunt, Shelby D.	40	29.47
9	Dawande, Milind	39	11.02
10	Green, Paul E.	39	19.85
11	Bagozzi, Richard P.	38	21.09
12	Chen, Ying-Ju	36	15.12
13	Hauser, John R.	36	16.81
14	Ruyter, Ko de	35	8.81
15	Janiszewski, Chris	35	16.46
16	Srinivasan, V.	34	15.76
17	Chintagunta, Pradeep K.	34	14.33
18	Whitt, Ward	34	23.54
19	Staelin, Richard	34	14.37
20	Bearden, William O.	33	12.69
21	Kouvelis, Panos	33	13.56
22	Shugan, Steven M.	31	23.01
23	Krishna, Aradhna	31	13.49
24	Monroe, Kent B.	30	13.46
25	Dekimpe, Marnik G.	29	11.79
26	Hoyer, Wayne D.	29	12.23
27	Rust, Roland T.	29	12.92
28	Parasuraman, A.	29	11.99
29	Netessine, Serguei	28	12.10
30	Janakiraman, Ganesh	28	9.08
31	Federgruen, A.	28	11.70

32	Sudhir, K.	28	14.18
33	Simchi-Levi, David	28	11.10
34	Lee, Hau L.	28	12.50
35	Shen, Zuo-Jun Max	28	12.15
36	Eliashberg, Jehoshua	27	12.16
37	Song, Jing-Sheng	27	11.99
38	Lilien, Gary L.	27	12.04
39	Mitchell, Will	27	12.01
40	Verhoef, Peter C.	27	8.34

Table 7.3. *Most published authors in absolute number of appearances versus adjusted appearance accounting for articles with multiple authors*

7.2.6. *Most influential papers*

Certain publications have played pivotal roles in the growth of every scientific subject. Due to their impact, these papers accelerate the field's growth [BER 93]. To better comprehend the future possibilities of the marketing discipline, it is necessary to determine the most prominent papers published in the 17 journals between 1936 and 2022. This research used the commonly accepted approach of summed citation counts to identify and assess the effect or impact of research work [TAH 99]; [RAM 04]; [BER 06].

Rank	Authors	Year	Journal	Citations	Citations per year
1	Fornell and Larcker	1981	Journal of Marketing Research	44,198	1,078
2	Verhoef et al.	2021	Journal of Consumer Research	773	773
3	Carpenter	2017	Academy of Marketing Science Review	3,844	769
4	Henseler et al.	2015	Journal of the Academy of Marketing Science	3,312	473
5	Zhao et al.	2010	Journal of Consumer Research	4,814	401
6	Morgan and Hunt	1994	Journal of Marketing	9,080	324
7	Bagozzi and Yi	1988	Journal of the Academy of Marketing Science	10,457	308

8	Bies et al.	2021	International Journal of Research in Marketing	278	278
9	Parasuraman et al.	1988	Journal of Retailing	8,702	256
10	Cassiman and Veugelers	2016	Academy of Marketing Science Review	1,398	233
11	Vargo and Lusch	2008	Journal of the Academy of Marketing Science	3,200	229
12	Hair et al.	2012	Journal of the Academy of Marketing Science	2,230	223
13	Armstrong and Overton	1977	Journal of Marketing Research	9,534	212
14	Keller	1993	Journal of Marketing	5,737	198
15	Parasuraman et al.	1985	Journal of Marketing	7,319	198
16	Zeithaml	1988	Journal of Marketing	6,543	192
17	Zeithaml et al.	1996	Journal of Marketing	4,879	188
18	Lemon and Verhoef	2016	Journal of Marketing	1,110	185
19	Thaler	2008	Marketing Science	2,550	182
20	Churchill	1979	Journal of Marketing Research	7,777	181
21	Zervas et al.	2017	Journal of Marketing Research	864	173
22	Vargo and Lusch	2016	Journal of the Academy of Marketing Science	1,011	169
23	Chevalier and Mayzlin	2006	Journal of Marketing Research	2,676	167
24	Oliver	1999	Journal of Marketing	3,819	166
25	Teece	1992	Journal of Marketing	4,493	150
26	Huselid	1995	Academy of Management Journal	4,038	150
27	Jarvis et al.	2003	Journal of Consumer Research	2,804	148
28	Cronin et al.	2000	Journal of Retailing	3,063	139
29	Norton	2021	Journal of Consumer Research	138	138
30	Fournier	1998	Journal of Consumer Research	3,249	135

31	Fiss	2011	Academy of Management Journal	1,468	133
32	Tsai and Ghoshal	1998	Academy of Management Journal	3,156	132
33	Cohen and Whang	2021	Operations Research	131	131
34	Narver and Slater	1990	Journal of Marketing	4,170	130
35	Ganesan	2021	Journal of Marketing	130	130
36	Doney and Cannon	1997	Journal of Marketing	3,228	129
37	Debo et al.	2018	Academy of Marketing Science Review	508	127
38	Bertsimas and Sim	2004	Operations Research	2,270	126
39	Kempf and Smith	2020	Academy of Marketing Science Review	252	126
40	Rich et al.	2010	Academy of Management Journal	1,499	125

Table 7.4. *Most influential papers ordered by citations per year to account for the time horizon of long-established researchers*

The most significant papers published in the 17 journals were determined based on the number of times they appeared in the *Web of Science* and *CrossRef* databases, accessed on the *Springer Verlag* website. Because a paper published earlier has a greater chance of receiving more citations than one published later, the articles were sorted according to the number of citations divided by the number of years they were published, thereby following the procedure displayed in [FUR 08]. The most influential works are listed in Table 7.4 with their number of citations and citations per year.

7.3. Conclusions

This chapter aimed at illustrating the structure and past evolution of the content of the marketing field as well as to find possible future directions for the development of marketing thought deriving from the author's in-depth analysis of all of the publications, based on 17 top journals relating to the field. Within this chapter, the author performed various unsupervised machine learning algorithms such as LDA, MCA and CA to detect major topics in the marketing research literature and then present the findings dynamically. In this manner, it is expected to have obtained a less biased, i.e. less subjective, view on marketing discipline. Indeed, a history of almost 80 years is a sufficient time to reflect on the maturity of the marketing discipline constructed here. Within contemporary technological advances, this research methodology was also a unique combination of manual scholarly research combined with the support of AI-driven capability (machine-learning using KNIME Analytics Platform) to embrace much larger sets of articles for more advanced and rigorous research findings. Based on the analysis of 28,253 articles for this study, the theoretical contributions could be reported in the following manner:

The research identified the major research streams within the top journals during their lifetime. The author furthermore studied the relationships between the subfields of marketing, therefore accounting for the structural evolution of the field since its foundation over time.

Changes pertaining to the evolution of the field based on this research indicate the essential timelines, wherein the rise and decline of the individual stream of research were embedded. While the author also illustrates the essential peaks and, furthermore, the decline and rise of the themes, the distinguished reader is referred to the essential topics established, based on the multiple plots illustrated within the diverse topics established above. To answer the second question, the top five list of the most influential papers in absolute number of appearances versus adjusted appearance accounting for articles with multiple authors were occupied by:

1) Kumar Viswanthan; 2) Gewal Dhruv; 3) Christopher Tang; 4) Christian Homburg and 5) Donald Lehman. The top five most essential papers published based on total citation and per year citation index are: 1) [FOR 81]; 2) [VER 21]; 3) [CAR 17]; 4) [HEN 15] and 5) [ZHA 10], which encompass a total of 56,941 citations and 3,494 citations per year. [VER 21] may be an outlier here, and this may change as this publication ages.

7.4. Discussion

While this research confirms [YAD 10, 20]'s observations on the nature of the decline of the conceptual articles and the lack of indigenous themes and theory development in marketing [HUN 18], this chapter still provides evidence that the contributions of conceptual-only articles are essential in addition to the quantitative papers. Here, a call for action for the peers could be shown, which is that there is an apparent necessity for additional theoretical contributions which would deliver solid impulses to the quantitative-based and consumer-behavior-driven research to expand their themes in the image of foundational marketing aspirations such as the dimensions of establishing sustainability. [LUN 18] and mindful consumption

[SHE 11], coping and framing an AI-value definition of marketing, and defining solid organizational theories for managing international companies so that effective marketing responses could be designed and that the theoretical limitations of the marketing field [HUN 20a]; [KAM 21]; [SHE 21] are overcome. The model developed by [YAD 14] is a possible path going forward, wherein the role of the doctoral programs has been observed as pivotal and topologically positioned at the locus of innovation and frontier of viability for a discipline. Figure 7.60 illustrates that evolution of knowledge development is rather a time-consuming facture within a discipline. This process of a careful and deliberative modus operandi is the essence of disciplines because it ensures that short-lived approaches and developments are integrated with care or discarded [YAD 14].

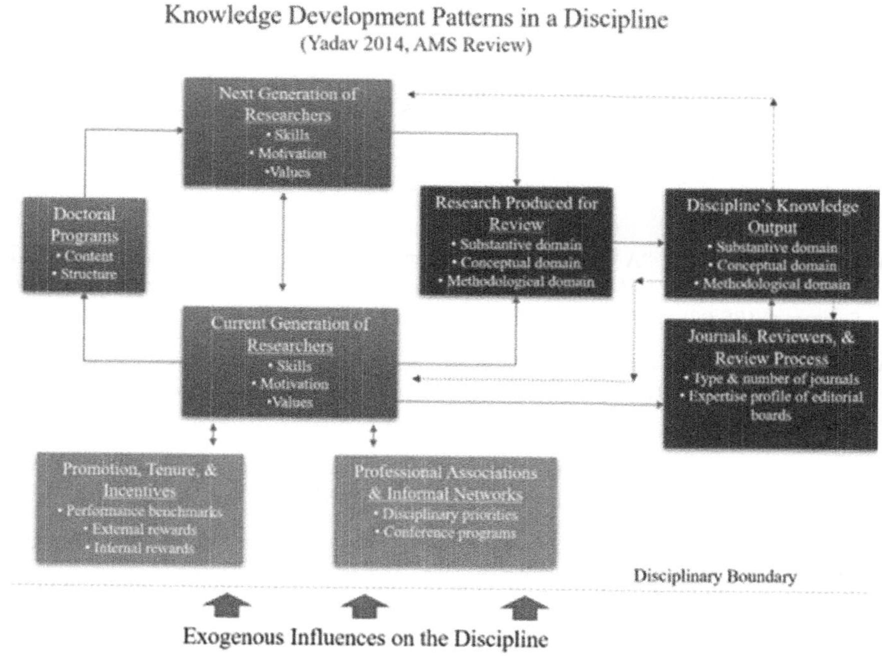

Figure 7.61. *Determinants of the type of knowledge produced by a discipline (source: an interpretation of [YAD 14]. For a color version of this figure, see www.iste.co.uk/machado/industry.zip*

These are sufficient reasons why it is essential that marketing establishes itself as a field of autopoetic sciences, thus to deliver a foundational and integrated framework for the field and to cope with contemporaneous technological development and societally shifting forces by developing more foundational grounds toward more viable indigenous theory developing capacities within the marketing doctoral schools.

7.5. References

[ALD 06a] ALDERSON W., "Marketing efficiency and the principle of postponement", in WOOLISCROFT B. (ed.), *A Twenty-First Century Guide to Aldersonian Marketing Thought*, Springer, New York, Heidelberg, pp. 109–113, 2006.

[ALD 06b] ALDERSON W., "The analytical framework for marketing", in WOOLISCROFT B. (ed.), *A Twenty-First Century Guide to Aldersonian Marketing Thought*, Springer, New York, Heidelberg, pp. 61–73, 2006.

[ARU 10] ARUN R., SURESH V., VENI MADHAVAN C.E. et al., "On finding the natural number of topics with latent Dirichlet allocation: Some observations", in ZAKI M., YU J.X., RAVINDRAN B., PUDI V. (eds), *Pacific-Asia Conference on Knowledge Discovery and Data Mining*, Springer, Berlin, Heidelberg, pp. 391–402, available at: https://link.springer.com/chapter/10.1007/978-3-642-13657-3_43, 2010.

[ASH 56] ASHBY W.R., *An Introduction to Cybernetics*, John Wiley & Sons, New York, 1956.

[ATT 99] ATTIAS H., "A variational Bayesian framework for graphical models", in *Proceedings of the 12th International Conference on Neural Information Processing Systems (NIPS'99)*, MIT Press, Cambridge, pp. 209–215, 1999.

[ATT 00] ATTIAS H., PLATT J., ACERO A. et al., "Speech denoising and dereverberation using probabilistic models", *Advances in Neural Information Processing Systems*, vol. 13, 2000.

[BAG 88] BAGOZZI R.P., YI Y., "On the evaluation of structural equation models", *Journal of the Academy of Marketing Science*, vol. 16, pp. 74–94, 1988.

[BAR 62] BARTELS R., SHAW E.H., TAMILIA R.D., "The development of marketing thought", *Journal of Marketing*, vol. 26, no. 2, pp. 29–33, 1962.

[BAR 67] BARTELS R., "A model for ethics in marketing", *Journal of Marketing*, vol. 31, no. 1, pp. 20-26, 1967.

[BAR 74] BARTELS R., "The identity crisis in marketing: Is marketing a specific function with general applicability or a general function that is specifically applied?", *Journal of Marketing*, vol. 38, no. 4, pp. 73–76, 1974.

[BEE 59] BEER S., *Cybernetics and Management*, John Wiley & Sons, New York, 1959.

[BEE 95] BEER S. (ed.), "Brain of the firm", in *The Managerial Cybernetics of Organization*, John Wiley & Sons, Chichester, 1995.

[BEL 20] BELK R., "Resurrecting marketing", *AMS Review*, vol. 10, nos 3–4, pp. 168–171, available at: https://link.springer.com/article/10.1007/s13162-020-00182-9, 2020.

[BEN 82] BENZECRI J.-P., *L'analyse des données. Tome 2 : l'analyse des correspondances*, Dunod, Paris, 1982.

[BER 69] BERTALANFFY L.V., "General systems theory and psychiatry—An overview", *Psychiatry*, vol. 32, no. 4, pp. 314–329, 1969.

[BER 93] BERRY L.L., PARASURAMAN A., "Building a new academic field—The case of services marketing", *Journal of Retailing*, vol. 69, no. 1, pp. 13–60, 1993.

[BER 06] BERGH D.D., PERRY J., HANKE R., "Some predictors of SMJ article impact", *Strategic Management Journal*, vol. 27, no. 1, pp. 81–100, 2006.

[BIS 12] BISCHOF J., AIROLDI E.M., "Summarizing topical content with word frequency and exclusivity", *ArXiv*, abs/1206.4631, 2012.

[BLE 03] BLEI D.M., NG A.Y., JORDAN M.I., "Latent Dirichlet allocation", *Journal of Machine Learning Research*, vol. 3, pp. 993–1022, 2003.

[BLE 12] BLEI D.M., "Probabilistic topic models", *Communications of the ACM*, vol. 55, no. 4, pp. 77–84, 2012.

[BOO 78] BOORSTIN D.J., *Republic of Technology: Reflections on our Future Community*, Harper and Row, Publishers, Inc., New York, 1978.

[BOR 64] BORDEN N.H., "The concept of the marketing mix", *Journal of Advertising Research*, vol. 4, no. 2, available at: https://motamem.org/wp-content/uploads/2019/07/borden-1984_the-concept-of-marketing-mix.pdf, 1964.

[BRI 14] BRINKER S., MCLELLAN L., "The rise of the chief marketing technologist", *Harvard Business Review*, vol. 92, no. 7, pp. 82–85, available at: https://dialnet.unirioja.es/servlet/articulo?codigo=4748978, 2014.

[CAO 09] CAO J., XIA T., LI J. et al., "A density-based method for adaptive LDA model selection", *Neurocomputing*, vol. 72, nos 7–9, pp. 1775–1781, 2009.

[CAR 17] CARPENTER B., GELMAN A., HOFFMAN M.D. et al., "Stan: A probabilistic programming language", *Journal of Statistical Software*, vol. 76, no. 1, pp. 1–32, doi: 10.18637/jss.v076.i01, 2017.

[CHA 22] CHALMERS D.J., *Reality+. Virtual Worlds and the Problems of Philosophy*, 1st edition, W.W. Norton & Company, New York, 2022.

[CHR 08] CHRISTENSEN C.M., *The Innovator's Dilemma. When New Technologies Cause Great Firms to Fail*, Harvard Business School Press, Boston, 2008.

[CHR 16a] CHRISTENSEN C.M., *Competing Against Luck. The Story of Innovation and Customer Choice*, HarperCollins Publishers, New York, 2016.

[CHR 16b] CHRISTENSEN C.M., "Know your customers' 'jobs to be done'", *Harvard Business Review*, available at: https://hbr.org/2016/09/know-your-customers-jobs-to-be-done [Accessed 01 August 2022], 2016.

[CLA 14] CLARK T., KEY T., HODIS M. et al., "The intellectual ecology of mainstream marketing research: An inquiry into the place of marketing in the family of business disciplines", *Journal of the Academy of Marketing Science*, vol. 42, no. 3, pp. 223–241, 2014.

[DIC 83] DICKEY J.M., "Multiple hypergeometric functions: Probabilistic interpretations and statistical uses", *Journal of the American Statistical Association*, vol. 78, pp. 628–637, doi: 10.1080/01621459.1983.10478022, 1983.

[EIG 76] EIGLIER P., LANGEARD E., Principes de politique marketing pour les entreprises de services, Centre d'Études et de Recherche sur les Organisations, Université d'Aix-Marseille, 1976.

[EIS 15] EISEND M., "Have we progressed marketing knowledge? A meta-meta-analysis of effect sizes in marketing research", *Journal of Marketing*, vol. 79, no. 3, pp. 23–40, 2015.

[ELA 18] EL-ANSARY A., SHAW E.H., LAZER W., "Marketing's identity crisis: Insights from the history of marketing thought", *AMS Review*, vol. 8, nos 1–2, pp. 5–17, available at: https://link.springer.com/article/10.1007/s13162-017-0102-y, 2018.

[FOR 81] FORNELL C., LARCKER D.F., "Evaluating structural equation models with unobservable variables and measurement error", *Journal of Marketing Research*, vol. 18, no. 1, pp. 39–50, 1981.

[FUR 08] FURRER O., THOMAS H., GOUSSEVSKAIA A., "The structure and evolution of the strategic management field: A content analysis of 26 years of strategic management research", *International Journal of Management Reviews*, vol. 10, no. 1, pp. 1–23, 2008.

[GAR 55] GARDNER B.B., LEVY S.J., "The product and the brand", *Harvard Business Review*, vol. 33, no. 2, pp. 33–39, 1955.

[GEE 73] GEERTZ C., *The Interpretation of Cultures*, Basic Books, New York, 1973.

[GRA 11] GRANT A.M., POLLOCK T.G., "Publishing in AMJ — Part 3: Setting the Hook", *Academy of Management Journal*, vol. 54, no. 5, pp. 873–879, 2011.

[GRE 76] GRETHER E.T., "The first forty years", *Journal of Marketing*, vol. 40, no. 3, pp. 63–69, 1976.

[GRÖ 82] GRÖNROOS C., "An applied service marketing theory", *European Journal of Marketing*, vol. 16 no. 7, pp. 30–41, available at: https://www.emerald.com/insight/content/doi/10.1108/eum0000000004859/full/pdf, 1982.

[GRÖ 06] GRÖNROOS C., "Adopting a service logic for marketing", *Marketing Theory*, vol. 6, no. 3, pp. 317–333, 2006.

[GUM 79] GUMMESSON E., "The marketing of professional services — An organizational dilemma", *European Journal of Marketing*, vol. 13, no. 5, pp. 308–318, available at: https://www.emerald.com/insight/content/doi/10.1108/eum0000000004951/full/pdf, 1979.

[GUM 08] GUMMESSON E., "Extending the service-dominant logic: From customer centricity to balanced centricity", *Journal of the Academy of Marketing Science*, vol. 36, no. 1, pp. 15–17, 2008.

[HEC 88] HECK J.L., COOLEY P.L., "Most frequent contributors to the finance literature", *Financial Management*, vol. 17, no. 3, p. 100, 1988.

[HEL 22] HELKKULA A., ARNOULD E., "Developing and renewing marketing as a scientific discipline through reflexive cocreation", *AMS Review*, vol. 12, no. 3, pp. 168–173, doi: 10.1007/s13162-022-00244-0, 2022.

[HEN 15] HENSELER J., RINGLE C.M., SARSTEDT M., "A new criterion for assessing discriminant validity in variance-based structural equation modeling", *Journal of the Academy of Marketing Science*, vol. 43, pp. 115–135, 2015.

[HOF 86] HOFFMAN D.L., FRANKE G.R., "Correspondence analysis: Graphical representation of categorical data in marketing research", *Journal of Marketing Research*, vol. 23, no. 3, pp. 213–227, 1986.

[HOF 94] HOFFMAN D.J., *Advanced Methods of Marketing Research*, Blackwell Publishers, Cambridge, 1994.

[HUN 15] HUNT S.D., "The theoretical foundations of strategic marketing and marketing strategy: foundational premises, R-A theory, three fundamental strategies, and societal welfare", *AMS Review*, vol. 5, nos 3–4, pp. 61–77, available at: https://link.springer.com/article/10.1007/s13162-015-0069-5, 2015.

[HUN 18] HUNT S.D., "Advancing marketing strategy in the marketing discipline and beyond: From promise, to neglect, to prominence, to fragment (to promise?)", *Journal of Marketing Management*, vol. 34, nos 1–2, pp. 16–51, 2018.

[HUN 20a] HUNT S.D., "For re-institutionalizing the marketing discipline in Era V", *AMS Review*, vol. 10, nos 3–4, pp. 189–198, 2020.

[HUN 20b] HUNT S.D., "Indigenous theory development in marketing: The foundational premises approach", *AMS Review*, vol. 10, nos 1–2, pp. 8–17, 2020.

[IAN 14] IANSITI M., LAKHANI K.R., "Digital ubiquity: How connections, sensors, and data are revolutionizing business", *Harvard Business Review*, vol. 92, no. 11, p. 19, available at: https://dialnet.unirioja.es/servlet/articulo?codigo=5544176, 2014.

[KAC 90] KACIAK E., LOUVIERE J., "Multiple correspondence analysis of multiple choice experiment data", *Journal of Marketing Research*, vol. 27, no. 4, p. 455, 1990.

[KAM 21] KAMRAN Q., TOPP S., HENSELER J., "Towards the co-evolution of food experience search spaces based on the design Weltanschauung model in food marketing", *Cogent Business & Management*, vol. 8, no. 1, article 1901643, 2021.

[KAM 22] KAMRAN Q., RESHANI A., TOPP S., "Towards a unified theory of the new market realities in cyber-physical design search spaces", in MACHADO C. (ed.), *Technological Challenges*, Springer, Cham, pp. 25–59, available at: https://link.springer.com/chapter/10.1007/978-3-030-98040-5_2, 2022.

[KEY 20] KEY T.M., CLARK T., FERRELL O.C. et al., "Marketing's theoretical and conceptual value proposition: Opportunities to address marketing's influence", *AMS Review*, vol. 10, nos 3–4, pp. 151–167, available at: https://link.springer.com/article/10.1007/s13162-020-00176-7, 2020.

[KEY 21] KEY T.M., CLARK T., FERRELL O.C. et al., "Re-institutionalizing marketing", *AMS Review*, vol. 11, nos 3–4, pp. 446–453, available at: https://link.springer.com/article/10.1007/s13162-021-00220-0, 2021.

[KOT 67] KOTLER P., *Marketing Management. Analysis, Planning, Implementation, and Control*, Prentice-Hall, Upper Saddle River, 1967.

[KOT 69] KOTLER P., LEVY S.J., "Broadening the concept of marketing", *Journal of Marketing*, vol. 33, no. 1, pp. 10–15, 1969.

[KOT 72] KOTLER P., "A generic concept of marketing", *Journal of Marketing*, vol. 36, no. 2, pp. 46–54, 1972.

[KOT 05] KOTLER P., "The role played by the broadening of marketing movement in the history of marketing thought", *Journal of Public Policy & Marketing*, vol. 24, no. 1, pp. 114–116, 2005.

[LEB 84] LEBART L., *Multivariate Descriptive Statistical Analysis: Correspondence Analysis and Related Techniques for Large Matrices*, Wiley, New York, 1984.

[LEV 79] LEVY S.J., KOTLER P., "Toward a broader concept of marketing's role in social order", *Journal of the Academy of Marketing Science*, vol. 7, no. 2, pp. 233–238, available at: https://journals.sagepub.com/doi/pdf/10.1177/009207037900700207, 1979.

[LEV 83] LEVITT T., "The globalization of markets", *Harvard Business Review*, vol. 61, no. 3, pp. 92–102, 1983.

[LEV 84] LEVITT T., "Marketing myopia", in FORD G.T. (ed.), *Marketing and the Library*, 1st edition, Haworth Press, New York, pp. 59–80, 1984.

[LEV 18] LEVY, S.J., "A marketing identity is flourishing," *AMS Review*, vol. 8, no. 1, pp. 18–19, 2018.

[LUC 69] LUCK D.J., "Marketing notes and communications", *Journal of Marketing*, vol. 33, no. 3, pp. 53–55, 1969.

[LUN 18] LUNDE M.B., "Sustainability in marketing: A systematic review unifying 20 years of theoretical and substantive contributions (1997–2016)", *AMS Review*, vol. 8, nos 3–4, pp. 85–110, 2018.

[MCC 78] MCCARTHY E.J., *Basic Marketing: A Managerial Approach*, 6th edition, Irwin, Homewood, 1978.

[MCC 02] MCCALLUM A., MALLET. A machine learning for language toolkit, available at: http://mallet.cs.umass.edu/ [Accessed 21 May 2021], 2002.

[MCK 57] MCKITTERICK J.B., "What is the marketing management concept?", in *Conference Proceedings of the American Marketing Association*, 1957.

[MED 21] MEDIUM, "Crypto is the New America?", *Very Brief Takes*, *Medium*, available at: https://verybrieftakes.medium.com/crypto-is-the-new-america-982972c1daea [Accessed 1 August 2022], 28 May 2021.

[MIM 11] MIMNO D., WALLACH H., TALLEY E. et al., "Optimizing semantic coherence in topic models", in *Proceedings of the 2011 Conference on Empirical Methods in Natural Language Processing*, pp. 262–272, available at: https://aclanthology.org/D11-1024/, 2011.

[MOO 19] MOORMAN C., HEERDE H.J., MOREAU C.P. et al., "Challenging the boundaries of marketing", *Journal of Marketing*, vol. 83, no. 5, pp. 1–4, doi: 10.1177/0022242919875305, 2019.

[MOR 94] MORGAN R.M., HUNT S.D., "The commitment-trust theory of relationship marketing", *Journal of Marketing*, vol. 58, no. 3, pp. 20–38, 1994.

[MOR 14] MORLEY J., LEIGH O., FLEMING W.F., *The Works of Voltaire: A Contemporary Version with Notes, Volume 12*, Legare Street Press, Paris, 2014.

[NEL 17] NELSON J., Definitions of marketing, available at: https://www.ama.org/the-definition-of-marketing-what-is-marketing/ [Accessed 01 August 2022], 2017.

[NEW 09] NEWMAN D., ASUNCION A., SMYTH P. et al., "Distributed algorithms for topic models", *Journal of Machine Learning Research*, vol. 10, no. 8, available at: https://www.jmlr.org/papers/volume10/newman09a/newman09a.pdf, 2009.

[PAR 21] PARVATIYAR A., SHETH J.N., "Toward an integrative theory of marketing", *AMS Review*, vol. 1, pp. 1–14, 2021.

[POS 93] POSTMAN N., *Technopoly*, Jim McBurney, Sunnyvale, 1993.

[POW 12] POWERS T.L., "Early schools of marketing thought and marketplace evolution", *Journal of Historical Research in Marketing*, vol. 4, no. 1, pp. 190–206, 2012.

[RAM 04] RAMOS-RODRÍGUEZ A.-R., RUÍZ-NAVARRO J., "Changes in the intellectual structure of strategic management research: A bibliometric study of the Strategic Management Journal, 1980–2000", *Strategic Management Journal*, vol. 25, no. 10, pp. 981–1004, 2004.

[REI 09] REIBSTEIN D.J., DAY G., WIND J., "Guest editorial: Is marketing losing its way?", *Journal of Marketing*, vol. 73, no. 1, pp. 1–3, 2009.

[RÖD 15] RÖDER M., BOTH A., HINNEBURG A., "Exploring the space of topic coherence measures", in *Proceedings of the Eighth ACM International Conference on Web Search and Data Mining (WSDM'15)*, ACM, pp. 399–408, doi: 10.1145/2684822.2685324, 2015.

[SHA 05] SHAW E.H., JONES D.B., "A history of schools of marketing thought", *Marketing Theory*, vol. 5, no. 3, pp. 239–281, 2005.

[SHA 09] SHAW E.H., "Reflections on the history of marketing thought", *Journal of Historical Research in Marketing*, vol. 1, no. 2, pp. 330–345, 2009.

[SHE 82] SHETH J.N., GARDNER D.M., *History of Marketing Thought: An Update*, American Marketing Association, Chicago, 1982.

[SHE 90] SHETH J.N., YOUNKIN N., HILLS G.E. et al., "Marketing theory: Evolution and evaluation", *Journal of Marketing*, vol. 54, no. 1, p. 138, 1990.

[SHE 11] SHETH J.N., SETHIA N.K., SRINIVAS S., "Mindful consumption: A customer-centric approach to sustainability", *Journal of the Academy of Marketing Science*, vol. 39, no. 1, pp. 21–39, 2011.

[SHE 21] SHETH J.N., "New areas of research in marketing strategy, consumer behavior, and marketing analytics: The future is bright", *Journal of Marketing Theory and Practice*, vol. 29, no. 1, pp. 3–12, 2021.

[SHO 77] SHOSTACK G.L., "Breaking free from product marketing", *Journal of Marketing*, vol. 41, no. 2, pp. 73–80, available at: https://journals.sagepub.com/doi/full/10.1177/002224297704100219, 1977.

[SIE 14] SIEVERT C., SHIRLEY K., "LDAvis: A method for visualizing and interpreting topics", in *Proceedings of the Workshop on Interactive Language Learning, Visualization, and Interfaces*, Association for Computational Linguistics, Stroudsburg, 2014.

[SKÅ 15] SKÅLÉN P., GUMMERUS J., KOSKULL C.V. et al., "Exploring value propositions and service innovation: A service-dominant logic study", *Journal of the Academy of Marketing Science*, vol. 43, no. 2, pp. 137–158, 2015.

[SMI 56] SMITH W.R., "Product differentiation and market segmentation as alternative marketing strategies", *Journal of Marketing*, vol. 21, no. 1, pp. 3–8, 1956.

[TAH 99] TAHAI A., MEYER M.J., "A revealed preference study of management journals' direct influences", *Strategic Management Journal*, vol. 20, no. 3, pp. 279–296, 1999.

[TAM 11] TAMILIA R.D., "Reflections on the history of marketing thought and theory development", *Marketing Theory*, vol. 11, no. 4, pp. 507–512, 2011.

[TEH 06] TEH Y.W., JORDAN M.I., BEAL M.J. et al., "Hierarchical Dirichlet processes", *Journal of the American Statistical Association*, vol. 101, no. 476, pp. 1566–1581, 2006.

[TEH 21] TEH C., "China is tempting customers with its flawless AI idols—Virtual influencers who don't gain weight, never age, and keep their computer-generated noses out of controversy", *Insider*, available at: https://www.insider.com/chinas-flawless-ai-influencers-the-hot-new-queens-of-advertising-2021-8 [Accessed 1 August 2022], 13 August 2021.

[VAR 04] VARGO S.L., LUSCH R.F., "Evolving to a new dominant logic for marketing", *Journal of Marketing*, vol. 68, no. 1, pp. 1–17, 2004.

[VAR 05] VARGO S.L., MORGAN F.W., "Services in society and academic thought: An historical analysis", *Journal of Macromarketing*, vol. 25, no. 1, pp. 42–53, 2005.

[VAR 08] VARGO S.L., LUSCH R.F., "Service-dominant logic: Continuing the evolution", *Journal of the Academy of Marketing Science*, vol. 36, no. 1, pp. 1–10, 2008.

[VAR 22] VARGO S.L., "Fostering scholarly discussion in marketing", *AMS Review*, vol. 12, pp. 137–138, doi: 10.1007/s13162-022-00250-2, 2022.

[VER 21] VERHOEF P.A., KANNAN S., STURGILL J.L. et al., "Severe acute respiratory syndrome–associated coronavirus 2 infection and organ dysfunction in the ICU: Opportunities for translational research", *Critical Care Explorations*, vol. 3, no. 3, e0374, 2021.

[WEB 13] WEBSTER F.E., LUSCH R.F., "Elevating marketing: Marketing is dead! Long live marketing!", *Journal of the Academy of Marketing Science*, vol. 41, no. 4, pp. 389–399, 2013.

[WIE 48] WIENER N., *Cybernetics or Control and Communication in the Animal and the Machine*, MIT Press, Cambridge, 1948.

[WIL 03] WILKIE W.L., MOORE E.S., "Scholarly research in marketing: Exploring the '4 eras' of thought development", *Journal of Public Policy & Marketing*, vol. 22, no. 2, pp. 116–146, 2003.

[YAD 10] YADAV M.S., "The decline of conceptual articles and implications for knowledge development", *Journal of Marketing*, vol. 74, no. 1, pp. 1–19, 2010.

[YAD 14] YADAV M.S., "Enhancing theory development in marketing", *AMS Review*, vol. 4, nos 1–2, pp. 1–4, 2014.

[YAD 20] YADAV M.S., "Reimagining marketing doctoral programs", *AMS Review*, vol. 10, nos 1–2, pp. 56–64, available at: https://link.springer.com/article/10.1007/s13162-020-00169-6, 2020.

[YAO 09] YAO L., MIMNO D., MCCALLAM A., "Efficient methods for topic model inference on streaming document collections", in *Knowledge Discovery and Data Mining*, 2009.

[ZHA 10] ZHAO X., LYNCH J.G. Jr, CHEN Q., "Reconsidering Baron and Kenny: Myths and truths about mediation analysis", *Journal of Consumer Research*, vol. 37, no. 2, pp. 197–206, 2010.

8

Industry 6.0: Why Talk About It Now?

The evolution of industrial revolutions has significantly shaped society and the economy, culminating in the transition from Industry 5.0 to the emerging Industry 6.0. This chapter addresses this historical path, highlighting some of the differences that Industry 6.0 has in relation to its predecessors. Focusing on integrating advanced technologies, such as artificial intelligence and hyperconnectivity, Industry 6.0 goes beyond human–machine collaboration, promoting environmental sustainability, human well-being and social responsibility as central pillars. This chapter seeks to take a brief look at how technological innovation drives this transformation, creating an industrial ecosystem that balances efficiency, ethics and sustainability to meet future demands.

8.1. Introduction

Industrial revolutions have catalyzed significant societal transformations, redefining how we produce, work and interact with technology. From the First Industrial Revolution, marked by mechanization and the use of steam in the 18th century, to the Fifth Industrial Revolution, which brought collaboration between humans and robots into the spotlight, each stage represented a crucial advance in the balance between innovation and human needs.

Currently, the transition from Industry 5.0 to Industry 6.0 ushers in a new era, where digital interconnection, advanced artificial intelligence and automation are directed to respond to even more ambitious objectives, namely, promoting sustainability, valuing human well-being and developing technologies centered on ethical and ecological values [CHO 22].

In this sense, Industry 6.0 stands out for going beyond the "simple" integration of humans and machines, typical of Industry 5.0, by seeking to incorporate broader

Chapter written by Carolina MACHADO and Catarina da Silva NOGUEIRA.

concepts of collective intelligence, hyper-connectivity and social responsibility. This model not only improves the efficiency and customization of processes but also prioritizes the creation of an industrial ecosystem that respects the environment and strengthens the quality of life.

In this chapter, we seek to address how Industry 6.0 differs from its predecessors, exploring the role of technological innovation as a driver of this transformation while highlighting the impact of sustainable practices centered on human well-being as fundamental pillars for industrial and societal evolution.

8.2. Historical context: from the First to the Fifth Industrial Revolution

Over the last few years, industry has faced numerous transformations caused by humans. The production of goods, such as clothing, food and weapons, as well as services was performed manually by humans. But everything changed with the appearance of new resources, which led to them reinventing existing techniques. These profound changes and their effects were especially striking for industry, so they were noted in history as revolutions [FRE 96; PIL 20].

According to Figure 8.1 (adapted from [DEM 19]; and [GEO 20]), the Industrial Revolutions were divided into five historical periods.

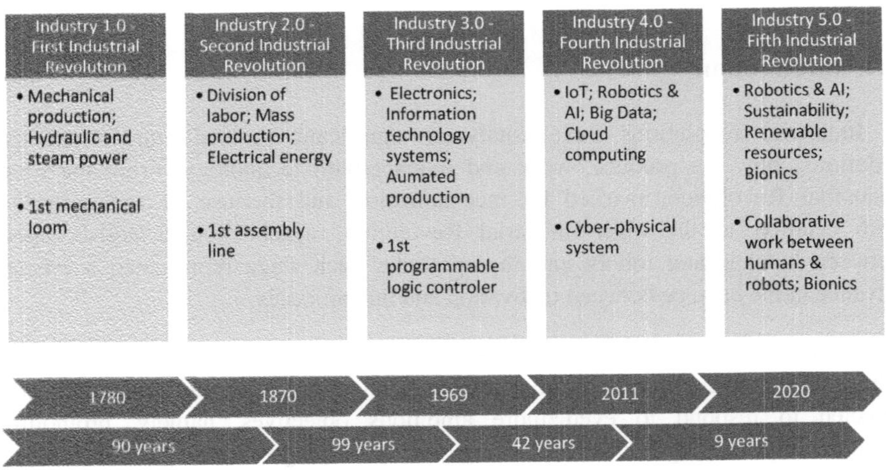

Figure 8.1. *From Industry 1.0 to Industry 5.0. (Adapted from [DEM 19] and [GEO 20])*

Starting in the 18th century, Industry 1.0 has become one of the most significant milestones in modern history. During this period, much of the production was done using manual force; i.e. it was manufactured and used a unitary production. However, with the beginning of the industrial revolution, this production began to be mechanized and in series, using steam and hydraulic energy, which led to, according to [GRO 21], eight times more volume simultaneously. This is considered by the author the most significant advance in increasing human productivity.

Furthermore, during this phase, the first steamships and locomotives appeared, allowing a reduction in the travel time of people, as well as goods over long distances, thus speeding up commercial exchanges [GRO 21].

The beginning of Industry 2.0, in the 19th century, was marked by the use of electricity instead of steam and hydraulic energy, as it was simpler and allowed the concentration of supply to a specific machine [GEO 20].

One of the pillars of Industry 2.0 is efficiency, as well as productivity and, in light of this, techniques were created, such as the division of labor, carried out en masse, in a production line. It should be noted that Frederick Taylor was one of the main people responsible for these changes when carrying out a study on work and productivity, which led to the introduction of the concepts of fast production, with quality and quantity [GEO 20].

In the 20th century and after the end of World War II, Industry 3.0 began, also known as the digital age and the information age. During this period, electronic components, computer technologies and even the Internet were introduced [FRE 96; ROB 04]. From these creations, the automated production process emerged, without the need for human assistance, which, by employing programmable logical controls, allowed the potential for great advances in the field of robotics, such as robots that execute programmed sequences without human intervention, as well as the automatic landing of an airplane without the assistance of the pilot [GRO 21]. The use of technology has made it much easier to carry out the most difficult tasks on a daily basis, which has enabled an increase in mass production [FRE 96; ROB 04].

Industry 4.0 began in 2011 and is characterized by the use of information and communication technologies. The use of the Internet, which had already arrived from Industry 3.0, allowed for communication between companies and for them to gather data about themselves. This connection led to the emergence of cyber-physical systems, i.e. a system that connects physical components such as the machine with digital components and software, among others, in a network. Using this system, it was possible to automate, enhance and better monitor production [GRO 21; VIL 21].

In the Fourth Industrial Revolution, changes were introduced in the production system, namely, instruments linked to robotics, artificial intelligence (AI), the Internet of Things and Big Data. In this context, according to [GRO 21], these advances have enabled the emergence of machines that can predict failures and trigger maintenance processes autonomously or self-organized logistics that react to unexpected changes in production. Using these elements ensured that companies became progressive and worked more efficiently, as the loss of time due to lack of information and communication was reduced [GRO 21].

Less than 10 years after the start of Industry 4.0, Industry 5.0 appeared. This industry seeks not only to support the pillars of efficiency and productivity but also to prioritize human health and production that considers the limits of the environment [LON 20; RAJ 23].

With regards to Industry 5.0 trends, it is worth highlighting that they focus on robotics, artificial intelligence (AI), the Internet of Things, Big Data, digital transformation and the Internet, concepts that are not new, as they have been applied in Industry 4.0 [AKU 22]. The use of artificial intelligence (AI), both in Industry 4.0 and Industry 5.0, is dominant. However, it is used in very different ways. In Industry 4.0, the "AI black box" is used, in which humans are almost not needed, and the machine carries out the majority of the production process. However, in Industry 5.0, a "weak AI" is used. This is because it values collaboration between man and machine. In this way, both perform the task and depend on each other [VOG 21]. According to [ROM 16], Industry 5.0 was introduced to humanize Industry 4.0.

From Figure 8.1, and based on what was cited by [DEM 19], it appears that when following a temporal sequence, the intervals between revolutions have been decreasing, as the first three revolutions have an interval of around 200 years, while the next one is 40 years. The interval is expected to become increasingly shorter between the following revolutions.

As mentioned, the digital age or the information age began in the historical period of the Third Industrial Revolution. However, all the terminologies associated with them have emerged over the years and evolved in Industries 4.0 and 5.0 [KEE 23].

In the literature, there is no explicit definition that includes all terms linked to the digital era, which is why several terminologies are considered, which are related to each other, particularly (1) digital transformation; (2) digitalization and (3) digital workers, which is why it is pertinent to make a distinction between them [ZHA 23].

In a study by [GON 21], based on 53 definitions of digital transformation, these authors arrived at the unified definition that digital transformation consists of: "[...] A fundamental change process, enabled by the innovative use of digital technologies accompanied by the strategic leverage of key resources and capabilities, aiming to radically improve an entity and redefine its value proposition for its stakeholders" (p. 12).

Digitalization, in turn, is distinguished from digital transformation, as it comprises: "[...] transformation of all information types (texts, sounds, visuals, video and other data from various sources) into the digital language" [MAC 17, p. 2].

For [VER 21], digitalization should be perceived as a digital tool and as an innovative business model, with ample opportunities to bring strategic benefits to the company in the long term.

Finally, as for digital workers, they are considered a "[...] group of individuals that are influenced by new digital technologies affecting their attitudes, competencies and actions" [ZHA 23, p. 4].

Currently, in Industry 5.0, it is essential to define the key technological components predominantly used in the world of work, these being [FAR 18; HUA 18; MAD 15; PIL 20; SAG 13; SCI 00]:

1. Internet of Things: a network of intelligent objects that are connected and that allow sharing, collecting and exchanging data.

2. Big Data: a method of analyzing and storing a large amount of data remotely.

3. Artificial intelligence (AI): the ability of a machine, robot or system to imitate/simulate human problem-solving and decision-making characteristics.

4. Robots: programmable machines that are intended to perform repetitive and dangerous tasks, at a continuous speed.

5. Virtual reality (VR): three-dimensional reproductions that enable interaction in a virtual environment.

6. Augmented reality (AR): transforms enhanced real elements into an interactive experience produced by a computer.

8.3. The transition from Industry 5.0 to Industry 6.0

With the implementation of technologies, life as we know it will be constantly changing, from the way we work to the way we live. Therefore, we can only guess how it will be [YAD 22].

According to predictions, by 2050, it is expected that a level of development will have been reached in which the technology is self-sufficient [YAD 22]. We are thus faced with the Industrial Revolution 6.0.

Industry 6.0 is characterized as a futuristic revolution, expected to begin in 2050, with an implementation period of between 10 and 15 years [CHO 22].

The first studies on the Industrial Revolution 6.0 have recently begun to appear in the literature, predicting that it may be closer than expected [GOL 23].

The concept of Industry 6.0 is constantly changing, as it will only happen in the near future, being subject to different interpretations [GOL 23].

As mentioned by [GOL 23], Industry 6.0 can be understood as something "[...] which enables customer-centric hyper-connected multi-domain factories, with dynamic supply chains, where human (as a production worker) becomes a part of the interconnected, digitalized, and optimized environment" (p. 105). In turn, [ANN 21] summarized the concept into ubiquitous, customer-oriented, virtualized and antifragile manufacturing.

According to [DAS 22], some of the main technologies prevalent in Industry 6.0 will be:

1. autonomous robots;
2. quantum computing;
3. Internet of Things;
4. augmented/virtual reality;
5. 3D printing;
6. cloud computing;
7. artificial intelligence (AI).

Just like its predecessors, Industry 6.0 will also transform the world of work, as a new culture at work will be established, in which workers, in addition to working with other humans, will become accustomed to production assisted by machines/robots, in an environment of high development, with agile production [ALP 19; DEM 19].

In this industry, it is expected that the worker will be able to establish a co-working relationship between humans and robots, in which they will use their intelligence in conjunction with the machine's potential, in order to achieve resilient, error-free production [ALP 19; DEM 19].

As in any of the previous Industrial Revolutions, 6.0 also raises concerns regarding the replacement of human labor by machines. In order to mitigate the conflict, it is suggested by the authors that when organizations introduce new technologies, which imply the replacement of human work, they should always create a higher number of jobs that are equivalent to those lost [CHO 22].

8.4. Industry 6.0

Industry 6.0 is an emerging concept that refers to the evolution of industrial practices after Industry 4.0 (focused on automation, connectivity and data) and Industry 5.0 (which emphasizes human–machine collaboration and sustainability) [CHO 22]. While Industry 4.0 and Industry 5.0 are still being implemented in many areas, the term Industry 6.0 is beginning to be discussed in advanced technology and innovation contexts.

Based on current trends, several aspects of Industry 6.0 can be highlighted. Among the main ones, we can highlight the increasing focus that has been given to *augmented intelligence (Advanced AI)*. More specifically, it is observed that Industry 6.0 goes beyond artificial intelligence used to automate processes, integrating augmented intelligence, where AI and humans work in symbiosis to achieve levels of efficiency, creativity and decision-making that are impossible to achieve when we are faced with machines or humans alone [HEI 23b; DAV 23]. This is the case of systems that not only suggest solutions but learn from human decisions to improve processes continually. *Biological* and *technological integration* [BER 20] has also gained particular prominence. Technologies such as brain–computer interfaces and biological sensors will allow for a more fluid interaction between machines and humans.

While Industry 5.0 prioritizes sustainable practices, Industry 6.0 emphasizes the concept of regeneration, where industrial systems not only minimize environmental impacts but also contribute to ecological restoration, which leads us to *regenerative sustainability* [CHO 22]. This is the case, for example, of factories that operate in zero (or even negative) carbon cycles, returning more resources to the environment than they consume. *Hyperconnected networks* and *advanced edge computing* also have potential in Industry 6.0 [DAS 22]. Expanding global connectivity between devices, machines and systems, using advanced technologies such as 6G and distributed computing, contributes to obtaining real-time responses and total autonomy in complex systems.

As the circular economy is a reality in Industry 5.0, it is observed that Industry 6.0 will enhance the implementation of production chains based on a *circular (digitized) economy* [ZOC 24], using blockchain and digital twins to track materials throughout of its life cycle. This is the case of intelligent products that inform when and how they can be recycled or repaired, ensuring zero waste.

As ethical issues are a growing concern, Industry 6.0 advocates automation based on ethical principles, thus ensuring that technological advancement respects privacy, equality and human rights [CHO 22]. In addition to *conscious* and *ethical automation*, *continuous education* and *learning* are also a concern. Effectively, in an Industry 6.0 context, workers must be constantly updated to interact with new technologies through *adaptive learning* and *gamification platforms* [HEI 23a].

In summary, although the concept of Industry 6.0 is still under development, it reflects the search for a balance between cutting-edge technology, sustainability and human well-being. This is an ambitious vision for the global industry's future [CHO 22].

8.5. Fundamental differences between Industry 5.0 and Industry 6.0

The criticism that it is premature to talk about Industry 6.0, given that Industry 5.0 is still being implemented, is understandable and valid to a certain extent. Indeed, many countries and sectors are still struggling to fully adopt Industry 4.0 concepts and technologies, while Industry 5.0, with its emphasis on human–machine collaboration and sustainability, is still in its early stages in several regions. However, the debate about Industry 6.0 is not just a matter of anticipation. It reflects a need to think further, to respond to emerging technological and social changes [GRO 21]. Therefore, it is important to highlight some of the main differences that are felt between Industry 5.0 and Industry 6.0 (Table 8.1) [CHO 22; HEI 23b].

	Industry 5.0	**Industry 6.0**
Focus on collaboration vs. full integration	Prioritizes collaboration between humans and machines, bringing back the role of the human worker as a creative and critical actor in the production process. Marked by mass customization and sustainability as its main objective.	Aims to deeply integrate humans, machines and even biological systems, creating an almost organic symbiosis. The relationship between humans and technology transcends collaboration to become a full integration with brain–machine interfaces, augmented intelligence and advanced ethical automation.

Sustainability vs. regeneration	Seeks to minimize environmental impact, promoting sustainability practices, energy efficiency and waste reduction.	Adopts the concept of regenerative sustainability, where industrial systems not only minimize damage but contribute positively to the restoration of the environment.
Advanced technologies	Uses technologies such as artificial intelligence, advanced automation, collaborative robotics and the Internet of Things (IoT) to optimize processes and promote personalization.	Explores emerging technologies such as 6G networks, quantum computing, decentralized blockchain, global-scale digital twins, biotechnology and brain–computer interfaces, enabling autonomous and hyperconnected systems.
Sustainability culture vs. integrated ethics	Based on an ethical model that places human beings and nature as priorities but still depends on external regulations and adaptations of practices.	Embeds ethics directly into technological systems, such as bias-free AI and conscious automation, to ensure that industrial decisions respect human rights, privacy and social equality.
Scope of innovation	Mainly focuses on improving industrial processes and a more human and sustainable impact.	Expands the scope to include a holistic approach to social, technological and environmental transformation, with a strong focus on regeneration and general well-being.

Table 8.1. *Main differences between Industry 5.0 and Industry 6.0. Source: Own elaboration (based on [CHO 22; HEI 23b]).*

8.6. Industry 6.0: technological innovation, sustainability and human well-being

Even though it is a concept in development, Industry 6.0 appears as a visionary response to the challenges and opportunities that technological and social evolution presents. It is not limited to advancing the use of technology but seeks to harmoniously integrate technological innovation, environmental sustainability and human well-being [CHO 22]. Rooted in the understanding that technological progress cannot be dissociated from social and environmental impacts, Industry 6.0, unlike previous revolutions, which often prioritized efficiency and profit over social or environmental considerations, seeks to align these advances with a greater purpose, namely, creating a more balanced, sustainable world oriented towards collective well-being. A balance is proposed here between these three fundamental pillars, resulting in the industrial solutions of the future being designed to meet human needs without compromising natural resources or creating social inequalities. This balance is achieved by placing technology as a facilitator of sustainable coexistence between society and the environment, and not just as an instrument of production [CHO 22].

In technological terms, Industry 6.0 goes far beyond the advanced automation and digitalization that characterize Industry 4.0. It introduces a model in which technology is not just a tool, but a partner in the decision-making process [GRO 21]. Industry 6.0 will be marked by the presence of intelligent systems that not only automate tasks but work in partnership with humans, promoting a symbiotic relationship [DAV 23]. Machines and algorithms will be attuned to ethical values, with the ability to learn and adapt to human decisions, while ensuring responsible practices. This will allow the creation of more inclusive, safe and worker-centered work environments. This results in ethical and responsible integration, ensuring that privacy and human dignity are protected.

Furthermore, sustainability will no longer be just a goal or an ideal, becoming intrinsic to all production processes. The notion of sustainability will be expanded to incorporate regenerative practices, where industries will actively contribute to environmental restoration, promoting production cycles that not only minimize ecological impact but also return resources to the planet [CHO 22]. This implies, for example, that factories and industrial processes are designed not only to consume less energy, water and resources but also to return more to the environment than they take out. Technologies such as blockchain, digital twins and hyperconnected networks will be essential to guarantee the transparency and traceability of production chains, enabling a true circular economy. Strongly driven by digitalization, the circular economy will play a crucial role in this context [ZOC 24]. Waste will become a rarity, and reuse will become a rule.

In social terms, Industry 6.0 will be transformative. It recognizes that technology only has real value if it improves people's lives. Human well-being will be at the center of industrial decisions. This implies that technology will be used to improve the quality of life, not only in work environments but also in the impact of the products and services generated. More specifically, workplaces will be redesigned to be safer, less physically demanding and more intellectually enriching. At the same time, the focus on personalization will allow products and services to be shaped to meet people's individual needs, promoting a more inclusive and human experience. Exoskeletons, brain–computer interfaces and other advances will create conditions for people to perform tasks with less physical or mental effort, enhancing human capabilities without overload [DAV 23].

Additionally, there will be a revolution in education and continuous training. Workers will not just be trained to operate machines. More than that, they will have access to continuous learning platforms adapted to their skills and interests, allowing them to keep up with technological developments without feeling excluded [HEI 23a]. Gamification and virtual reality will transform training into something dynamic and accessible.

It is also important to highlight that Industry 6.0 represents an effort to shape an intrinsically ethical economic and industrial system. This means that business decisions will be made taking into account human rights, equal access and the preservation of cultures and communities. Major technological advances like AI will be regulated to prevent bias and discrimination, ensuring everyone benefits fairly.

In short, Industry 6.0 reflects a collective effort to build a future where technology is a powerful ally [HEI 23b] but one that is humanized, and where economic growth is aligned with the preservation of the environment and the promotion of fairer societies [CHO 22]. It is an ambitious but deeply necessary vision that invites us to reimagine the way we interact with the world and the technologies we create.

8.7. By way of conclusion

8.7.1. *Why talk about Industry 6.0 now?*

The debate on Industry 6.0 serves to anticipate the directions that technology and society are taking. The accelerated pace of technological innovation requires industries, governments and institutions to be prepared for changes that could happen in only a few decades or even sooner.

Many of the technologies that underpin the Industry 6.0 vision are already in development. Brain–computer interfaces, advanced biotechnology, quantum computing, blockchain for value chains and 6G networks are moving from laboratories to practical applications. This creates a foundation to begin designing the future of the industry.

The world we live in faces complex global challenges, such as climate change, social inequality and ethical tensions in the application of advanced technologies. It follows that Industry 6.0 is not just an evolution of production, but a model of systemic response that combines technological progress with regenerative solutions centered on human well-being.

Finally, it is important to highlight that society is increasingly demanding that companies be responsible, transparent and focused on positive impact. Industry 6.0 reflects this demand, projecting a future where industrial innovation is completely aligned with social and environmental values.

In short

As mentioned previously, although talking about Industry 6.0 seems extemporaneous, it is important to realize that technological and industrial evolution does not happen in perfectly defined linear stages. Concepts like Industry 6.0 emerge not as an attempt to invalidate the efforts of Industry 5.0 but as an exercise in imagination and planning for what will come next. Differentiating oneself from Industry 5.0 is not just a matter of new technologies but of an approach that places environmental regeneration, biological and technological integration and ethics as the foundations of the industrial future. In this context, discussing Industry 6.0 now is preparing the ground for a more integrated, sustainable and humane future.

8.8. References

[AKU 22] AKUNDI A., EURESTI D., LUNA S. et al., "State of Industry 5.0—Analysis and identification of current research trends", *Applied System Innovation*, vol. 5, no. 1, p. 27. doi: 10.3390/asi5010027, 2022.

[ALP 19] ALPASLAN DEMIR K., CICIBAŞ H., The next industrial revolution: Industry 5.0 and discussions on Industry 4.0, available at: https://www.researchgate.net/publication/336653504, 2019.

[ANN 21] ANNANPERÄ, E., JURMU, M., KAIVO-OJA, J. et al., From Industry X to Industry 6.0: Antifragile manufacturing for people, planet, and profit with passion, White Paper, Issue 5/2021, pp. 1–38, Allied ICT Finland, available at: https://cris.vtt.fi/en/publications/from-industry-x-to-industry-60-antifragile-manufacturing-for-peop [Accessed 20 December 2024], 2021.

[BER 20] BERGS T. et al., "Application cases of biological transformation in manufacturing technology", *CIRP Journal of Manufacturing Science and Technology*, vol. 31, pp. 68–77, doi: 10.1016/j.cirpj.2020.09.010, 2020.

[CHO 22] CHOURASIA S., TYAGI A., PANDEY S.M. et al., "Sustainability of Industry 6.0 in global perspective: Benefits and challenges", *MAPAN*, vol. 37, no. 2, pp. 443–452. doi: 10.1007/s12647-022-00541-w, 2022.

[DAS 22] DAS S., PAN T., "A strategic outline of Industry 6.0: Exploring the future", *SSRN Electronic Journal*, pp. 1–10, doi: 10.6084/m9.figshare.19738909.v1, 2022.

[DAV 23] DAVE D.M., MANDVIKAR S., "Augmented intelligence: Human-AI collaboration in the era of digital transformation", *International Journal of Engineering Applied Sciences and Technology*, vol. 8, no. 6, pp. 24–33, doi: 10.33564/ijeast.2023.v08i06.003, 2023.

[DEM 19] DEMIR K.A., DÖVEN G., SEZEN B., "Industry 5.0 and human-robot co-working", *Procedia Computer Science*, vol. 158, pp. 688–695, doi: 10.1016/j.procs.2019.09.104, 2019.

[FAR 18] FARSHID M., PASCHEN J., ERIKSSON T. et al., "Go boldly!: Explore augmented reality (AR), virtual reality (VR), and mixed reality (MR) for business", *Business Horizons*, vol. 61, no. 5, pp. 657–663, doi: 10.1016/j.bushor.2018.05.009, 2018.

[FRE 96] FREMDLING R., Industrial revolution and scientific and technological progress, Working Paper, University of Groningen, available at: https://research.rug.nl/en/publications/industrial-revolution-and-scientific-and-technological-progress-3, 1996.

[GEO 20] GEORGE S.A., GEORGE H.A.S., "Industrial Revolution 5.0: The transformation of the modern manufacturing process to enable man and machine to work hand in hand", *The Seybold Report Journal*, vol. 15, no. 9, pp. 214–234, doi: 10.5281/zenodo.6548092, 2020.

[GOL 23] GOLOVIANKO M., TERZIYAN V., BRANYTSKYI V. et al., "Industry 4.0 vs. Industry 5.0: Co-existence, transition, or a hybrid", *Procedia Computer Science*, vol. 217, pp. 102–113, doi: 10.1016/j.procs.2022.12.206, 2023.

[GON 21] GONG C., RIBIERE V., "Developing a unified definition of digital transformation", *Technovation*, vol. 102, doi: 10.1016/j.technovation.2020.102217, 2021.

[GRO 21] GROUMPOS P.P., "A critical historical and scientific overview of all industrial revolutions", *IFAC-PapersOnLine*, vol. 54, no. 13, pp. 464–471, doi: 10.1016/j.ifacol.2021.10.492, 2021.

[HEI 23a] HEILALA J., SAEID P., "Superhuman-centered design for manufacturing metaverse", *CDEUR Workshop Proceedings, 6th Conference on Technology Ethics, Tethics 2023*, vol. 3582, pp. 156–161, Turku, 18 October 2023.

[HEI 23b] HEILALA J., SINGH K., "Evaluation planning for artificial intelligence-based Industry 6.0 Metaverse Integration", in AHRAM T., KARWOWSKI W., BUCCHIANICO P.D. et al. (eds), *Intelligent Human Systems Integration (IHSI 2023): Integrating People and Intelligent Systems*, vol. 69. AHFE International, doi: 10.54941/ahfe1002892, 2023.

[HUA 18] HUANG M.-H., RUST R.T., "Artificial Intelligence in service", *Journal of Service Research*, vol. 21, no. 2, pp. 155–172, doi: 10.1177/1094670517752459, 2018.

[KEE 23] KEENGWE J., GIKANDI J.W., "Critical issues surrounding competency-based education in the digital age: The 'Wickedness Narrative of the Problem'", in KEENGWE J., GIKANDI J.W. (eds), *Competence-Based Curriculum and E-Learning in Higher Education*, IGI Global, Hershey, pp. 199–215, 2023.

[LON 20] LONGO F., PADOVANO A., UMBRELLO S., "Value-oriented and ethical technology engineering in Industry 5.0: A human-centric perspective for the design of the factory of the future", *Applied Sciences*, vol. 10, no. 12, p. 4182, doi: 10.3390/app10124182, 2020.

[MAC 17] MACHEKHINA O.N., "Digitalization of education as a trend of its modernization and reforming", *Revista Espacios*, vol. 38, no. 40, pp. 1–26, 2017.

[MAD 15] MADAKAM S., RAMASWAMY R., TRIPATHI S., "Internet of Things (IoT): A literature review", *Journal of Computer and Communications*, vol. 3, no. 5, pp. 164–173, doi: 10.4236/jcc.2015.35021, 2015.

[PIL 20] PILEVARI N., YAVARI F., "Industry revolutions development from Industry 1.0 to Industry 5.0 in manufacturing", *Journal of Industrial Strategic Management*, vol. 5, no. 2, pp. 44–63, 2020.

[RAJ 23] RAJUMESH S., "Promoting sustainable and human-centric industry 5.0: A thematic analysis of emerging research topics and opportunities", *Journal of Business and Socio-Economic Development*, doi: 10.1108/JBSED-10-2022-0116, 2023.

[ROB 04] ROBERT L.E., "The Industrial revolution: Past and future", *American Institute for Economic Research*, vol. 44, no. 8, pp. 1–8, available at: www.aier.org, 2004.

[ROM 16] ROMERO D., BERNUS P., NORAN O. et al., "The Operator 4.0: Human cyber-physical systems & adaptive automation towards human-automation symbiosis work systems", in NÄÄS I., VENDRAMETTO O., REIS J.M. (eds), *Advances in Production Management Systems. Initiatives for a Sustainable World*. Springer, Cham, pp. 677–686, doi: 10.1007/978-3-319-51133-7_80, 2016.

[SAG 13] SAGIROGLU S., SINANC D., "Big data: A review", in *2013 International Conference on Collaboration Technologies and Systems (CTS)*, pp. 42–47, doi: 10.1109/CTS.2013.6567202, 2013.

[SCI 00] SCIAVICCO L., SICILIANO B., *Modelling and Control of Robot Manipulators*, 2nd edition, Springer, London, doi: 10.1007/978-1-4471-0449-0, 2000.

[VER 21] VERHOEF P.C., BROEKHUIZEN T., BART Y. et al., "Digital transformation: A multidisciplinary reflection and research agenda", *Journal of Business Research*, vol. 122, pp. 889–901, doi: 10.1016/j.jbusres.2019.09.022, 2021.

[VIL 21] VILLALONGA A., NEGRI E., BISCARDO G. et al., "A decision-making framework for dynamic scheduling of cyber-physical production systems based on digital twins", *Annual Reviews in Control*, vol. 51, pp. 357–373, doi: 10.1016/j.arcontrol.2021.04.008, 2021.

[VOG 21] VOGT J., "Where is the human got to go? Artificial intelligence, machine learning, big data, digitalisation, and human–robot interaction in Industry 4.0 and 5.0", *AI & Society*, vol. 36, no. 3, pp. 1083–1087, doi: 10.1007/s00146-020-01123-7, 2021.

[YAD 22] YADAV R., ARORA S., DHULL S., "A path way to Industrial Revolution 6.0", *International Journal of Mechanical Engineering*, vol. 7, no. 1, pp. 1452–1459, 2022.

[ZHA 23] ZHANG J., CHEN Z., "Exploring human resource management digital transformation in the digital age", *Journal of the Knowledge Economy*, doi: 10.1007/s13132-023-01214-y, 2023.

[ZOC 24] ZOCCO F., HADDAD W.M., CORTI A. et al., "A unification between deep-learning vision, compartmental dynamical thermodynamics, and robotic manipulation for a circular economy", *IEEE Access*, vol. 12, pp. 173502–173516, doi: 10.1109/ACCESS.2024.3493755, 2024.

List of Authors

J. Paulo DAVIM
Department of Mechanical Engineering
University of Aveiro
Portugal

Qeis KAMRAN
Weiden Business School
Ostbayerische Technische
Hochschule (OTH)
Amberg-Weiden
Germany

Carolina MACHADO
Department of Management
School of Economics,
Management and Political Science
and
Interdisciplinary Centre of Social
Sciences
CICS.NOVA
University of Minho
Braga
Portugal

António Brandão MONIZ
Faculty of Science and Technology
Department of Applied Social Sciences
Nova University of Lisbon
Portugal

Catarina da Silva NOGUEIRA
Department of Management
School of Economics,
Management and Political Science
University of Minho
Braga
Portugal

Lurdes PEDRO
Setúbal Polytechnic Institute
Higher School of Business Sciences
ESCE/IPS
Setúbal
Portugal

Célia QUINTAS
Setúbal Polytechnic Institute
Higher School of Business Sciences
ESCE/IPS
Setúbal
Portugal

Cosmina-Mihaela ROSCA
Department of Automatic Control,
Computers, and Electronics
Faculty of Mechanical and Electrical
Engineering
Petroleum-Gas University of Ploiești
Romania

José Rebelo dos SANTOS
Setúbal Polytechnic Institute
Higher School of Business Sciences
ESCE/IPS
Setúbal
Portugal

Adrian STANCU
Department of Business Administration
Faculty of Economic Sciences
Petroleum-Gas University of Ploiești
Romania

Ismail W.R. TAIFA
Department of Mechanical and Industrial Engineering
College of Engineering and Technology
University of Dar Es Salaam
Tanzania

Index

C, D

N, O

P, Q

R, S

T, U

V, W

Other titles from

iSTE

in

Systems and Industrial Engineering – Robotics

2024

ANDRÉ Jean-Claude
Knowledge Production Modes between Science and Applications 1: Concepts
Knowledge Production Modes between Science and Applications 2: Applications

BRIFFAUT Jean-Pierre
Complexities 2: Various Approaches in the Field of Social and Human Sciences
(Systems of Systems Complexity Set – Volume 6)

OUSTALOUP Alain
Control in System Dynamics: Comparative Analysis of Feedback Strategies

YALAOUI Farouk, NGUYEN Nhan-Quy, OUAZENE Yassine, ZEMZAMI Maria, MERGHEM-BOULAHIA Leïla
Connected Innovation and Technology X.0 1: Advancing Digital Transformation: Innovational Strategies for Smart Industry and Sustainable Innovation

2023

BERRAH Lamia, CLIVILLÉ Vincent
Decision and Decision-maker in an Industrial Environment

BRIFFAUT Jean-Pierre
Complexities 1: Various Approaches in the Field of Techno-scientific Knowledge
(Systems of Systems Complexity Set – Volume 5)

CHAINEAUX Jacques
The Mechanisms of Explosions: 27 Case Studies for their Understanding

DIALLO Saikou, BOGUI Jean-Jacques Maomra, ATCHOUA Julien
Technologies and African Societies in Pandemic Times: Using Technology to Survive and Thrive During the Covid-19 Era
(Africa Digital Age Set – Volume 2)

JAEGLER Anicia, ROQUES Thierry
Sustainable Supply Chain Management: Sustainable Decision Support

PRINTZ Jacques
Organization and Pedagogy of Complexity: Systemic Case Studies and Prospects
(Systems of Systems Complexity Set – Volume 4)

2022

AMARA Yacine, BEN AHMED Hamid, GABSI Mohamed
Hybrid Excited Synchronous Machines: Topologies, Design and Analysis

BOURRIÈRES Jean-Paul, PINÈDE Nathalie, TRAORÉ Mamadou Kaba, ZACHAREWICZ Grégory
From Logistic Networks to Social Networks: Similarities, Specificities, Modeling, Evaluation

DEMOLY Frédéric, ANDRÉ Jean-Claude
4D Printing 1: Between Disruptive Research and Industrial Applications
4D Printing 2: Between Science and Technology

HAJJI Rafika, JARAR OULIDI Hassane
Building Information Modeling for a Smart and Sustainable Urban Space

KROB Daniel
Model-based Systems Architecting: Using CESAM to Architect Complex Systems
(Systems of Systems Complexity Set – Volume 3)

LOUIS Gilles
Dynamics of Aircraft Flight

2020

BRON Jean-Yves
System Requirements Engineering

KRYSINSKI TOMASZ, MALBURET FRANÇOIS
Energy and Motorization in the Automotive and Aeronautics Industries

PRINTZ Jacques
System Architecture and Complexity: Contribution of Systems of Systems to Systems Thinking

2019

ANDRÉ Jean-Claude
Industry 4.0: Paradoxes and Conflicts

BENSALAH Mounir, ELOUADI Abdelmajid, MHARZI Hassan
Railway Information Modeling RIM: The Track to Rail Modernization

BLUA Philippe, YALAOU Farouk, AMODEO Lionel, DE BLOCK Michaël, LAPLANCHE David
Hospital Logistics and e-Management: Digital Transition and Revolution

BRIFFAUT Jean-Pierre
From Complexity in the Natural Sciences to Complexity in Operations Management Systems
(Systems of Systems Complexity Set – Volume 1)

BUDINGER Marc, HAZYUK Ion, COÏC Clément
Multi-Physics Modeling of Technological Systems

FLAUS Jean-Marie
Cybersecurity of Industrial Systems

JAULIN Luc
Mobile Robotics – Second Edition Revised and Updated

KUMAR Kaushik, DAVIM Paulo J.
Optimization for Engineering Problems

TRIGEASSOU Jean-Claude, MAAMRI Nezha
Analysis, Modeling and Stability of Fractional Order Differential Systems 1: The Infinite State Approach
Analysis, Modeling and Stability of Fractional Order Differential Systems 2: The Infinite State Approach

VANDERHAEGEN Frédéric, MAAOUI Choubeila, SALLAK Mohamed, BERDJAG Denis
Automation Challenges of Socio-technical Systems

2018

BERRAH Lamia, CLIVILLÉ Vincent, FOULLOY Laurent
Industrial Objectives and Industrial Performance: Concepts and Fuzzy Handling

GONZALEZ-FELIU Jesus
Sustainable Urban Logistics: Planning and Evaluation

GROUS Ammar
Applied Mechanical Design

LEROY Alain
Production Availability and Reliability: Use in the Oil and Gas Industry

MARÉ Jean-Charles
Aerospace Actuators 3: European Commercial Aircraft and Tiltrotor Aircraft

MAXA Jean-Aimé, BEN MAHMOUD Mohamed Slim, LARRIEU Nicolas
Model-driven Development for Embedded Software: Application to Communications for Drone Swarm

MBIHI Jean
Analog Automation and Digital Feedback Control Techniques
Advanced Techniques and Technology of Computer-Aided Feedback Control

MORANA Joëlle
Logistics

SIMON Christophe, WEBER Philippe, SALLAK Mohamed
Data Uncertainty and Important Measures
(Systems Dependability Assessment Set – Volume 3)

TANIGUCHI Eiichi, THOMPSON Russell G.
City Logistics 1: New Opportunities and Challenges
City Logistics 2: Modeling and Planning Initiatives
City Logistics 3: Towards Sustainable and Liveable Cities

ZELM Martin, JAEKEL Frank-Walter, DOUMEINGTS Guy, WOLLSCHLAEGER Martin
Enterprise Interoperability: Smart Services and Business Impact of Enterprise Interoperability

2017

ANDRÉ Jean-Claude
From Additive Manufacturing to 3D/4D Printing 1: From Concepts to Achievements
From Additive Manufacturing to 3D/4D Printing 2: Current Techniques, Improvements and their Limitations
From Additive Manufacturing to 3D/4D Printing 3: Breakthrough Innovations: Programmable Material, 4D Printing and Bio-printing

ARCHIMÈDE Bernard, VALLESPIR Bruno
Enterprise Interoperability: INTEROP-PGSO Vision

CAMMAN Christelle, FIORE Claude, LIVOLSI Laurent, QUERRO Pascal
Supply Chain Management and Business Performance: The VASC Model

FEYEL Philippe
Robust Control, Optimization with Metaheuristics

MARÉ Jean-Charles
Aerospace Actuators 2: Signal-by-Wire and Power-by-Wire

POPESCU Dumitru, AMIRA Gharbi, STEFANOIU Dan, BORNE Pierre
Process Control Design for Industrial Applications

RÉVEILLAC Jean-Michel
Modeling and Simulation of Logistics Flows 1: Theory and Fundamentals
Modeling and Simulation of Logistics Flows 2: Dashboards, Traffic Planning and Management
Modeling and Simulation of Logistics Flows 3: Discrete and Continuous Flows in 2D/3D

2016

ANDRÉ Michel, SAMARAS Zissis
Energy and Environment
(Research for Innovative Transports Set – Volume 1)

AUBRY Jean-François, BRINZEI Nicolae, MAZOUNI Mohammed-Habib
Systems Dependability Assessment: Benefits of Petri Net Models
(Systems Dependability Assessment Set – Volume 1)

BLANQUART Corinne, CLAUSEN Uwe, JACOB Bernard
Towards Innovative Freight and Logistics
(Research for Innovative Transports Set – Volume 2)

COHEN Simon, YANNIS George
Traffic Management
(Research for Innovative Transports Set – Volume 3)

MARÉ Jean-Charles
Aerospace Actuators 1: Needs, Reliability and Hydraulic Power Solutions

REZG Nidhal, HAJEJ Zied, BOSCHIAN-CAMPANER Valerio
Production and Maintenance Optimization Problems: Logistic Constraints and Leasing Warranty Services

TORRENTI Jean-Michel, LA TORRE Francesca
Materials and Infrastructures 1
(Research for Innovative Transports Set – Volume 5A)
Materials and Infrastructures 2
(Research for Innovative Transports Set – Volume 5B)

WEBER Philippe, SIMON Christophe
Benefits of Bayesian Network Models
(Systems Dependability Assessment Set – Volume 2)

YANNIS George, COHEN Simon
Traffic Safety
(Research for Innovative Transports Set – Volume 4)

2015

AUBRY Jean-François, BRINZEI Nicolae
Systems Dependability Assessment: Modeling with Graphs and Finite State Automata

BOULANGER Jean-Louis
CENELEC 50128 and IEC 62279 Standards

BRIFFAUT Jean-Pierre
E-Enabled Operations Management

MISSIKOFF Michele, CANDUCCI Massimo, MAIDEN Neil
Enterprise Innovation

2014

CHETTO Maryline
Real-time Systems Scheduling
Volume 1 – Fundamentals
Volume 2 – Focuses

DAVIM J. Paulo
Machinability of Advanced Materials

ESTAMPE Dominique
Supply Chain Performance and Evaluation Models

FAVRE Bernard
Introduction to Sustainable Transports

GAUTHIER Michaël, ANDREFF Nicolas, DOMBRE Etienne
Intracorporeal Robotics: From Milliscale to Nanoscale

MICOUIN Patrice
Model Based Systems Engineering: Fundamentals and Methods

MILLOT Patrick
Designing Human–Machine Cooperation Systems

NI Zhenjiang, PACORET Céline, BENOSMAN Ryad, RÉGNIER Stéphane
Haptic Feedback Teleoperation of Optical Tweezers

OUSTALOUP Alain
Diversity and Non-integer Differentiation for System Dynamics

REZG Nidhal, DELLAGI Sofien, KHATAD Abdelhakim
Joint Optimization of Maintenance and Production Policies

STEFANOIU Dan, BORNE Pierre, POPESCU Dumitru, FILIP Florin Gh.,
EL KAMEL Abdelkader
Optimization in Engineering Sciences: Metaheuristics, Stochastic Methods and Decision Support

2013

ALAZARD Daniel
Reverse Engineering in Control Design

ARIOUI Hichem, NEHAOUA Lamri
Driving Simulation

CHADLI Mohammed, COPPIER Hervé
Command-control for Real-time Systems

DAAFOUZ Jamal, TARBOURIECH Sophie, SIGALOTTI Mario
Hybrid Systems with Constraints

FEYEL Philippe
Loop-shaping Robust Control

FLAUS Jean-Marie
Risk Analysis: Socio-technical and Industrial Systems

FRIBOURG Laurent, SOULAT Romain
Control of Switching Systems by Invariance Analysis: Application to Power Electronics

GROSSARD Mathieu, REGNIER Stéphane, CHAILLET Nicolas
Flexible Robotics: Applications to Multiscale Manipulations

GRUNN Emmanuel, PHAM Anh Tuan
Modeling of Complex Systems: Application to Aeronautical Dynamics

HABIB Maki K., DAVIM J. Paulo
Interdisciplinary Mechatronics: Engineering Science and Research Development

HAMMADI Slim, KSOURI Mekki
Multimodal Transport Systems

JARBOUI Bassem, SIARRY Patrick, TEGHEM Jacques
Metaheuristics for Production Scheduling

KIRILLOV Oleg N., PELINOVSKY Dmitry E.
Nonlinear Physical Systems

LE Vu Tuan Hieu, STOICA Cristina, ALAMO Teodoro,
CAMACHO Eduardo F., DUMUR Didier
Zonotopes: From Guaranteed State-estimation to Control

MACHADO Carolina, DAVIM J. Paulo
Management and Engineering Innovation

MORANA Joëlle
Sustainable Supply Chain Management

SANDOU Guillaume
Metaheuristic Optimization for the Design of Automatic Control Laws

STOICAN Florin, OLARU Sorin
Set-theoretic Fault Detection in Multisensor Systems

2012

AÏT-KADI Daoud, CHOUINARD Marc, MARCOTTE Suzanne, RIOPEL Diane
Sustainable Reverse Logistics Network: Engineering and Management

BORNE Pierre, POPESCU Dumitru, FILIP Florin G., STEFANOIU Dan
Optimization in Engineering Sciences: Exact Methods

CHADLI Mohammed, BORNE Pierre
Multiple Models Approach in Automation: Takagi-Sugeno Fuzzy Systems

DAVIM J. Paulo
Lasers in Manufacturing

DECLERCK Philippe
Discrete Event Systems in Dioid Algebra and Conventional Algebra

DOUMIATI Moustapha, CHARARA Ali, VICTORINO Alessandro, LECHNER Daniel
Vehicle Dynamics Estimation using Kalman Filtering: Experimental Validation

GUERRERO José A, LOZANO Rogelio
Flight Formation Control

HAMMADI Slim, KSOURI Mekki
Advanced Mobility and Transport Engineering

MAILLARD Pierre
Competitive Quality Strategies

MATTA Nada, VANDENBOOMGAERDE Yves, ARLAT Jean
Supervision and Safety of Complex Systems

POLER Raúl, CARNEIRO Luis Maia, JASINSKI Thomas, ZOLGHADRI Marc, PEDRAZZOLI Paolo
Intelligent Non-hierarchical Manufacturing Networks

TROCCAZ Jocelyne
Medical Robotics

YALAOUI Alice, CHEHADE Hicham, YALAOUI Farouk, AMODEO Lionel
Optimization of Logistics

ZELM Martin, SANCHIS Raquel, POLER Raúl, DOUMEINGTS Guy
Enterprise Interoperability: I-ESA'12 Proceedings

2011

CANTOT Pascal, LUZEAUX Dominique
Simulation and Modeling of Systems of Systems

DAVIM J. Paulo
Mechatronics

DAVIM J. Paulo
Wood Machining

GROUS Ammar
Applied Metrology for Manufacturing Engineering

KOLSKI Christophe
Human–Computer Interactions in Transport

LUZEAUX Dominique, RUAULT Jean-René, WIPPLER Jean-Luc
Complex Systems and Systems of Systems Engineering

ZELM Martin, VAN SINDEREN Marten, DOUMEINGTS Guy, JOHNSON Pontus
Enterprise Interoperability: IWEI2011 Proceedings

2010

BOTTA-GENOULAZ Valérie, CAMPAGNE Jean-Pierre, LLERENA Daniel, PELLEGRIN Claude
Supply Chain Performance: Collaboration, Alignment and Coordination

BOURLÈS Henri, GODFREY K.C. Kwan
Linear Systems

BOURRIÈRES Jean-Paul
Proceedings of CEISIE'09

CHAILLET Nicolas, REGNIER Stéphane
Microrobotics for Micromanipulation

DAVIM J. Paulo
Sustainable Manufacturing

GIORDANO Max, MATHIEU Luc, VILLENEUVE François
Product Life-Cycle Management: Geometric Variations

LOZANO Rogelio
Unmanned Aerial Vehicles: Embedded Control

LUZEAUX Dominique, RUAULT Jean-René
Systems of Systems

VILLENEUVE François, MATHIEU Luc
Geometric Tolerancing of Products

2009

DIAZ Michel
Petri Nets: Fundamental Models, Verification and Applications

OZEL Tugrul, DAVIM J. Paulo
Intelligent Machining

PITRAT Jacques
Artificial Beings

2008

ARTIGUES Christian, DEMASSEY Sophie, NERON Emmanuel
Resources–Constrained Project Scheduling

BILLAUT Jean-Charles, MOUKRIM Aziz, SANLAVILLE Eric
Flexibility and Robustness in Scheduling

DOCHAIN Denis
Bioprocess Control

LOPEZ Pierre, ROUBELLAT François
Production Scheduling

THIERRY Caroline, THOMAS André, BEL Gérard
Supply Chain Simulation and Management

2007

DE LARMINAT Philippe
Analysis and Control of Linear Systems

DOMBRE Etienne, KHALIL Wisama
Robot Manipulators

LAMNABHI-LAGARRIGUE Françoise, LAGHROUCHE Salah, LORIA Antonio, PANTELEY Elena
Taming Heterogeneity and Complexity of Embedded Control

LIMNIOS Nikolaos
Fault Trees

2006

FRENCH COLLEGE OF METROLOGY
Metrology in Industry

NAJIM Kaddour
Control of Continuous Linear Systems

Printed and bound by CPI Group (UK) Ltd, Croydon, CR0 4YY

13/09/2026

14961012-0007